QUANTITATIVE GEOLOGY OF LATE JURASSIC EPICONTINENTAL SEDIMENTS IN THE JURA MOUNTAINS OF SWITZERLAND

 Birkhäuser

Reinhart A. Gygi

QUANTITATIVE GEOLOGY OF LATE JURASSIC EPICONTINENTAL SEDIMENTS IN THE JURA MOUNTAINS OF SWITZERLAND

Reinhart A. Gygi
Carl Spitteler-Strasse 70/B112
8053 Zürich
Switzerland

Supported by "Jura-Stiftung für Sediment- und Strukturgeologie"

ISBN 978-3-0348-0135-5   ISBN 978-3-0348-0136-2 (eBook)
DOI 10.1007/978-3-0348-0136-2
Springer Basel Heidelberg New York Dordrecht London

Library of Congress Control Number: 2012953473

Cover illustrations: Photography by W. Etter

Polished cross-section of a thickly-branching, reef-building coral, Cryptocoenia limbata GOLDFUSS*, with bands of annual accretion, out of a patch reef in the upper St-Ursanne Formation, Transversarium Chron, middle Oxfordian. The coral is a facies fossil diagnostic of the environment. The ammonite in the lower left is a Perisphinctes (Perisphinctes) alatus ÉNAY**, a wholly septate nucleus out of the lower part of the Mumienkalk Bed, Antecedens Chron, middle Oxfordian. The ammonite is a guide fossil diagnostic of relative age.

*) Specimen Gy 4323, section RG 370, unit no. 11, Courtételle, Canton Jura. Scale ca. ×1.4.
**) Specimen J24575, section RG 81b, bed no. 14a, Gächlingen, Canton Schaffhausen. Scale ×1. Both specimens are kept in the Museum of Natural History, Basel, Switzerland.

Printed on acid-free paper

Springer Basel is part of Springer Science+Business Media (www.springer.com)

To the honor of GOD, Our
CREATOR and LORD

Introduction

This study is a summary and a critical review of what the author has published since 1966 about sedimentary geology in an epicontinental (shelf) sea during the Late Jurassic in northern Switzerland. Most of the results that are presented in this study were previously documented in numerous papers and in four books published in Switzerland and abroad. A synopsis of the interplay of varying paleoclimate, sea level variations, varying water depth, sea floor topography, vertical and lateral facies changes, processes of sedimentation such as aggradation and progradation, compaction, the great regional differences in rates of sedimentation and in isostatic equilibration of the lithosphere under load, and concomitant synsedimentary tectonics is presented here. Regional variation in isostatic adjustment of the lithosphere to the increasing load of sediments is analyzed by means of time correlations based on a detailed biochronology of ammonites in combination with mineral stratigraphy using the comparatively stable clay mineral kaolinite, and with sequence stratigraphy.

Differential vertical displacement of the basement can be the effect either of processes deep in the earth interior (endogenic), or of lateral variation in rates of sedimentation and thereby of regional differences in loading and in isostatic equilibration of the lithosphere (exogenic). An epicontinental basin can accommodate partially compacted sediments with a thickness that is about three times the initial depth of the basin, until the basin is filled up to close below sea level. Cauliflower pellets of glauconite were formed in deeper water, and iron ooids were accreted in shallower water. Both kinds of particles evolved during nondeposition at the surface of mud-grade sediments, or somewhat below the sediment surface in times of a minimal rate of deposition. The bathymetrically controlled, vertical and lateral boundaries between facies where cauliflower pellets of glauconite or iron ooids were formed are clear-cut. The pertinent facies boundaries were at a paleodepth of approximately 100 meters *in the basin investigated*. Rapid relative sea level rises which could be documented to have occurred in Oxfordian and in Kimmeridgian time were essentially eustatic. They were of small scale. Ammonites collected from *in situ* in sections measured bed by bed have a great potential in the investigation of sedimentary geology, provided that they are well preserved and numerous enough, and that they are used in combination with detailed lithostratigraphy, sedimentology, mineral stratigraphy, and sequence stratigraphy.

The entire material which was collected by the author since 1962 and that is cited in this study is stored in the Museum of Natural History at Basel, Switzerland, together with the originals of the fieldbooks which are occasionally cited in this study. Copies of the fieldbooks are in the Landesgeologie at Wabern near Bern. The exact location with coordinates of all of the cited sections with the suffix RG can be read from Table 1 in GYGI (2000a). Names of localities are preceded in the text by the name of the township the locality belongs to. The names of all of the townships cited in the text of the present study can be found in the "Road map of Switzerland" 1:200,000 by swisstopo (Swiss Federal Office of Topography). An alphabetic list of townships including all of the sections measured and all of the localities studied by the author is kept at the Museum of Natural History at Basel and in the section "Landesgeologie" of swisstopo at 3084 Wabern.

The names of all of the lithostratigraphic units in the Upper Jurassic Series in northern Switzerland mentioned in this study are quoted in the stratigraphic dictionary by GYGI (2000b).

Each entry in this annotated dictionary includes the name(s) of the author(s) of the unit, the reference of publication, and, where possible, the type locality with coordinates and reference of the type section. The following text in the dictionary specifies the name and the sheet number of the topographic map, Landeskarte der Schweiz 1:25,000, in which the type locality of a unit can be found. Then the lithology, the macrofossil assemblage, and the age of a unit is described. Description of the lithostratigraphic units in the Central Jura Mts. by GYGI (2000a) is preceded by a synonymy list of older names for each unit.

Contents

List of Figures

1.1 From the Beginnings to 1962

The first modern scientific account of marine sediments preserved on land was given in France by the priest J. L. Giraud-Soulavié in 1780, as cited from Geikie (1905, p. 338 of an unabridged, undated reprint, second edition by Dover, New York). According to Geikie, on p. 344 of this reprint, A. L. Lavoisier, the founder of modern chemistry, distinguished on p. 350, Pl. 7 of a study published in 1789 on marine strata preserved on land, "littoral banks" from "pelagic banks" which were deposited in a sea at different distances from land. Lavoisier also concluded that sea level must have varied in the past. W. Smith pioneered the modern concept of stratigraphic paleontology in England, first in 1816, and then in an extended version in 1817.

Merian (1821) published the first systematic study in geology of the Swiss Jura Mountains. He described the main stratigraphic units of the region around Basel briefly and appropriately. Merian probably was, like many earth scientists of the time, impressed by A. G. Werner, who was a very influential teacher at the Mining Academy of Freiberg in Saxony, Germany, almost to his death in 1817. Werner taught, in the words of Geikie (1905, p. 212 in the Dover reprint cited above), that the formations labeled by him were to be recognized all over the world in the same order and with the same characters. Merian correlated what is the modern Bärschwil Formation with the younger, modern Wildegg Formation, because both formations have a high content of clay minerals. Merian must have been aware of the emerging method of biochronology which was initiated by W. Smith in England, because he emphasized on p. 102 that not only lithology, but fossils as well must be used in time correlation. He added on p. 103 that coeval strata which were laid down in different environments include unequal fossil assemblages according to variation in the environment. He arrived at this conclusion by comparison with sediments of the Recent. Merian's remark on p. 103 that the state of fossil taxonomy at his time was as yet insufficient

for reliable time correlation of sediments held true for several decades to come.

A. Gressly investigated the geology of the Jura Mountains mainly in Canton Solothurn. Gressly (1838–1841, p. 91) followed Merian (1821) in that he correlated his terrain à chailles, which is the equivalent of the modern Sornetan Member, with the modern Wildegg Formation. Most of the Wildegg Formation is younger than Gressly's terrain à chailles (Fig. 1.5 in this study). Gressly stated on p. 95 that the terrain à chailles, which included at the time the modern Liesberg Member, grades upward into limestone with coral banks (biostromes) and reefs. This limestone with corals is the modern St-Ursanne Formation. Gressly continued on pp. 97–98 with a description of the *lateral* transition from the coral-dominated assemblage indicating shallow water in his upper terrain à chailles (now Liesberg Member) to intermediate assemblages with bivalves of the genus *Pholadomya* and ammonites, and gradually to very abundant siliceous sponges from relatively deep water like in the modern Birmenstorf Member, mainly in Canton Aargau. Large perisphinctid ammonites and siliceous sponges occur in the lowermost bed of the more distal part of the Pichoux Formation, which Gressly regarded to be an intercalation in his terrain à chailles. Gressly's time correlation between what is now the Liesberg Member (with coral biostromes) with the basal Pichoux Formation and with the Birmenstorf Member with biostromes of siliceous sponges and abundant ammonites in Canton Aargau is correct. The inconsistency that the Birmenstorf Member is *below* the Effingen Member, which he thought to be the time equivalent of his upper terrain à chailles, apparently did not occur to him. Gressly noted on p. 102 that ammonites were rare in "littoral regions". He added that they are very abundant in "pelagic regions", and meant by this deeper water in an epicontinental sea. Gressly thereby showed that ammonites were sensitive to facies, and mainly to water depth in their habitat.

Gressly (1838–1841, p. 13) compared the "coral banks in littoral deposits", which he had studied in the Jura

Mountains, with Recent reefs in the tropics. He observed that the coral assemblages in the Jura Mountains were often associated with calcareous oolite. This is the case in the modern Tiergarten Member of the upper St-Ursanne Formation. GRESSLY introduced on the same p. 13 the term facies into the Swiss literature on earth science. Facies is the Latin word for the English face. CROSS and HOMEWOOD (1997, p. 1618) drew attention to references testifying that GRESSLY was not the first earth scientist to use the term facies. But it was GRESSLY (1838–1841) who formulated for the first time two of what he called laws of facies. First (p. 20): The facies of a given, regionally restricted stratigraphic unit is characterized by its lithology and by the macrofossil assemblage which is preserved in the rock. This facies is in lateral contrast to adjacent facies in coeval stratigraphic units. Second (p. 21): A given facies can ascend with time through successively younger stratigraphic units. GRESSLY probably meant by this, but did not plainly state, that a facies could shift *laterally* with time, and thereby anticipated progradation. GRESSLY used the term facies loosely. His "region" and "zone" were synonyms of facies. He is probably the inventor of the paleogeographic map. GRESSLY (1838–1841) presented such a map in Pl. 6.

WALTHER (1893–1894) extended the theory of facies. He stated on p. 994 that the depth of the sea varied continuously (at a given location). As a consequence, a distinct littoral fauna was forced to shift landward during a sea level rise, or seaward in the course of a sea level fall (p. 993). WALTHER probably envisaged with this eustatic fluctuations of sea level. He wrote on p. 979 that only such facies can become to be placed into *vertical* succession during geologic history, which are adjacent in *lateral* succession at the present time. This conclusion received much attention from subsequent, mainly Anglo-Saxon authors. It became known as "WALTHER's law" (see for instance COE, ed., 2003, p. 14, or NEUENDORF, K. K. E. et al., eds., Glossary of Geology, 5th edition 2005, p. 714). The proximal, thick part of shallowing-upward succession no. 1 in northwestern Switzerland is an instructive example of WALTHER's law. This part of the succession is explained in main conclusion no. 36.

A. OPPEL is the founder of zonal biostratigraphy. He subdivided the stages of D'ORBIGNY into zones. OPPEL's zone is a stack of marine strata which is documented at a type locality along with the assemblage of macrofossils embedded in the strata. An important example of this is OPPEL's revised and final version of the *Transversarium* (ammonite) Zone which was published after OPPEL's death by W. WAAGEN (OPPEL and WAAGEN 1866). OPPEL designated the type locality of the *Transversarium* Zone to be at Birmensdorf, Canton Aargau (now: spelled Birmenstorf) in northern Switzerland. This is the first biozone named after an ammonite taxon which was ever documented in detail. OPPEL

closely cooperated with C. MOESCH, who was geologically mapping at that time in Canton Aargau. MOESCH gave many of the ammonites that he had collected to OPPEL for publication. MOESCH (1863, 1867) as well as WÜRTENBERGER and WÜRTENBERGER (1866) pioneered modern stratigraphy in that they defined lithostratigraphic units in a way that the units could be mapped. These authors named most of their lithostratigraphic units after a township or a locality, and they added a separate list of fossils to the description of each of the lithostratigraphic units. They thereby kept lithostratigraphy separate from biostratigraphy, maybe for the first time.

ROLLIER (1888) followed GRESSLY in correlating the Liesberg Member with the Birmenstorf Member in Canton Aargau. ROLLIER (1888) correlated what is now the St-Ursanne Formation and the Günsberg Formation combined with the Effingen Member. It is likely that ROLLIER arrived at this opinion judging from the semicircular cliff above Combe des Geais on the south slope of Mt. Raimeux. This cirque is north of the village of Grandval east of Moutier. The Günsberg Formation is in the western part of the outcrop mostly massive limestone and forms a vertical cliff together with well-bedded limestone of the Pichoux Formation below. The Günsberg Formation becomes progressively marly eastward in the outcrop and concomitantly weathers back. This conspicuous lateral facies transition within a single outcrop was imaged in Figs. 6–8 by GYGI in GYGI and PERSOZ (1986). ROLLIER's correlation remained to be unchallenged until BOLLIGER and BURRI (1967) proposed another solution.

SCOTT (1940) gave a comprehensive account of the paleoecology of ammonites in sediments from the vast Cretaceous epicontinental seas in Texas, USA. He pointed out the interrelation between the occurrence of these cephalopods and facies of the sediments in which he found the ammonites. SCOTT (1940, p. 308 and 315) stated that ammonites are rare or absent in sediments laid down in very shallow water, and he distinguished in his Fig. 8 different ranges of water paleodepth which were occupied by distinct assemblages of ammonites. SCOTT concluded from this on p. 320 that the mode of life of the ammonites investigated by him was "necto-benthonic".

1.2   From 1962 to the Present Time

A turning point in the investigation of Jurassic successions was when the International Congress "Colloque du Jurassique" was convened at Luxembourg in 1962. The majority of participants in the congress voted to accept the conception of the modern Oxfordian Stage following ARKELL (1956, p. 84), who abandoned the stages Argovien, Rauracien, and Séquanien. The Oxfordian Stage *sensu lato*,

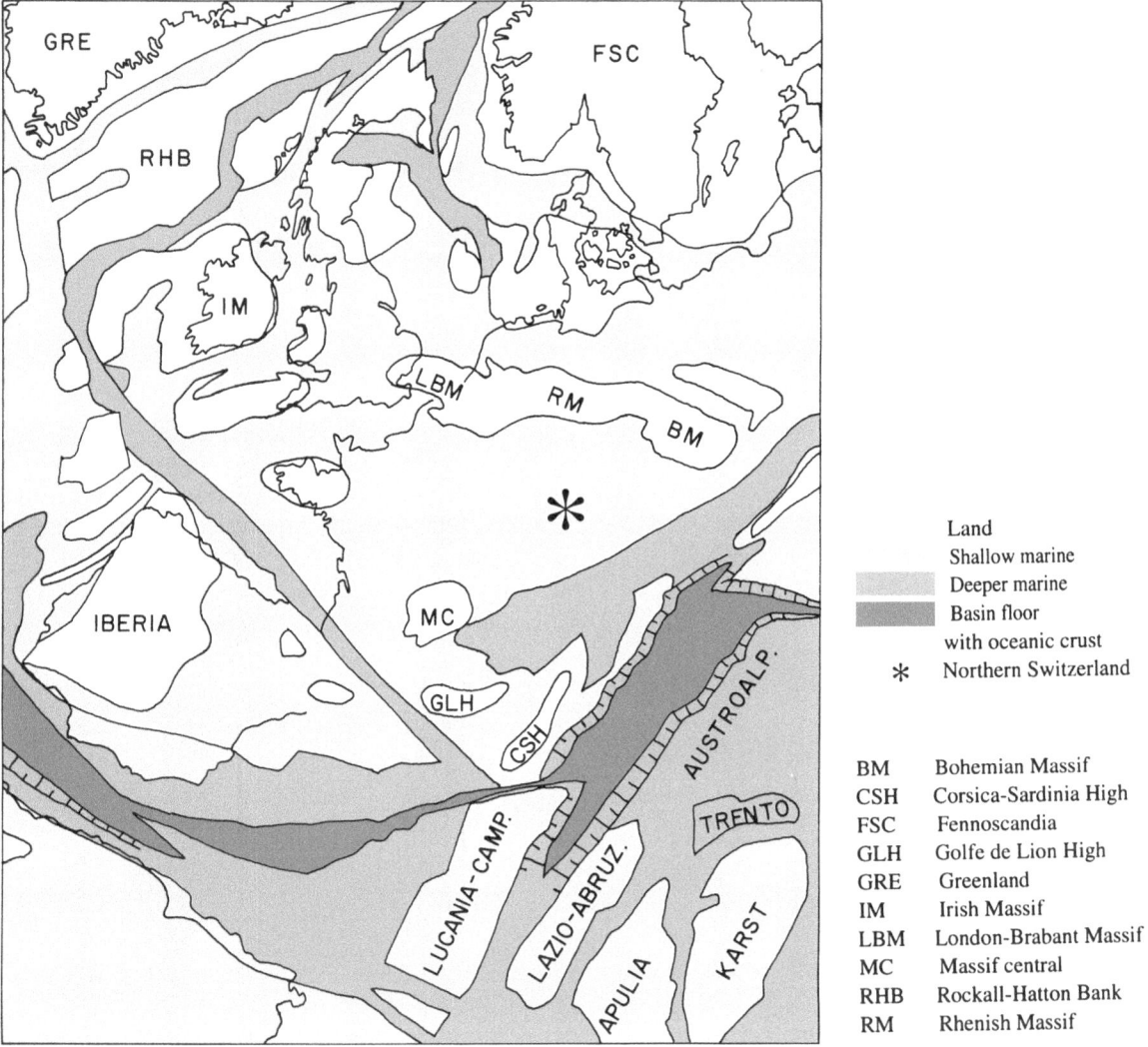

Fig. 1.1 Paleogeographic map of Europe in the Late Jurassic. After Ziegler (1988, Pl. 13). Refigured from Gygi (2000a, Fig. 2)

as it was conceived by Arkell and was then sanctioned by the Colloque du Jurassique, is now adopted worldwide. The congress decided on 4th August 1962 to make the following recommendations, which were printed in 1964 on p. 85 in the volume of the proceedings of the congress, Colloque du Jurassique Luxembourg 1962 (1964): The top of the Callovian Stage (and of the Middle Jurassic Series) is at the top of the ammonite zone of *Quenstedtoceras lamberti*. The base of the Oxfordian Stage (and of the Upper Jurassic Series) is at the base of the ammonite zone of *Quenstedtoceras mariae*. The top of the Oxfordian (in southern England) is at the top of the ammonite zone of *Ringsteadia pseudocordata*, and (in southern Germany) at the top of the ammonite zone of *Idoceras planula*. The base of the Kimmeridgian (in southern England) is at the base of the ammonite zone of *Pictonia baylei*, and (in southern Germany) at the base of the ammonite zone of *Sutneria platynota* (see Fig. 1.6).

Ziegler (1963, 1967) investigated ammonites and the composition of the associated macrofossil assemblages in sediments of the Late Jurassic in Europe. Ziegler arrived at the same conclusions as Gressly (1838–1841) and Scott (1940), that ammonites were facies-dependent, and that the mode of life of most of the ammonites was consequently necto-benthic. Ziegler (1967, Fig. 11) made broad estimates of the paleobathymetric ranges of distinct macrofossil assemblages in European sediments of the Late Jurassic.

Research by the author of this study on strata of Late Jurassic age in northern Switzerland began, after preliminary work since 1960, in the spring of 1962, after R. Trümpy at the Geological Institute of the Swiss Federal Institute of Technology ETH Zürich proposed to investigate such sediments in Canton Aargau for a Ph. D. thesis. The Geological Institute at Zürich is mutually run by the ETH and the University of Zürich. The author's Ph. D. thesis was accepted by the Philosophical Faculty II of the University

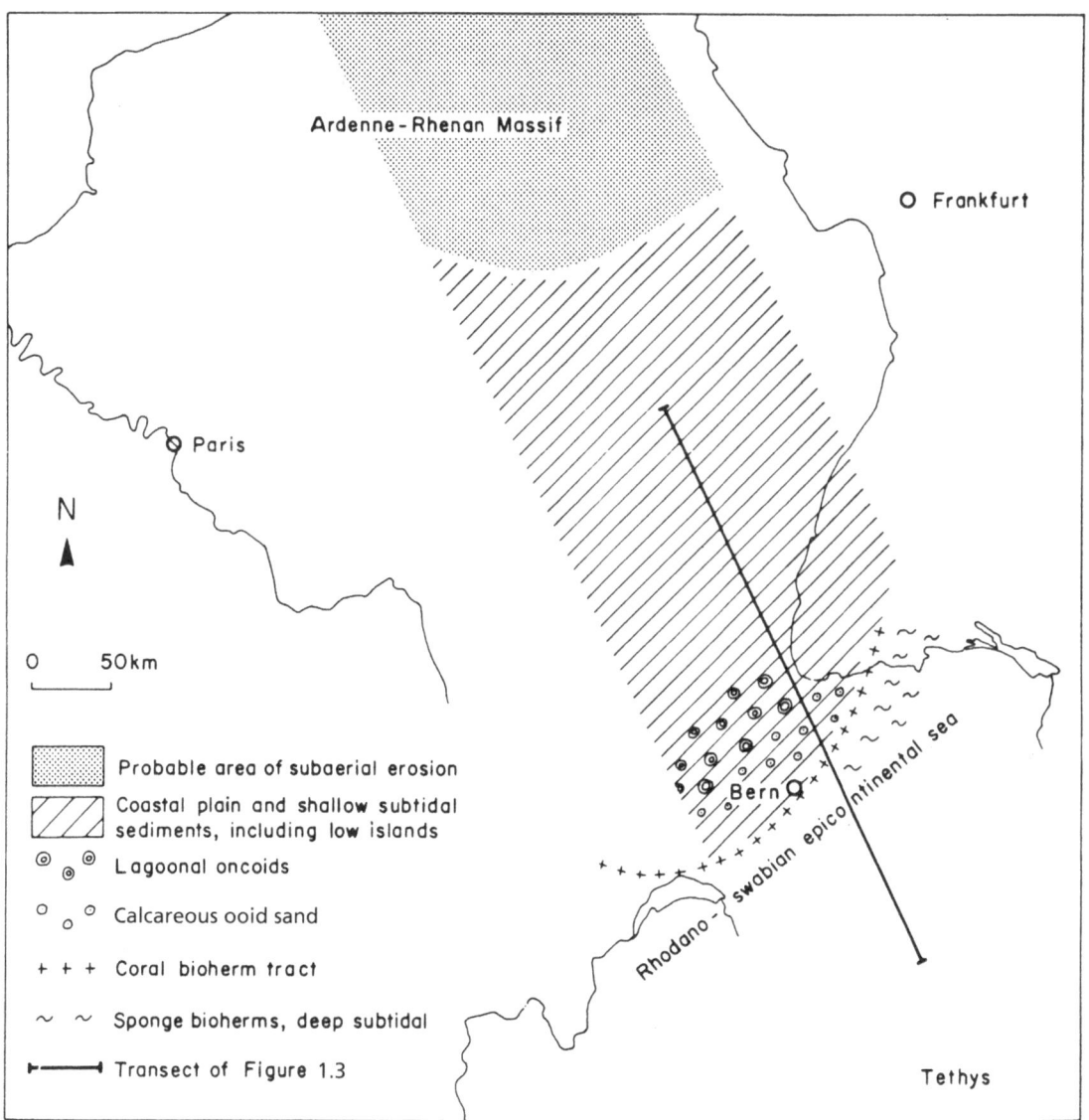

Fig. 1.2 Paleogeographic map of land in the north and of the Rhodano-Swabian Epicontinental Basin. Refigured from GYGI (1992, Fig. 1)

of Zürich on November 2, 1967, and it was published early in 1969. B. ZIEGLER, then at the Paleontological Institute of the University of Zürich, gave the author an introduction to biostratigraphic practice with ammonites. It was thanks to a recommendation by K. HSÜ at the ETH Zürich that the author could participate in the Organism to Sediment Seminar which was held by R. N. GINSBURG at the Bermuda Biological Station in 1968. The results of the author's investigation of coral reefs in Bermuda, which was begun during GINSBURG's seminar in 1968, were published in GYGI (1969b, 1970, 1975). Participation in the Third International Coral Reef Symposium at Miami, Florida, in May 1977 and in the field trips to the coral reefs in Florida, the Bahamas, and Belize, which were organized by the congress, was the author's opportunity to augment his knowledge of Recent

coral reefs and of shallow-marine sediments in the tropical West Atlantic. The author's wife SYLVIA helped to collect macrofossils mainly in systematic, bed by bed excavations from 1970 and prepared the major part of the well preserved ammonites that were incorporated into the author's collection in the Museum of Natural History Basel. She typed manuscripts and drew some of the figures in ink for publication. F. PERSOZ of the University of Neuchâtel proposed to the author to prepare a joint paper on mineral stratigraphy and ammonite biostratigraphy combined, which was published by GYGI and PERSOZ (1986). P. R. VAIL of Rice University at Houston, Texas, spent a sabbatical year at the Ecole des Mines at Paris. He invited the author to Paris in order to prepare a joint paper on sequence stratigraphy of sediments of the Late Jurassic in northern Switzerland.

Fig. 1.3 Cross-section from land in the north to the Rhodano-Swabian Epicontinental Basin and to the Helvetic facies belt. Refigured from GYGI (1992, Fig. 2). Inset enlarged as Fig. 1.5

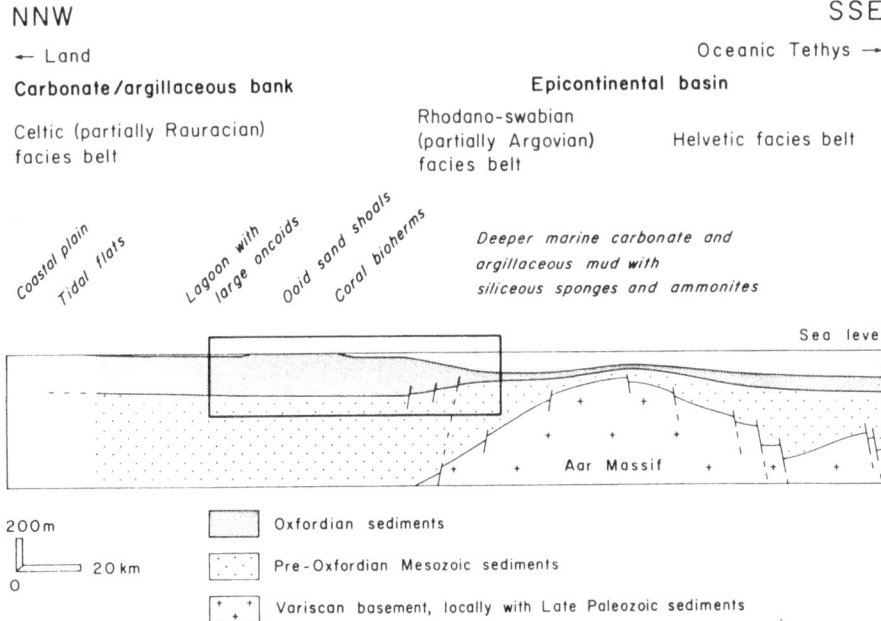

A. L. COE of the Open University at Milton Keynes, England, finished the part on sequence stratigraphy in this paper after P. R. VAIL fell very seriously ill at Paris. The paper was published by GYGI et al. (1998).

The author of this study could investigate sediments from the Bathonian Stage upward to the Kimmeridgian Stage in excellent outcrops, either natural or artificial, from 1960 to 1995. This study is based on 221 lithologic sections which were measured in detail, most of them in northern Switzerland. The remainder of sections was measured in adjacent France and in southern Germany. The location of all of the sections measured is shown in the map of Fig. 1.4. Each section has an individual number and is subdivided into numbered beds. A complete list of the sections with coordinates was published by GYGI (2000a, Tab. 1). The individual sections were first projected into the partial transects that are arranged perpendicularly to depositional strike and which are shown in the map of Fig. 1 in GYGI and PERSOZ (1986). Then the partial transects were assembled into the section that is represented in Fig. 1.5 of this study. This composite section is the result of the author's work mainly on lithostratigraphy (GYGI 1969a, 2000a; GYGI and PERSOZ 1986), on sedimentology of deep-water iron oolites and sea floor topography (GYGI 1981), calculations of water paleodepth and basement subsidence (GYGI 1986), and ammonite biochronology in GYGI (1977, 1990a, 1991b, 1995, 2000a, 2001, 2003).

The between 9,000 and 10,000 ammonites which were collected mainly from sediments in northern Switzerland were taken from *in situ* with a few exceptions. Cardioceratid ammonites typical of the Boreal Faunal Province are in northern Switzerland abundant in sediments of the early

Oxfordian. *Gregoryceras* are ammonites that first appeared in the *Cordatum* Chron, the last chron of the early Oxfordian. *Gregoryceras* were most abundant in low paleolatitudes of the Tethyan Faunal Province (see mainly BERT et al. 2009). *Gregoryceras* are important in intercontinental time correlation (GYGI and VON HILLEBRANDT 1991). Ammonites of the Subtethyan Faunal Province are abundant in northern Switzerland in the whole of the Oxfordian and in the Kimmeridgian Stage (Fig. 1.6). Time correlation within sediments laid down during the Late Jurassic in northern Switzerland could be made in detail thanks to ammonites, lithologic marker beds like the Knollen Bed, mineral stratigraphy, and sequence stratigraphy. The formations, members, and beds shown in Fig. 1.5 of this study are a frame of reference for ammonite biochronology. A curve of eustatic sea level variation in Oxfordian time was semiquantitatively calculated by GYGI (1986, Fig. 4). FISCHER and GYGI (1989) assigned their own radiometric ages to a choice of particularly well-preserved ammonites of the early and middle Oxfordian from northern Switzerland.

A selection of ammonite taxa which are essential in biochronology of the Late Jurassic in northern Switzerland was published by GYGI (2000a). The detailed succession of ammonite taxa in the Oxfordian and in the Kimmeridgian in northern Switzerland was documented with figured specimens in several papers by the author and by some co-authors, as well as in the two books by GYGI (2001, 2003). GYGI and MARCHAND (1982) figured numerous Cardioceratinae. Photographs of specimens of the successive genus *Amoeboceras* from northern Switzerland were published by ATROPS et al. (1993). The ammonite genus *Gregoryceras* was revised by GYGI (1977) based mainly on

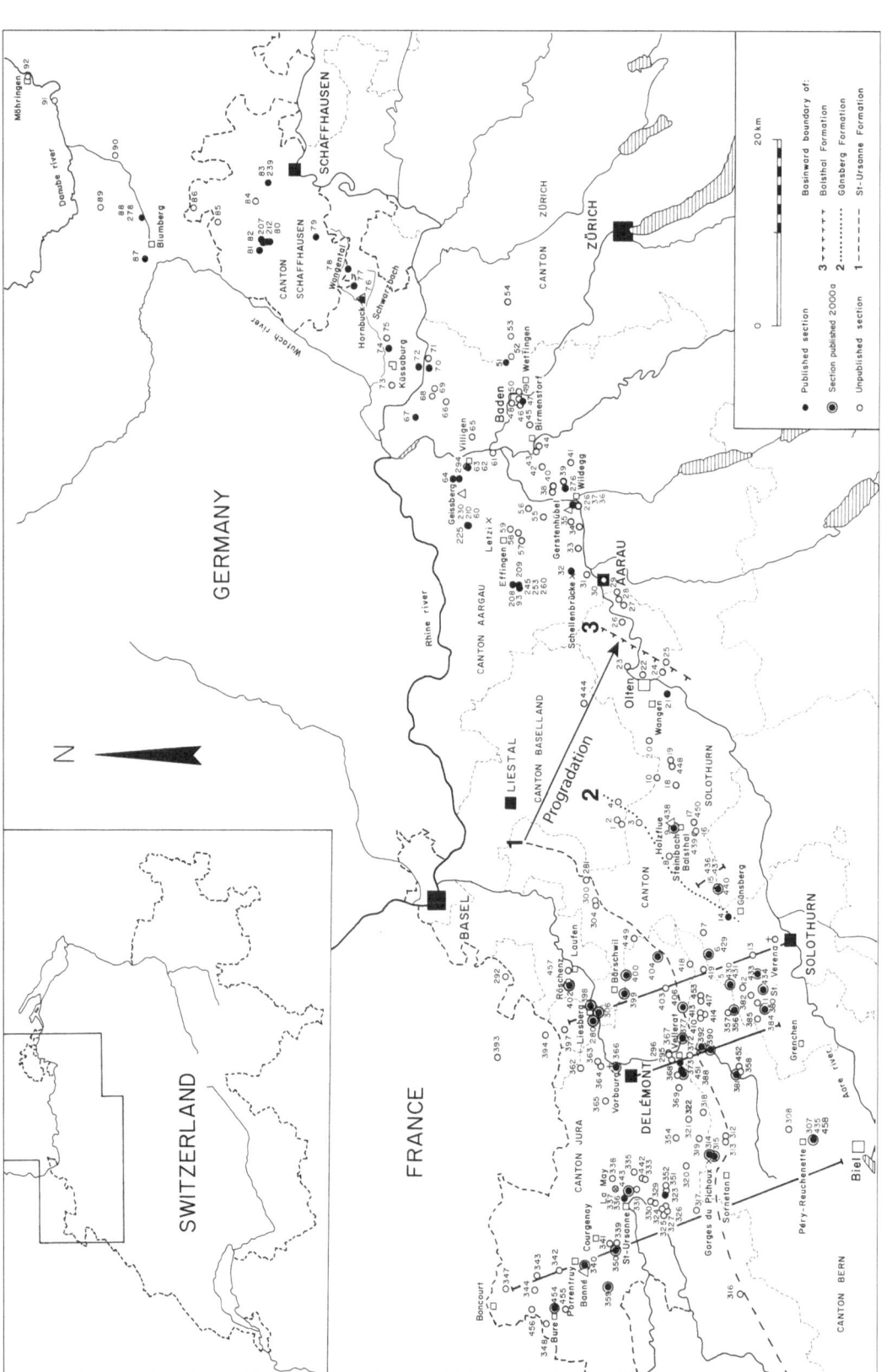

Fig. 1.4 Map with location and number of all of the sections which were measured in detail by the author between 1960 and 1995 mainly in northern Switzerland. The sections were projected into transects which are represented as bold, straight lines in the left part of the map. These transects and those further east, which are shown in the map of Fig. 1.5. Each of the transects runs as closely as possible in the direction of the maximal gradient of facies change, perpendicularly to depositional strike. Depositional strike east of Baden, in the eastern half of the map, runs approximately north-south (see map of Fig. 1 in GYGI et al. 1998). The belt of outcrops in that region runs from southwest to northeast, obliquely in relation to depositional strike. Accordingly, the short transects (unrepresented) which were drawn in that region and that were projected into Fig. 1.5 are en echelon. The curved, basinward margins of three subsequent carbonate platforms, no. 1 of the St-Ursanne Formation, no. 2 of the Günsberg Formation, and no. 3 of the Balsthal Formation (see inset in lower right of this map and Fig. 1.5) indicate the course of depositional strike. The direction in which the successive platform margins shifted is the direction of progradation. Refigured from Fig. 1 in GYGI (2000a).

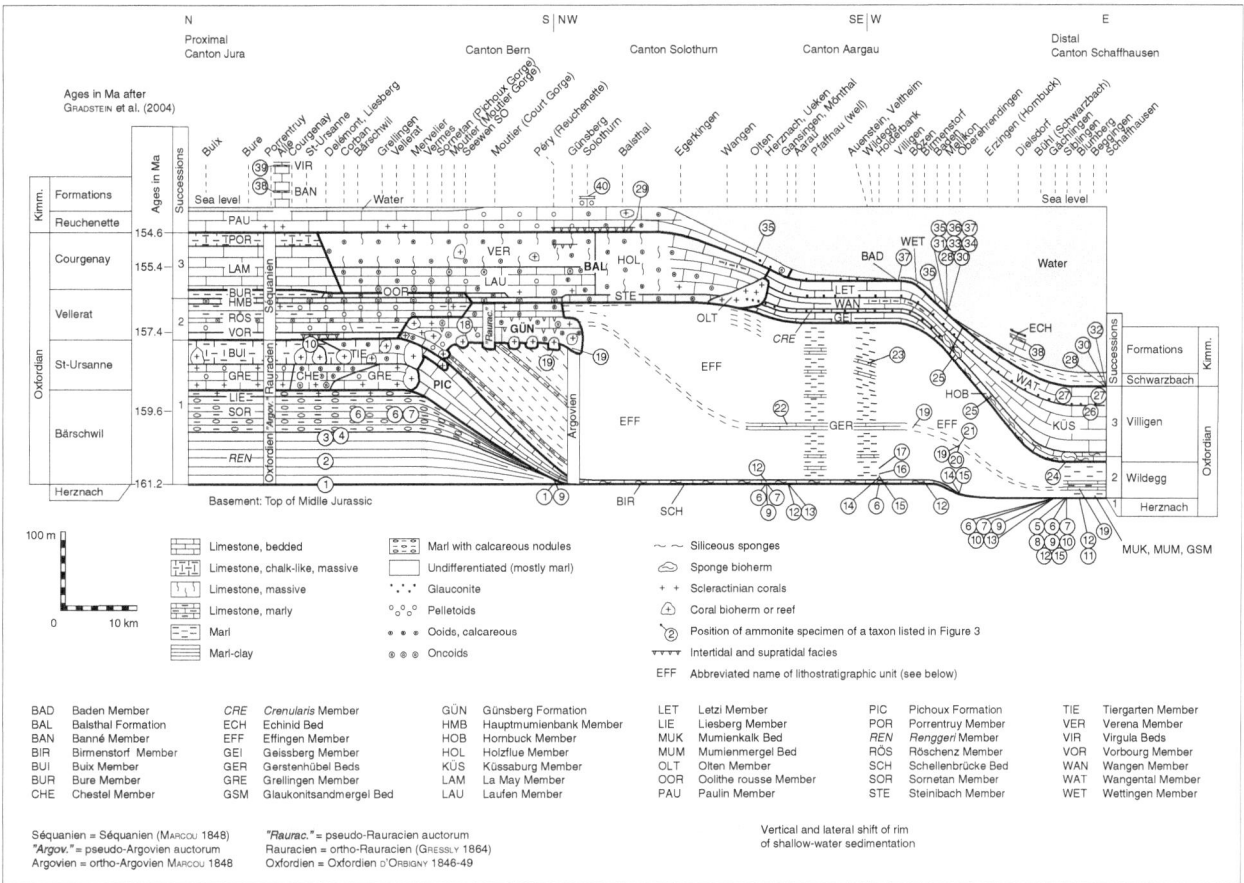

Fig. 1.5 Assembled, palinspastic cross-section through formations, members, and beds of the Upper Jurassic Series in northern Switzerland. Each of the three parts of the assembled section runs in a different direction which is indicated at the top. The direction of every one of the three parts of the section is arranged as closely as possible perpendicularly to the curved lines of depositional strike. Such lines are the basinward margins of the three successive carbonate platforms of the St-Ursanne Formation (1 in Fig. 1.4), Günsberg Formation (2 in Fig. 1.4), and Balsthal Formation (3 in Fig. 1.4), which are listed in the inset in the lower right of Fig. 1.4. Progradation of the base of the carbonate platforms of the Günsberg Formation (GÜN) and the Balsthal Formation (BAL) over the basinal Effingen Member (EFF) of the Wildegg Formation is outlined in this assembled cross-section by a gray band. The encircled numbers in the section mark the position of ammonites. The numbers encircled in this figure correspond to the numbers listed on the left side of Fig. 1.6. The numbered ammonites were collected from *in situ* with the following exceptions mentioned in the text. Ammonite no. 23, *Euaspidoceras hypselum* (OPPEL), and ammonite no. 24, *Epipeltoceras berrense* (FAVRE), were collected from talus below a known unit. Sediments in the upper part of the Kimmeridgian Stage are very incompletely preserved in most of northern Switzerland. The position of ammonites no. 38, 39, and 40 is therefore indicated in schematic columns near Alle in Canton Jura, near the city of Solothurn, and near Dielsdorf in northwestern Canton Zürich. The succession of ammonite chrons is shown in Fig. 1.6. The base of the assembled cross-section of Fig. 1.5 corresponds to the sea floor topography as it probably existed at the beginning of the Late Jurassic (compare with Fig. 4.15A). Thicknesses of formations and members shown in Fig. 1.5 were averaged from the corresponding thicknesses which were measured in the 221 sections shown in Fig. 1.4. The measured and averaged thicknesses of successive units

were then added up in Fig. 1.5, as if basement subsidence were equal everywhere in the investigated region during deposition of succession nos. 1–3. The water depth shown on the right hand side of Fig. 1.5 above the Schwarzbach Formation in Canton Schaffhausen is therefore much greater than it was in fact in the *Lussasense* and in the *Evolutum* Chron. The presumably real water depth at that time is represented in Fig. 4.15D. For the calculated history of sedimentation and differential basement subsidence see Fig. 4.15A-D. Lines delimiting formations and members as well as lines within these units in Fig. 1.5 intersect time planes. Note the extreme thinning of succession no. 1 from the proximal, left to the distal part of the section on the right hand side. The *vertical* scale in the lower left of the figure is exaggerated one hundredfold in order to visualize depositional slopes, which in fact were very gentle. Ages in million years are according to GRADSTEIN et al. (2004, p. 310). After GYGI (2003, Fig. 173), supplemented. Formations, members, and beds mentioned in Fig. 1.5 are described in the dictionary of lithostratigraphic units of Late Jurassic age in northern Switzerland by GYGI (2000b). Synonymy lists of these units were published by GYGI (2000a). The Steinibach Member in Fig. 1.5 was inappropriately spelled Steinebach Member in some of the earlier publications by the author. Steinibach, as the name is pronounced by the inhabitants of Balsthal, is the name of a creek, after which the member was named. The creek is shown, but its name is not recorded in sheet no. 1107 of the Landeskarte (Federal Map) at the scale of 1:25,000. The creek flows southward through the gorge north of the old church of Balsthal between Bisecht Hill on the western flank of the gorge and rocky Holzflue east of Steinibach Gorge. Bisecht Hill and the cliffs of Holzflue are visible in the left part of the upper photograph on Pl. 6 in ARKELL (1956). The position of Steinibach Gorge is indicated in the photograph of Fig. 4 on p. 84 in GYGI (1969a). An enlarged foldout of this Fig.1.5 is to be found at the end of the book.

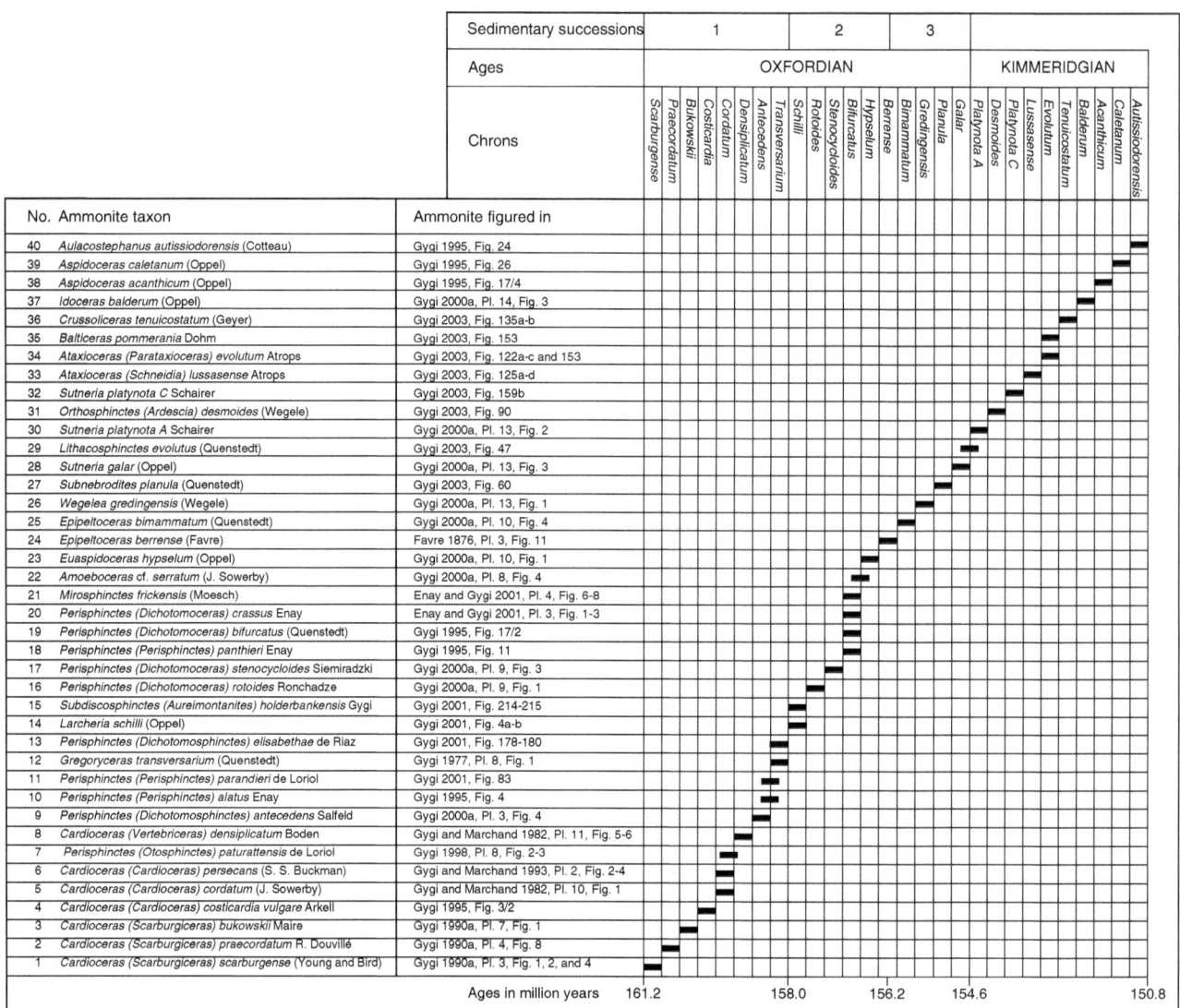

Fig. 1.6 Temporal succession of ammonite taxa which were *found* in northern Switzerland, and that are essential in biochronology of the Oxfordian and of the Kimmeridgian Age *sensu* GRADSTEIN et al. (2004). The chrons of the two ages or the ammonite *zones*, respectively, are as far as possible the same as the *subzones* in CARIOU and HANTZPERGUE, coord. (1997). Zonal indices which occur in France, but not in northern Switzerland, are replaced in this study by ammonite taxa which are shown in Figs. 7.1–7.6. The temporal succession of index taxa of new material found in northern Switzerland. The Swiss representatives of Euaspidoceratinae are figured in BONNOT and GYGI (1998, 2001). The *Paraspidoceras* which were published by GYGI et al. (1979) are unequivocal evidence that the pertinent ammonite taxa were rare, and some of them even very rare, but that these taxa probably represent perfectly distinct *species* in ammonites. *Glochiceras* as well as *Creniceras* and *Bukowskites* which were recently found in northern Switzerland were figured by GYGI (1991b). Photographs of early Oxfordian Perisphinctaceae were published in GYGI (1990a) and are shown in GYGI (1998).

ammonite chrons, which is listed in Fig. 1.6, does not represent an evolutive lineage. Therefore there may be some overlaps or gaps between the zones distinguished in this study. This is inevitable in a zonation, in which some zones are documented by a single index specimen. The numbers of taxa listed in this figure 1.6 correspond to the encircled numbers in Fig. 1.5. Ages in million years are according to GRADSTEIN et al. (2004, p. 310). Some of the ages shown in Fig. 1.6 are interpolated

Sediments of Late Jurassic age from the shallow-water realm in northwestern Switzerland could be dated with ammonites which were figured by GYGI (1995). An inventory and taxonomic description of the perisphinctacean ammonites of the type *Transversarium* Zone in northern Switzerland was made by GYGI (2001). ÉNAY and GYGI (2001) figured ammonites of the *Bifurcatus* Zone in northern Switzerland which are kept in the ROLLIER collection at the ETH Zürich. The Oxfordian perisphinctaceans from the *Berrense* Zone on upward and those of the earlier part of the Kimmeridgian were published by GYGI (2003). Ammonite ecology was investigated by

GYGI (1999). The diagnostic value of ammonites in sedimentary geology was stressed by GYGI (2003).

The boundary beds between the Middle and the Upper Jurassic Series were sedimented in the whole of northern Switzerland at a very low rate and in relatively deep water. The beds are thin and widespread iron oolites between Canton Jura and Canton Aargau. The iron oolites include a macrofossil assemblage with predominantly ammonites and with a well-represented benthos. Ammonites which were carefully excavated bed by bed from iron oolites of the earliest Oxfordian Age are evidence that some of these iron oolites are not stratigraphically condensed in spite of their slight thickness. The term stratigraphic condensation in the sense of how it is used in this study is explained below in Sect. 8.2. Iron oolitic beds of the earliest Oxfordian which are *not stratigraphically condensed* were excavated in section RG 280 near Liesberg in Canton Basel-Landschaft, in section RG 307 near Péry in Canton Bern, and in Canton Aargau in section RG 208 at Ueken north of Herznach. Section RG 208 at Ueken is refigured in this study in Figs. 4.14 and 8.2. A *stratigraphically condensed* iron oolite is bed no. 7 at the base of the Oxfordian in section RG 210 in the natural cleft called Eisengraben (in English: iron ditch) southwest of Gansingen in Canton Aargau. Section RG 210 is shown in Fig. 8.3. The condensed iron oolite of bed no. 7 in section RG 210 includes ammonites from the middle Callovian to the *Cordatum* Chron of the early Oxfordian, which are mixed in the same level. Thin iron oolites which are stratigraphically condensed, because ammonites of varying age occur side by side on the same level in the sediment, must therefore be carefully distinguished from equally thin iron oolites which are not stratigraphically condensed, because coeval ammonites within them occur on separate levels which could be distinguished.

MARCHAND (1979) introduced a *Paucicostatum* Horizon which he based on what he then interpreted to be the ammonite taxon *Cardioceras (Scarburgiceras) paucicostatum* LANGE. MARCHAND (1979) assigned his new horizon to the very base of the *Mariae* Zone in the lowermost Oxfordian. MARCHAND et al. (2000, Pl. 1, Figs. 2–4) re-interpreted the taxon *paucicostatum* LANGE to belong to the genus *Quenstedtoceras,* and therefore re-declared the *Paucicostatum* Horizon to be the uppermost part of the *Lamberti* Zone in the uppermost Callovian Stage. This was done based on ammonites which were collected in the iron mine at Herznach mainly during World War II. Section RG 208 at Ueken was excavated in 1971 a few hundred meters north of the Herznach mine. The lower, Callovian part of section RG 208 with beds no. 5 and 6 is represented in Fig. 4.14 of this study. The contiguous, upper part of section RG 208 with the Oxfordian beds no. 7a-9 is shown in this study in Fig. 8.2A–E. Every bed which was distinguished in

section RG 208 at Ueken can be correlated exactly with the corresponding section in the neighboring iron mine of Herznach, which was published for the first time by JEANNET (1951, Fig. 2). The bed by bed correlation between the Herznach mine and excavation RG 208 at Ueken is represented in Fig. 1 by MARCHAND et al. (2000). Two *Cardioceras scarburgense* (YOUNG and BIRD) from probably bed E2 in the Herznach mine were figured by MARCHAND et al. (2000, Pl. 1, Figs. 11 and 12). The base of the Oxfordian Stage in section RG 208 in Canton Aargau is therefore at the base of bed no. 7 in Fig. 8.2 in this study.

The base of the Oxfordian Stage is in iron oolitic facies as well in the two very similar boundary sections RG 307 in the quarry of La Charuque at Péry (GYGI 1990a, Fig. 4) and RG 280 in the clay pit of Andil at Liesberg (compare Pl. 22 with Pl. 30 in GYGI 2000a). Bed no. 20 in section RG 307 at Péry is brownish, clayey marl with a thickness of 25 cm. The iron ooids fade away completely towards the top of the bed. *Quenstedtoceras* of the *lamberti* group and *Quenstedtoceras leachi* (J. SOWERBY) occur in the lowermost part of the bed, and *Cardioceras (Scarburgiceras)* cf. *scarburgense* was found near the top. The boundary beds between the Callovian and the Oxfordian Stage near Liesberg are shown in an enlarged part of section RG 280 in GYGI (1990a, Fig. 2). Only *Quenstedtoceras lamberti* (J. SOWERBY) and *Quenstedtoceras paucicostatum* LANGE were found in the lower part of the iron oolitic clayey marl of bed no. 6 in this section. Bed no. 6 of section RG 280 is the time equivalent of bed no. 20 in section RG 307 at Péry. Both ammonite specimens mentioned above from bed no. 6 in section RG 280 were figured by GYGI (1990a, Pl. 1, Fig. 1, and Pl. 3, Fig. 14). The Callovian/Oxfordian boundary in section RG 280 near Liesberg is therefore within bed no. 6, and in section RG 307 near Péry within bed no. 20. It can be read from Fig. 2 in GYGI (1990a), that well-preserved *Cardioceras (Scarburgiceras) scarburgense* (YOUNG and BIRD) first appear in section RG 280 about 1.2 m above the base of the blue-gray clayey marl of the *Renggeri* Member. According to Fig. 3 in the same paper, the first *Creniceras renggeri* (OPPEL) coexist in section RG 280 with the last *Cardioceras (Scarburgiceras) scarburgense.* The specific name *renggeri* was by modern standards insufficiently defined by OPPEL, but it was widely used since OPPEL's time. This is why GYGI (1991a) proposed to validate the name. The name of the *Renggeri* Member refers to the ammonite taxon *Creniceras renggeri* which is abundant in the lowermost part of the member. Several well-preserved specimens of *Cardioceras (Scarburgiceras) scarburgense* and of *Creniceras renggeri,* which were collected from *in situ* in the lower *Renggeri* Member at Liesberg, were figured on the photographic plates in GYGI (1990a). Ammonites are evidence that the boundary beds between the Middle and the

Upper Jurassic Series are *not* stratigraphically condensed in section RG 307 at Péry in Canton Bern and in section RG 208 at Ueken in Canton Aargau.

Sedimentation at a normal rate and in ammonite facies began in the deep subtidal zone in northwestern Switzerland in the lowermost part of the proximal, thick part of succession no. 1 above the thin, iron-oolitic bed at the base of the succession. Normal sedimentation continued in this area until the latest part of the *Cordatum* Chron. At this time a thin, very fossiliferous bed was sedimented in the distal part of the Sornetan Member. This marker bed separates the lower half from the upper half of the Sornetan Member everywhere in the pertinent paleogeographic position. The ammonites from this fossil bed were figured by GYGI and MARCHAND (1993). Sedimentation at a normal rate resumed in this paleogeographic position during the following *Densiplicatum* Chron and continued to the end of the *Transversarium* Chron, when sedimentation of succession no. 1 was completed. The strata above the marker bed in the middle of the distal Sornetan Member were sedimented in water which was so shallow, that ammonites were few during sedimentation of the uppermost part of the Bärschwil Formation. Ammonites are rare or absent in most of the St-Ursanne Formation above. Sedimentation in ammonite facies and at a normal rate began in the relatively deep water of the adjacent epicontinental basin, in the area between eastern Canton Solothurn and northern Canton Aargau, in the *Transversarium* Chron. Succession no. 2 in eastern Canton Solothurn filled in the basin up to sea level. Ammonites therefore do not occur in the upper part of succession no. 2 in that region. A complete ammonite succession in sedimentary succession no. 2 exists in Canton Aargau. The ammonite succession in depositional succession no. 3 is complete in Canton Schaffhausen.

The boundary beds between the Oxfordian and the Kimmeridgian Stage in the sense of the International Congress "Colloque du Jurassique" at Luxembourg in 1962 were excavated in 1974 as section RG 239 in an old quarry which is easily accessible beside the road leading from the city of Schaffhausen to the adjacent village of Hemmental. The beds excavated in that quarry encompass the uppermost *Planula* Zone, the *Galar* Zone, and probably the entire *Platynota* Zone A–C (Fig. 1.6). Section RG 239 was for the first time published informally by GYGI (1990b) on p. 69 of the guidebook for the 2nd Oxfordian Working Group Meeting, which was held in Switzerland from September 10–15, 1990. A formal publication of this section is Fig. 160 on p. 140 in GYGI (2003). More than 1,000 ammonite specimens (see GYGI 1986, Fig. 6B) were excavated bed by bed at this site near Schaffhausen, first by GYGI (1969a, section RG 83 on Pl. 16), and then by R. and S. GYGI in 1974. Most of these ammonites are prepared. Out of these, only a very well-preserved *Sutneria galar* (OPPEL) and an equally well-preserved *Sutneria platynota* (REINECKE), morphotype A SCHAIRER, both specimens with peristome and lappets, are shown in this study in Fig. 7.3E, F at the scale ×2. The youngest specimen of *Sutneria galar* was found in bed no. 20 of section RG 239. The first *Sutneria platynota* appears in the same bed of that section. This does not prove that the vertical ranges of the two successive taxa overlap. The conclusion is more probable that *Sutneria galar* evolved to the morphologically clearly distinct *Sutneria platynota* geologically speaking instantaneously, and that the vertical ranges of the two easily distinguishable taxa touch each other within bed no. 20. This is an indication that evolution in ammonites was not always gradual as it was in most cases, like for instance in *Gregoryceras* (GYGI 1977), or in *Epipeltoceras* (see below). The base of the Kimmeridgian Stage in northern Switzerland, as it was conceived at the latest since HEIM (1919), can be defined biochronologically with precision. The base of the stage at the base of the Platynota A Zone is in section RG 239 *within* bed no. 20. The stage boundary thus defined is in agreement with the decision made by the International Congress COLLOQUE DU JURASSIQUE LUXEMBOURG 1962 (1964), which was mentioned above.

The biochronologically well-defined boundary between the vertical ranges of *Sutneria galar* (OPPEL) and *Sutneria platynota* (REINECKE), morphotype A SCHAIRER, approximately coincides in Canton Schaffhausen with the mappable, lithologic boundary between the limestone of the Villigen Formation below and the predominantly marly Schwarzbach Formation above. The same biochronologic boundary coincides in section RG 70 at Mellikon in Canton Aargau with the lithologic boundary between the Villigen Formation below and the Baden Member above, according to *Sutneria galar* (OPPEL) from bed no. 114, and *Sutneria platynota* (REINECKE), morphotype A SCHAIRER, from bed no. 120 of the section (GYGI 1969a, Pl. 17). The pertinent ammonites from Mellikon were figured by GYGI (2003, Fig. 64 on p. 64 and 159a on p. 139, respectively). The only locality in Canton Aargau which is accessible to the public, and where the uppermost Letzi Member, the Baden Member, and the bedded lower part as well as the massive upper part of the Wettingen Member crop out, is along the footpath 150 m west of the ruin of Besserstein castle above the village of Villigen. This is the uppermost part of section RG 62 which is figured in GYGI (1969a, Pl. 17).

Correlation L by PERSOZ in GYGI and PERSOZ (1986, Fig. 10 and Pl. 1B) indicates a minimum of kaolinite content. This correlation was first calibrated biochronologically in section RG 70 at Mellikon, where the lowermost, glauconitic Baden Member was sedimented at a very low rate. The lowermost part of the member in section RG 70 is probably incomplete, because *Sutneria platynota* (REINECKE), morphotype A SCHAIRER was found only in bed no. 120, and no *Sutneria platynota* (REINECKE), morphotype B or C SCHAIRER were found in beds no. 121–124, in which other ammonites are abundant. Correlation L coincides in the type section of the Reuchenette Formation in the fraction of La Reuchenette of the township of Péry in Canton Bern, with the intertidal stromatolite in bed no. 236 of section RG 307 (GYGI 2000a, Pl. 22). The photograph of a polished cross-section of the stromatolite is shown in this study in Fig. 5.8. The stromatolite is 3.5 m above the base of the Reuchenette Formation. The formation's base is in section RG 307 at the base of bed no. 235. Another intertidal stromatolite is 3.5 m above the base of the Reuchenette Formation in bed no. 20 of section RG 440 east of the summit of Mt. Rüttelhorn north of the village of Rumisberg in Canton Bern (GYGI 2000a, Pl. 43).

The type section of the Balsthal Formation is RG 438 in Steinibach Gorge north of the village of Balsthal (GYGI 2000a, Pl. 44). The section is 8 km east-northeast of section RG 440 north of Rumisberg. The probable time equivalent of the intertidal stromatolite in section RG 307 at Péry and in RG 440 at Rumisberg is at Balsthal in section RG 438 the paleosol of bed no. 52 above a subaerial erosion surface shown in Fig. 5.9. Time equivalence of the paleosol at Balsthal with the stromatolites further west is indicated by the ammonite *Lithacosphinctes evolutus* (QUENSTEDT) which was found by B. MARTIN 2.6 m below the paleosol in what is now bed no. 9 of the unpublished section RG 439 at Innere Klus near Balsthal (MARTIN 1984). The ammonite was identified by F. ATROPS (Lyon, France), and it was figured by GYGI (1995, Fig. 19). ATROPS assigned the specimen to the earliest *Platynota* Chron (GYGI 1995, p. 42). The base of the *Platynota* A Zone in the sections near Balsthal RG 438 in Steinibach Gorge, RG 439 at Innere Klus, and RG 450 at Chluser Roggen, is therefore at least 2.6 m below the paleosol. Provided that the paleosol at Balsthal is coeval with the stromatolite 3.5 m above the base of the Reuchenette Formation in section RG 440 at Mt. Rüttelhorn 8 km to the west-southwest, and coeval with the stromatolite 3.5 m above the base of the Reuchenette Formation in the type section of the formation at Péry, then the boundary between the Balsthal Formation and the Reuchenette Formation at Péry is probably coeval with the boundary between the Villigen Formation and the Baden Member at Mellikon in Canton Aargau, which is dated with ammonites. The boundary between the Courgenay Formation and the Reuchenette Formation in Ajoie region of Canton Jura is coeval as well, according to Fig. 10 in GYGI and PERSOZ (1986).

The boundary between the massive Verena Member in the upper Balsthal Formation below and the thickly-bedded, lowermost member of the Reuchenette Formation above is conspicuous in many sections, even when seen from a distance. The formation boundary is mappable. The lower-most, well-bedded member in the Reuchenette Formation has a thickness of between 40 and 50 m and is now called Paulin Member (see below). The best place to see the

Fig. 1.7 Rock of Rouge Pertuis north of the village of Undervelier in Canton Jura, as seen from the west. Massive calcareous oolite of the Verena Member in the upper Balsthal Formation is on the left. On the right hand side is well-bedded, mostly micritic limestone of the Paulin Member in the lower Reuchenette Formation. The strata are in tectoni-cally vertical position. Refigured from GYGI (2000a, Fig. 38 on p. 65)

formation boundary is the road beside the dam of Sorne River north of the village of Undervelier in Canton Jura, south of locality Les Forges. The soaring rock of Rouge Pertuis east of Sorne River is a prominent landmark above the dam. The massive Verena Member and the thickly-bedded Paulin Member in the rock are in tectonically verti-cal position. The formation boundary between the members is clearly visible in the rock when viewed from the road. This view is shown in the photograph of Fig. 1.7. The *coeval* boundary near Balsthal is atypical and cannot be discerned in the massive calcareous oolite of bed no. 49 below the paleosol in the type section of the Balsthal Formation (RG 438, Pl. 44 in GYGI 2000a). The type section of the forma-tion was originally and inappropriately chosen in Steinibach Gorge north of Balsthal by GYGI (1969a, p. 83 and Pl. 18). The Balsthal Formation has, therefore, to be re-defined in this study in the following way. It extends in section RG 307 at Péry from the base of the very fossiliferous bed no. 197 to the top of bed no. 234 (GYGI 2000a, Pl. 22). The Balsthal Formation has to be corrected in section RG 315 in Pichoux Gorge near Sornetan, in Pl. 21 in GYGI (2000a). The top of the Balsthal Formation in section RG 315 is at the top of bed no. 128. The erosion surface at the base of the paleosol at Balsthal is now recommended to be used in the region around Balsthal, and only there, as mark of the base of the Reuchenette Formation. Provided that the paleosol near

Balsthal (Fig. 5.9) and the stromatolite in bed no. 236 in section RG 307 near Péry (Fig. 5.8) are coeval, the base of the Reuchenette Formation near Balsthal would be somewhat younger than the base of the Reuchenette Formation in its type section at Péry.

The Paulin Member in the lowermost Reuchenette Formation is in the northern folds of the Jura Mountains and in Ajoie region of Canton Jura a thickly-bedded succession of mostly micritic limestone between the Verena or the coeval Porrentruy Member below and the Banné Member above, where the Banné Member occurs in Ajoie region (Fig. 1.7, and GYGI 2000a, Pl. 19). The Paulin Member was for the first time distinguished and named "Couches du Vorbourg" by GREPPIN (1893, p. 16). GREPPIN named his member after the tower of a former castle called Vorbourg with an adjacent chapel northeast of Delémont in Canton Jura. The well-individualized member named by GREPPIN with its distinct, thick bedding was not adopted by subsequent authors, probably because Delémont is far from the village of Oberbuchsiten in Canton Solothurn where the strata are that GREPPIN (1893) dealt with. A different Vorbourg limestone unit was named by ZIEGLER (1962, p. 21), without mention and therefore probably without knowledge of GREPPIN (1893). ZIEGLER's Vorbourg Member crops out along the road leading from Delémont to Vorbourg, in the same section as GREPPIN's Couches du Vorbourg, but ZIEGLER's unit encompasses much older strata than those which were originally named by GREPPIN. A revised version of ZIEGLER's unit is used in mapping since 1962 and is named Vorbourg Member. The excellent section along the road beside Vorbourg Tower was published in its full length by GYGI (2000a, Pl. 23). GYGI (2000b, p. 142) drew attention to the name of GREPPIN's member which is now a forgotten term. The Paulin Member in the lowermost Reuchenette Formation was well exposed and almost complete in the quarry of La Rasse south of Porrentruy when it was measured by the author on the rope as section RG 340 in 1981. The section is shown in Pl. 17 in GYGI (2000a). JANK et al. (2006) re-drew the section in their Fig. 10 and published a photograph of it in Fig. 12. The member is accessible on foot and crops out almost completely along the forest road called Chemin paulin on the territory of the township of Courgenay in Canton Jura.

Time-equivalent with the oolitic, massive Verena Member is in landward, northwestern direction the equally massive, mostly micritic Porrentruy Member. The Porrentruy Member was measured in the three sections RG 443, an exploration well near St-Ursanne, RG 350 along Chemin paulin road near Courgenay, and RG 340 in the quarry at La Rasse south of Porrentruy. The three sections are all in Canton Jura. They are shown in Pl. 17, 19, and 20 in GYGI (2000a). The lowermost, well-bedded succession of mainly micritic limestone of the Reuchenette Formation is in Ajoie region of Canton Jura between the Porrentruy Member below and the Banné Member above. The forest road called **Chemin paulin** is the only locality in the region of Ajoie where the whole thickness of this succession is at the present time accessible on foot. It is therefore proposed here that beds no. 80–155 of section RG 350 in GYGI (2000a, Pl. 19), beginning 4 km southwest of the village of Courgenay at coordinates 573,920/247,400, be named **Paulin Member**, and to designate section RG 350 to be the type section of the member. The base of the member at the boundary between the Courgenay Formation and the Reuchenette Formation is near St-Ursanne and in Ajoie region just as clear-cut as it is in the Rock of Rouge Pertuis north of Undervelier which is shown in Fig. 1.7. The Paulin Member is entirely exposed in section RG 315 in Pichoux Gorge at Sornetan, beginning at the base of bed no. 129 and extending upward to the top of bed no. 153 (GYGI 2000a, Pl. 21). The best and complete section of the Paulin Member was provided by the exploration well which was drilled beside the farm La Coperie near St-Ursanne. The well log was measured by the author, and it is represented as section RG 443 in Pl. 20 in GYGI (2000a). The Paulin Member encompasses in this section beds no. 66–124.

No ammonites were found in the Paulin Member, probably because it was sedimented at a water depth of less than 10 m. The base of the member is according to the mineral stratigraphic correlation L of PERSOZ in GYGI and PERSOZ (1986, Fig. 10 and Pl. 1B) coeval with the base of the *Platynota* A Zone and thereby with the base of the Kimmeridgian Stage in the sense of Swiss geologists following HEIM (1919). The ammonite *Aspidoceras* cf. *acanthicum* (OPPEL) was found by A. and H. ZBINDEN in the Banné Member in the quarry at L'Alombre aux Vaches 1 km to the northeast of section RG 350. The specimen is figured in GYGI (1995, Fig. 17/4). The Paulin Member consequently includes, according to Fig. 1.6 in this study, the seven ammonite zones from the base of the *Platynota* A Zone to the top of the *Balderum* Zone.

It was stated above that the base of the Kimmeridgian Stage in northern Switzerland at the base of the *Reineckianus* Beds in the sense of HEIM (1919, in the stratigraphic table of the Upper Jurassic following p. 506) or at the base of the modern *Platynota* Zone A–C is in agreement with the recommendation by the International Congress COLLOQUE DU JURASSIQUE LUXEMBOURG 1962 (1964, p. 85). The boundary between the Oxfordian and the Kimmeridgian Stage in northern Switzerland can be biochronologically defined with precision according to figured, very well-preserved ammonites. The boundary thus defined coincides with a vertical change in lithology which is conspicuous in many sections and can be correlated from eastern France through northern Switzerland to southern Germany. The lithologic boundary was caused by a change from a relatively dry climate at the end of sedimentation of the Balsthal Formation to a time with more rainfall. Time correlation of the Oxfordian/Kimmeridgian boundary between England and Switzerland is difficult, primarily because there are very few ammonites in what ARKELL (1956, p. 22) called "uppermost Corallian Beds" in southern England, where the city of Oxford and the village of Kimmeridge are located. The base of the Kimmeridge Clay is, according to ARKELL (1947, Fig. 16), near Oxford above a non-sequence or an unconformity. The stage boundary cannot be biochronologically defined with accuracy as well near Kimmeridge on the British south coast for lack of ammonites at that locality.

OPPEL (1863, p. 201) named the ammonite taxon *Amoeboceras bauhini* and stated that it is coeval with *Taramelliceras hauffianum* (OPPEL). According to SCHWEIGERT and CALLOMON (1997, p. 30), the vertical range of *Amoeboceras bauhini* (OPPEL) begins in southern Germany in a glauconitic bed. This bed is locally very fossiliferous, and it is the time equivalent and the lateral continuation of the glauconitic Knollen Bed of northern Switzerland. The Knollen Bed was traced by the author from Schönenwerd, a village west of the city of Aarau in northern Switzerland, to the hill of Ortenberg near the village of Deilingen in southern Germany. Deilingen is 12 km south-southwest of Balingen, and it is 110 km to the northeast of Schönenwerd. The Knollen Bed is therefore an excellent marker bed in the uppermost *Gredingensis* Zone. The bed could probably be followed in southern Germany beyond Deilingen.

The question whether the first appearance of *Sutneria platynota* (REINECKE), morphotype A SCHAIRER in northern Switzerland is contemporaneous with the first appearance of *Pictonia baylei* in England cannot be answered directly because of the pronounced provinciality of the ammonite populations in the pertinent time span. The subboreal ammonite genus *Ringsteadia* of the Late Oxfordian in southern England is also well represented in northern Switzerland. This was documented for the first time by MOESCH (1867, Pl. 2, Fig. 1), and again by GYGI (2000a, 2003). The youngest specimen of *Ringsteadia* which was found to date in Central Europe is *Ringsteadia suebica* GYGI from the glauconitic Knollen Bed near Spaichingen in southern Germany. This taxon was described to be new on p. 96 and figured on Pl. 12, Fig. 1 by GYGI (2000a). No taxa of *Ringsteadia* occur both in England and Central Europe. Ammonites of the genus *Pictonia* were not found to date in northern Switzerland. Provided that SCHWEIGERT and CALLOMON (1997) are right, the Oxfordian/Kimmeridgian boundary as it is now drawn in southern England would be considerably older than the boundary between the *Galar* and the *Platynota* Zone in northern Switzerland.

The youngest specimen of *Amoeboceras bauhini* (OPPEL) which was found in northern Switzerland is from bed no. 9 in section RG 239 at locality Summerhalde near Schaffhausen (ATROPS et al. 1993, Pl. 1, Fig. 19). Bed no. 9 of section RG

239 is in the middle of the *Galar* Zone (GYGI 2003, Fig. 160). CALLOMON in SCHWEIGERT and CALLOMON (1997, Fig. 9) stated to have found an *Amoeboceras bauhini* (OPPEL) in a section at South Ferriby in Lincolnshire, England, in the lower *Baylei* Zone. This confirms the opinion of ATROPS et al. (1993, p. 224), that the base of the Kimmeridgian Stage in England is approximately time-equivalent with the middle or lower *Planula* Zone in Central Europe. Nevertheless, the conventional and mappable, possibly younger base of the Kimmeridgian Stage at the base of the *Platynota* A Zone (Fig. 1.6) is retained in this study for the time being, as long as time correlation of the Oxfordian/Kimmeridgian boundary between *southern* England and northern Switzerland cannot be substantially improved.

A pragmatic and above all a practicable solution would be to define the boundary between the Oxfordian and the Kimmeridgian Stage not in the British Isles, but rather in a section in the Subtethyan Faunal Province. Section RG 239 at locality Summerhalde beside the road leading from the city of Schaffhausen to the village of Hemmental would be a suitable boundary type section with the approximately 1,000 ammonites which were excavated bed by bed from *in situ* in this easily accessible section. All of these ammonites are kept in the Museum of Natural History at Basel, Switzerland. Most of them are prepared. The numerous *Amoeboceras* from section RG 239 could be used in correlation with the Boreal Faunal Province, and *Sutneria* in correlation with the Tethyan Faunal Province. Redefinition of this boundary in a remote section in the Boreal Faunal Province on the Isle of Skye in northern Scotland as it is now aimed at therefore appears to be unrealistic and unnecessary.

The boundary sections at the base of the Oxfordian and of the Kimmeridgian Stage in northern Switzerland can be summarized as follows. The boundary between the Middle and the Upper Jurassic Series and with it the base of the Oxfordian Stage was excavated in northern Switzerland at several localities, mainly in the large excavation RG 208 near Ueken in Canton Aargau which is figured in GYGI (2000a, Fig. 5). The boundary beds in section RG 208 are figured in this study in Fig. 4.14 and in contiguous Fig. 8.2. The Callovian/Oxfordian boundary could be fixed in section RG 208 according to ammonites which were figured by GYGI and MARCHAND (1982) as well as by MARCHAND et al. (2000). The base of the Oxfordian Stage in excavation RG 208 is at the base of bed no. 7 in stratigraphically *uncondensed* iron oolite. The boundary between the Oxfordian and the Kimmeridgian Stage was excavated in the large excavation RG 239 near the city of Schaffhausen. The boundary section is figured in GYGI (2003, Fig. 160 on p. 140). The base of the Kimmeridgian Stage is within bed no. 20 of section RG 239. The section is documented with a rich and well preserved ammonite assemblage. As stated above, *Sutneria galar*

(Oppel) and *Sutneria platynota* (Reinecke), morphotype A Schairer are shown in this study in Fig. 7.3E, F.

European ammonite specialists as well as geologists from European countries, USA, and Canada gave the author valuable advice and help, either when visiting sections in the field, or when being seen through the author's collection, as co-authors, or as reviewers of manuscripts. The uninterrupted succession of ammonite chrons represented in Fig. 1.6 of this study could only be documented, because several private collectors gave the author specimens of rare ammonite taxa which they had found as a present. The private collectors thereby filled in the gaps which remained in the author's collection in spite of the great amount of time spent in collecting ammonites, some from existing outcrops, but most of them from systematic, bed by bed excavations. A photograph made at an early stage of the largest of these excavations is shown as Fig. 5 in Gygi (2000a). The author's collection at Basel includes between 9,000 and 10,000 ammonites. More than 6,000 specimens out of these are prepared. The result of the joint effort, which was supplemented by contributions of many more persons who cannot all be mentioned here, is that several processes of sedimentary geology going on during the Late Jurassic in what is now the Swiss Jura Mountains could be documented in detail.

1.3 Obsolete Stage Names in Modern Geological Maps of the Jura Mountains

Some remarks must be made on the names of obsolete stages which are now, following the decision of the International Congress Colloque du Jurassique Luxembourg 1962 (1964), integrated into the modern Oxfordian Stage (see below). The position of the obsolete stages within the modern Oxfordian Stage is shown in a column in the left part of the assembled section of Fig. 1.5 in this study. The obsolete stage names cannot be passed over in silence even after the several decades that elapsed since the Luxembourg congress, because the stage names are still represented in geological maps which are now for sale, as well as in the younger literature. The stage names Oxfordien, Argovien, and Séquanien were coined in the French part, and Rauracien in the Swiss part of the Jura Mountains, from which the geological term Jurassic is derived. When geological maps of the Jura Mountains are being used, the facts which became evident from the work by Gygi and Persoz (1986) should be noted.

The Étage Oxfordien of D'Orbigny (1842–1849, p. 608) comprised the entire Bärschwil Formation of modern Swiss geologists. Subsequent French stratigraphers assigned to their Oxfordien only the time equivalent of the *Renggeri* Member of Swiss geologists. The "groupe Argovien" of Marcou (1848) was, according to Magné and Mascle

(1964), defined by Marcou in a succession which is close to the village of Andelot-en-Montagne near Salins, Département Jura, in the French part of the Jura Mountains. The paper by Magné and Mascle (1964) was read before the International Congress "Colloque du Jurassique" at Luxembourg in 1962, but it did not gain the audience's due attention. Marcou's "groupe" was after its publication in 1848 interpreted to be a stage. The stage name Argovien refers to Canton Aargau in northern Switzerland, where Marcou probably has seen the biostromes of siliceous sponges in the thin succession which was later named Birmensdorf Member by Moesch (1863), and that is now spelled Birmenstorf Member. The thin succession of biostromes of silicious sponges at the base of Marcou's type Argovien near Andelot in France has the same facies and a similar thickness as the Birmenstorf Member in Canton Aargau. It is now documented with ammonites that the sponge biostromes in the lowermost part of Marcou's Argovien in France are coeval with those in the Birmenstorf Member in Switzerland.

Subsequent authors were bewildered by the ambiguous text in Marcou (1848), where Marcou explained why he named his new "groupe" in France after a region in Switzerland. Stages are normally named after a township close to the type section of the stage or, unlike the Argovien, after the region in which the stage was defined. Most authors after Marcou's time, except Dreyfuss et al. (1964) and Magné and Mascle (1964), were therefore of the opinion that the Argovien was a stage that had been defined in Switzerland rather than in France. This is why Marcou's Argovien was omitted in the compilation of stages that were defined in France, in a book which was edited by Cavelier and Roger (1980). Above the biostromes with siliceous sponges with a total thickness of 2.5 m near Andelot in France, which R. Énay showed to the author in 1964, is a thick succession of blue-gray marl with intercalations of marly limestone. The time equivalent of this marly succession with few macrofossils near Andelot in France is in Switzerland *in the corresponding paleogeographic position* near Günsberg in Canton Solothurn the Effingen Member with few macrofossils above the Birmenstorf Member. According to Magné and Mascle (1964, Fig. 3), the thickness of the marly succession is roughly 115 m near Andelot, and according to Gygi (1969a, Pl. 18, section RG 14) about 110 m near Günsberg. Above the Effingen Member near Günsberg is the predominantly carbonate Günsberg Formation with small coral bioherms in its lower part (labeled "GÜN" within the assembled section of Fig. 1.5 in this study). The coral bioherms in the lower Günsberg Formation at the type locality of the formation, at the head of the western part of the landslide called Gschlief north of the village of Günsberg, can be perceived in the left part of the photograph of Fig. 3 on p. 79 in Gygi (1969a). A limestone formation with hermatypic corals similar to the Günsberg Formation is above MARCOU's type Argovien in France. This limestone with corals in France was erroneously called Rauracien by DREYFUSS et al. (1964, p. 305).

GRESSLY (1864, p. 96) replaced the stage name Corallien by Rauracien, because lithostratigraphic units with hermatypic corals of different age were at his time indiscriminately called "Corallien". The lowermost part of GRESSLY's étage Rauracien encompasses the coral biostromes within marl, which were subsequently called Liesberg Member by ROLLIER (1888). This member is represented by the abbreviation "LIE" in the left part of Fig. 1.5 in this study. The major, upper part of GRESSLY's Rauracien is the modern, all-carbonate St-Ursanne Formation with hermatypic corals which was named by BOLLIGER and BURRI (1970). French stratigraphers correlated GRESSLY's Rauracien, now mainly the St-Ursanne Formation, with the younger, predominantly carbonate succession with hermatypic corals above MARCOU's Argovien in France, which is the time equivalent of the mainly carbonate Günsberg Formation in Switzerland (GÜN in Fig. 1.5). The French also thought, like MERIAN (1821), that GRESSLY's Rauracien was time-equivalent with what is now the even younger, carbonate Villigen Formation in the upper Oxfordian in Canton Aargau. The Villigen Formation was named by GYGI (1969a, p. 68). The formation in its type section near Villigen in Canton Aargau is a well-bedded, micritic limestone succession with mostly bivalves and with few ammonites above the mixed calcareous and argillaceous Effingen Member. Ammonites prevail in the macrofossil assemblage of the Villigen Formation in Canton Schaffhausen, where the formation was sedimented in deeper water than in Canton Aargau (Fig. 1.5).

As a consequence, French stratigraphers correlated in France the proximal and older, blue-gray clayey marl with carbonate concretions *below* GRESSLY's Rauracien, which is in Switzerland equivalent to the modern Sornetan Member named by GYGI (2000a, p. 53), with the equally blue-gray, but basinal and younger, marly succession which is the bulk of MARCOU's Argovien *above* the sponge biostromes at the base. French stratigraphers thereby correlated time equivalents of the Sornetan Member with time equivalents of the Effingen Member. The Sornetan Member is represented by the abbreviation SOR in Fig. 1.5 of this study. The paleogeographic position of MARCOU's type section of the Argovien near Andelot in France is equivalent to that of section RG 14 near Günsberg in Switzerland (see Fig. 1.5 in this study, or the detailed section RG 14 in GYGI 1969a, Pl. 18). It can be read from Fig. 1.5 in this study that GRESSLY's Rauracien is time-equivalent with *only* the sponge biostromes of the Birmenstorf Member in Canton Aargau and with the sponge biostromes at the base of MARCOU's type Argovien in France. This was documented with ammonites which are figured in GYGI (1995, 2001). Moreover it is

evident from Fig. 1.5 why Swiss geologists, following ROLLIER (1888, 1892, p.292, and especially 1911, Fig. 55 on p. 177), interpreted Rauracien with calcareous oolite and hermatypic corals from shallow water and Argovien with siliceous sponges and abundant ammonites from deeper water to be *coeval facies*, whereas French stratigraphers regarded Argovien and Rauracien for a long time to be separate *stages successive in time*. French mapping geologists continued until late in the twentieth century to make the same correlations as were published by MERIAN (1821), who tentatively and with due reservation correlated the argillaceous unit which was later called Bärschwil Formation by GYGI and MARCHAND (1993, p. 998) with the modern Wildegg Formation that was named by GYGI (1969a).

French stratigraphers of the first half of the twentieth century used the stages which are labeled in the column in the left part of Fig. 1.5 in this study. They arrived at a superposition of the true Rauracien over a misinterpreted "Argovien", which is the time equivalent of the Swiss Sornetan Member. This was published for instance by DREYFUSS et al. (1964, p. 305). The pseudo-Argovien and the pseudo-Rauracien of French authors are represented in Fig. 1.5 of this study with a gray shade. The mixed-up "stages" are represented in several modern geologic maps of the French Jura as well as in the text by CHAUVE and PERRIAUX (1974, Fig. 212) which was contributed to a French treatise. This treatise was edited by DEBELMAS (1974) and was published no less than 110 years after the introduction of the Rauracien by GRESSLY (1864). French stratigraphers assigned the stage Séquanien of MARCOU (1848) to the Kimmeridgian Stage. Their Séquanien encompasses the modern Vellerat Formation ("combe séquanienne"), the Courgenay Formation directly above, and the Balsthal Formation of Swiss geologists.

ARKELL (1956, p. 84) initiated the gradual clearing up of the confusion by his suggestion to abandon the stages Oxfordien, Rauracien, Argovien, and Séquanien, because none of them except the Oxfordien were dated biochronologically with ammonites. The pseudo-stages were characterized by a distinct lithology and by macrofossils diagnostic of a particular facies: hermatypic corals for the Rauracien (*olim* Corallien), siliceous sponges for the Argovien, and bivalves for the Séquanien. ARKELL (1956) advocated uniting the four old stages mentioned above into a new Oxfordian Stage *sensu lato* which was to be defined exclusively according to ammonite biochronology. This was done after ARKELL's death by J. H. CALLOMON in a paper which he read before the International Congress "Colloque du Jurassique" at Luxembourg in 1962. CALLOMON's paper was approved by the congress, and it was published in the proceedings of the congress in 1964: COLLOQUE DU JURASSIQUE LUXEMBOURG 1962 (1964, pp. 269–291). The Oxfordian Stage in the sense of ARKELL (1956) with the ammonite zonation according to CALLOMON (1964) is now applied worldwide.

2.1   Sediment Thickness

The first step in quantifying sedimentation in this regional study was to measure and sample 221 detailed sections which are spaced as closely as possible. The longest of these sections is RG 307 at Péry in Canton Bern, which is represented in Pl. 22 in GYGI (2000a). The section has a length of 357 m, and it includes 246 individually numbered units. Exact thickness of a massive and very thick unit can be difficult to measure. For instance, the Balsthal Formation in its type section RG 438 in Steinibach Gorge north of Balsthal is tectonically in near-vertical position. The lower, mostly massive part of the Holzflue Member is intersected by the road leading through Steinibach Gorge at a very acute angle. Measurement of thicknesses in this part of the Holzflue Member was therefore prone to error.

This is why the thickness of the Balsthal Formation as measured in Steinibach Gorge was checked by measuring the cliff of Chluser Roggen south of Balsthal in the unpublished section RG 450 using a rope. A photograph of the cliff is shown in GYGI (2000a, Fig. 37 on p. 64). The massive limestone strata in this perfectly vertical cliff dip with 8° north. No correction of the measured thicknesses was necessary at this degree of dip. The total thickness of the Balsthal Formation at Balsthal was measured in section RG 438 in Steinibach Gorge to be 90.0 m (GYGI 2000a, Pl. 44), and 90.6 m in section RG 450 descending with a rope from Point 702 on Chluser Roggen. Massive strata of mainly the St-Ursanne Formation crop out in the cliff of Peute Roche 1 km southwest of the village of Vellerat in Canton Jura. The cliff was measured on the rope as section RG 451. The cliff is exactly vertical and 92.8 m high. The strata intersected dip with 13° north. The thicknesses measured are by 3% greater than real because of the dip, or by 2.8 m. The true thickness of the strata measured is consequently 90 m. Section RG 451 is drawn on Pl. 24 in GYGI (2000a). No correction of the measured thicknesses was made in the plate.

It is evident from Figs. 1.5 and 4.15 in this study that there are extreme regional discrepancies in thickness of coeval sediments, both in succession no. 1 and no. 2. The results of measurement of detailed sections and descriptions of lithostratigraphic units are to be found in GYGI (1969a, 2000a, b). The 221 sections measured are assembled in Fig. 1.5 of this study. This figure is the frame of reference for all further work, especially for ammonite biochronology. Thicknesses were measured as accurately as possible in order to quantify synsedimentary tectonics and to discern endogenic from exogenic basement subsidence. *Endogenic* basement subsidence included regional and equable subsidence and in some cases superimposed differential subsidence varying by tens of meters in a short time span over short lateral distances. Such differential endogenic subsidence is evident in carbonate members of the shallow water realm. The thickness of the St-Ursanne Formation varies by 60 m between approximately 95 m in section RG 306 near Liesberg (GYGI 2000a, Pl. 31) and 35.2 m in the unpublished section RG 397 near Kleinlützel over a palinspastic distance of 5 km (see below and GYGI 1990c, Fig. 5). Thicknesses of the Verena Member vary by as much as 27 m between approximately 59 m in section RG 431 in the quarry at Gänsbrunnen (GYGI 2000a, Pl. 40, supplemented on August 30, 2008), 57.2 m in section RG 404 near Mervelier (GYGI 2000a, Pl. 36), and 31.6 m in section RG 429 near Welschenrohr (GYGI 2000a, Pl. 39). Substantial variation in thickness in coeval lithostratigraphic units from shallow water over short horizontal distances was probably caused by endogenically driven normal faults. It is discussed in Sect. 11.2 that neither depth of sedimentation nor facies of carbonate units from very shallow water were affected to a significant degree directly above such synsedimentary faults.

Endogenic uplift amounted to less than 10 m, for instance in the latest *Hypselum* Chron when the top of the calcareous oolite of the Steinibach Member was raised above sea level in the unpublished section RG 4 near Waldenburg (see Fig. 1.4 and mainly 4.3) and south of Aedermannsdorf in the unpublished section RG 15. Local endogenic uplift above sea level of marine calcarenite at the top of the Balsthal Formation near Balsthal amounted to more than 4 m. *Exogenic subsidence* of adjacent basement blocks varied by amounts in excess of 100 m. It can be read from Fig. 4.15B, C that the effect of great regional variation in loading of the seafloor with coeval sediments and water caused the basement to subside differentially below the thickest part of the shallowing-upward succession no. 1 and later below the thickest part of shallowing-upward succession no. 2.

2.2 Direct Time Calibration of Sediments

2.2.1 Radiometric Ages

The first attempt at dating pure pellets of submicroscopic glauconite crystals in sediments of Late Jurassic age in northern Switzerland with the K–Ar method was made by Gygi and McDowell (1970). Fischer in Fischer and Gygi (1989) measured some of the same glauconite pellets again with an improved technique. These authors assigned the calculated ages to well-preserved, figured ammonites excavated from the same strata as glauconite. All of the numerical ages which were published in the two papers proved later to be too young. Smith et al. (1993) showed that dating of glauconite *can* be successful, but only if the ^{40}Ar-^{39}Ar method is used.

2.2.2 Geomagnetic Polarity Scale

An attempt at establishing an Oxfordian and early Kimmeridgian chronology of reversals of the Earth's magnetic field based on sediments on land was made by J. Ogg in 1987, when the author of this study guided him through sections in the Swiss Jura Mountains. Ogg then collected a complete set of oriented samples from fine-grained sediments. No results of this work were published.

2.3 Relative Time Calibration of Sediments

2.3.1 Biochronology of Ammonites and Ammonite Zones

Ammonites are the organisms with the highest rate of evolution which became fossilized in the sediments investigated. Establishing a reliable ammonite biochronology is an extremely time-consuming procedure. Almost 10,000 casts

of ammonites in a state of preservation sufficient for study were collected over the years from *in situ,* some in existing outcrops, but most of them in systematic, bed by bed excavations. The well-preserved specimens among these were prepared, measured, identified, and many of them were figured in several publications. The ammonites were organized into zones, as far as possible according to the comprehensive volume coordinated by Cariou and Hantzpergue (1997). The ammonite zone in the present study is the taxon-range zone of the International Stratigraphic Guide, second edition by Salvador (1994, p. 57). The ammonite zone in this study is equivalent to the subzone in earlier publications by the author of the present study. Since most of the ammonites lived preferentially in water deeper than about 30 m (see below), the 28 ammonite zones which are defined in Sect. 7.3 below were conceived in sediments that were laid down in the deeper part of the epicontinental Rhodano-Swabian Basin. The duration of ammonite chrons, or the time span which is represented by ammonite zones, is probably unequal. Only an average duration of the chrons in the Oxfordian and Kimmeridgian Age could therefore be calculated. This was done by dividing the aggregate duration of the two ages of 10.4 million years, according to Gradstein et al. (2004, p. 310), by the number of the 28 ammonite chrons which are listed in Fig. 1.6 of the present study. One of the listed chrons consequently lasted on average approximately 370,000 years. This is close to, for instance, the 380,000 years which were calculated by Weedon et al. (1999, p. 1800) to be the average time represented by what they rated to be a subzone of the Pliensbachian Stage in England. The similarity is remarkable, because such a subzone in England is temporally and geographically distant from northern Switzerland. There are possibly gaps or overlaps between ammonite zones used in this study, because ammonites in sediments laid down at a *normal* rate in deeper water are uncommon. For instance, the *Hypselum* Zone is represented in northern Switzerland by a single specimen of the zonal index taxon.

2.3.2 The Three Principal Marl-Limestone Successions in the Swiss Oxfordian

Gygi in Gygi and Persoz (1986, Pl. 1A) subdivided the Oxfordian sediments in northern Switzerland into three principal marl-limestone successions. The term marl was then, and is used in this study in the sense of Barth, Correns and Eskola (1939) in Pettijohn (1975, Fig. 10–41). The partially argillaceous sediments of the Bärschwil Formation in succession no. 1, of the Wildegg Formation in succession no. 2, and of the Bure Member in succession no. 3 were initially mixtures with widely varying percentages of clay minerals and of mud-grade carbonate minerals like aragonite and calcite. Such mixed rocks and all of the limestones are classified in the present study using the terminology in Fig. 10–41 in PETTIJOHN (1975). The sediment with the highest content of clay minerals is the clayey marl of the *Renggeri* Member in the lowermost part of the Bärschwil Formation. The Buix Member in the upper St-Ursanne Formation is a pure limestone with a carbonate content of up to more than 99%.

Succession no. 1 in Fig. 1.5 is thick in its proximal part. Below is clayey marl and marl of the Bärschwil Formation. Above is limestone of the St-Ursanne Formation, which can be extremely pure in the Buix Member. The Bärschwil Formation represents approximately 6.4 ammonite chrons of on average 370,000 years each, which amount to a total of almost 2.4 million years. The St-Ursanne Formation was sedimented during about 1.6 ammonite chrons. This is equivalent to close to 600,000 years on the assumption of an equal duration of ammonite chrons. The total time represented by succession no. 1 is 8 ammonite chrons or nearly 3 million years.

Succession no. 2 represents the 6 ammonite chrons from the *Schilli* Chron to the *Berrense* Chron. The Effingen Member is the thick, lower part, where the succession was sedimented in deeper water. The member is mostly marl with some intercalations of marly limestone or of limestone. The Effingen Member represents the *Schilli* to the *Hypselum* Chron, or 5 chrons. The limestone in the uppermost part of succession no. 2 in Canton Aargau is the Geissberg Member, which represents the greater part of the *Berrense* Chron. The Effingen Member was sedimented in a time span of about 1.8 million years and the Geissberg Member during probably approximately 300,000 years. Succession no. 2 then represents a time span which can be calculated at 2.1 million years.

Succession no. 3 in Canton Schaffhausen includes in its lower part the upper, partially marly Hornbuck Member of the *Bimammatum* Chron, and above the limestones of the Küssaburg Member (*Gredingensis* Chron) and of the Wangental Member (*Planula* and *Galar* Chron). The four ammonite chrons represented by succession no. 3 amount to a total time span of about 1.5 million years. Succession no. 3 in northwestern Switzerland includes the marl of the Bure Member below and the limestones of the La May and of the Porrentruy Member above (Fig. 1.5). The Bure Member probably represents the *Bimammatum* Chron or 370,000 years. No ammonites were found in the three members in northwestern Switzerland. Time correlation with Canton Aargau was made using the mineral-stratigraphic correlations by PERSOZ in GYGI and PERSOZ (1986, Pl. 1A). Time correlation between Canton Aargau and Canton Schaffhausen could be made with ammonites.

The times of sedimentation of successions no. 1–3 calculated above differ from those which can be read from chart 6 by HARDENBOL et al. (1998): 2.63 ma for succession no. 1, 1.58

ma for succession no. 2, and 1.17 ma for succession no. 3. The time represented by each of the three principal Oxfordian marl-limestone successions in northern Switzerland is unequal, whether according to the calculations mentioned above or following HARDENBOL et al. (1998, chart 6). The proportions of time of marl to limestone sedimentation diverge within successions no. 1–3 even more than the total time represented by each of the successions. The proportion of time between marl and limestone sedimentation can be calculated to be in succession no. 1 approximately 2.4:0.6 ma, or 1.8:0.3 ma in succession no. 2, and 0.3:1.2 ma in succession no. 3.

2.3.3   Elementary Marl-Limestone Cycles

The limestone member representing the *Planula* Zone and the *Galar* Zone in southern Germany is called Wohlgeschichtete Kalke (well-bedded limestones). SEIBOLD (1952) analyzed the carbonate content of the thick, micritic limestone tiers and of the thin marly intercalations in between. He found that the carbonate content of the marly intercalations separating the limestone tiers or beds was on average 13% less than that of the limestone layers. SEIBOLD concluded from this that the supply of clay minerals to the deeper part of the Rhodano-Swabian Basin in southern Germany was slight and continuous in the pertinent time span. He presumed that the carbonate mud was chemically precipitated intermittently within the basin. SEIBOLD (1952, p. 368) presented evidence that the variation in the carbonate content between limestone beds and thin, marly intercalations was primary, not caused by diagenesis. The bedding rhythm was constant and could be followed in southern Germany over a great horizontal distance.

Being inspired by SEIBOLD (1952), GYGI (1969a, Pl. 17) published the result of calcimetry in a bed by bed set of analyses through the entire micritic limestone succession of the Villigen Formation in section RG 62, which was measured in 1960 west of Villigen in Canton Aargau. The average content of $CaCO_3$ of the limestone tiers in the Letzi Member in section RG 62 is 96%. The $MgCO_3$ content of the limestone tiers analyzed for magnesium in the Villigen Formation does not exceed 3% in the section. The partings with an elevated content of clay minerals between limestone tiers in the Letzi Member are a few millimeters thick and were not analyzed. Marly intercalations with a substantial thickness between limestone beds were found in section RG 62 only in the Geissberg Member. The carbonate content of these intercalations, where analyzed, is at most 15% lower than that of the limestone beds. This confirms the results of SEIBOLD (1952, p. 368). Marly intercalations weather back and thereby produce the distinct bedding as it appears in weathered, natural cliffs of the Villigen Formation.

Section RG 62 was measured in the small valley west of the ruin of Besserstein castle located above the village of Villigen. The road from Villigen upward to the table mountain of Mt. Geissberg is leading through the valley. Section RG 62 was assembled from minor outcrops on both sides of the valley. The location of these partial sections was sketched from two viewpoints on the brink of the table mountain, each viewpoint overlooking the steep slope and cliff on the opposite side of the narrow valley. The sketches were drawn early in March, before the trees in the deciduous forest had sprouted their foliage. The minor sections in the valley partially overlap. Their overlapping parts could be correlated using the pattern of successive beds with varying thickness in a similar way as the succession of unequal growth rings in the trunk of trees is correlated in dendrochronology. Section RG 62 being thus assembled was checked against the overlapping part of the nearby, unpublished section RG 63 which was measured on the rope descending at a vertical cliff 300 m northwest of the ruin of Besserstein castle on the northeastern slope of Mt. Geissberg. The neighboring sections RG 62 and 63 are shown at a reduced scale in the upper composite section on Pl. 19 in GYGI (1969a).

The Letzi Member in Canton Aargau includes the *Planula* and the *Galar* ammonite Zone. The *Galar* Zone possibly represents significantly less time than the *Planula* Zone. The assumption can therefore be made that the Letzi Member was sedimented in a time span of around 600,000 years instead of the 740,000 years, which are the average duration of two entire ammonite chrons in the Late Jurassic. The Letzi Member in section RG 62 near Villigen includes 61 elementary marl-limestone cycles. Fifty-seven such cycles could be counted in the Letzi Member at Mellikon in section RG 70 (GYGI 1969a, Pl. 17). The fact that the numbers of elementary cycles are so close together over the considerable distance of 12 km between sections RG 62 and RG 70 in the Tabular Jura is an indication that the marl-limestone cycles were caused by short-term variations of climate. If this is so, then the average length of such an elementary climatic cycle would have been approximately 10,000 years during sedimentation of the Letzi Member.

The Pichoux Formation is a succession of well-bedded micritic limestone much like the Villigen Formation. It can be read from Fig. 1.5 that the formation is a wedge when imaged in cross-section perpendicularly to depositional strike. In the most proximal part of the wedge, in Pichoux Gorge near Sornetan in Canton Bern, the lower half of the formation includes 44 elementary marl to limestone cycles. The lower half of the Pichoux Formation at Sornetan ranges from the formation's base in section RG 314 to the marly boundary bed no. 22 of section RG 315 in GYGI (2000a, Pl. 21). The boundary beds between the top of the Sornetan Member and the base of the Pichoux Formation do not crop out in section RG 314 in Pichoux Gorge. They were excavated in a trench. The time span represented by the lower Pichoux Formation is about one third of the *Antecedens* Chron, and it can be estimated to be around 120,000 years. Provided that this estimate is of the right order of magnitude, the average duration of one of the 44 elementary marl-limestone cycles in the lower Pichoux Formation at Sornetan would be about 2,730 years. Notwithstanding the uncertainties of how much time is represented by the Letzi Member and mainly by the proximal part of the lower Pichoux Formation, the duration of elementary cycles in the lower Pichoux Formation at Sornetan is certainly less than that of elementary cycles in the Letzi Member of the Villigen Formation. Further down in the wedge of the Pichoux Formation perpendicularly to depositional strike, in section RG 307 at Péry in GYGI (2000a, Pl. 22), the coeval lower part of the Pichoux Formation includes only 17 marl-limestone couplets. The conclusion to be drawn from this is that long-distance time correlation using the bedding rhythm of micritic limestone is unfeasible.

The time equivalent of the lower Pichoux Formation in the adjacent carbonate platform of the St-Ursanne Formation is the lithostratigraphic unit that GYGI (2000a, p. 55) re-defined and named Delémont Member. The name of the member was challenged since and is therefore replaced here by the new name **Chestel Member**. Chestel is the name of the ridge which is formed by limestone of the St-Ursanne Formation south of the village of Liesberg in Canton Basel-Landschaft. A photograph of Chestel ridge is presented in Fig. 24 on p. 41 in GYGI (2000a). The name Chestel is shown on the topographic map Landeskarte 1:25,000, sheet no. 1086 Delémont. The type section of Chestel Member is bed nos. 107–110 in section RG 306 in Chestel ridge, which is represented in GYGI (2000a, Pl. 31). A photograph of Chestel Member near St-Ursanne is Fig. 4.19B in the present study. The member is at that locality massive calcareous oolite with only one parting in its upper part. The time equivalent of the micritic, well-bedded Letzi Member in the upper Villigen Formation in Canton Aargau is in northwestern Switzerland the Verena Member in the upper part of the carbonate platform of the Balsthal Formation. The Verena Member is massive calcareous oolite in most places, like for instance in the rock of Rouge Pertuis north of the village of Undervelier in Canton Jura. Strata are tectonically perpendicular in this landmark rock. No partings are visible in the Verena Member on this rock which is shown in the photograph of Fig. 1.7.

The Paulin Member in the lowermost Reuchenette Formation in the northern Jura Mountains includes 62 beds of pure or marly limestone in section RG 350 near Courgenay, 65 such beds in section RG 340 in the quarry of La Rasse south of Porrentruy where the top of the member does not crop out, and 56 limestone beds in section RG 443, where the member is complete in an exploration well near St-Ursanne.

The three sections are shown in detail in Pl. 17, 19, and 20 in GYGI (2000a). There are on average approximately 60 limestone beds or tiers in the Paulin Member of Ajoie region. The member represents a time of 1.4 million years according to HARDENBOL et al. (1998), or 1.5 million years following GRADSTEIN et al. (2004, p. 310). The thickness of the limestone tiers in the member is very unequal. The duration of an elementary cycle with a thin layer of marl below and a thick limestone tier above can be calculated, following the authors cited above, to have been on average approximately 25,000 years in this lagoonal sediment from very shallow water. The time represented per elementary cycle in the Paulin Member is probably about as unequal as the widely varying thicknesses of the individual cycles. When the average duration of one of the seven ammonite chrons represented by the Paulin Member is assumed to have been about 370,000 years as it was calculated above, then the average duration of an elementary cycle in the Paulin Member would have been as much as approximately 43,000 years, or almost twice as much as the time of around 25,000 years (see below) to be concluded from HARDENBOL et al. (1998).

2.3.4   Comparison of Marl-Limestone Cycles with MILANKOVITCH-Type Periodicity

The most probable cause of change between marl and limestone sedimentation in the region investigated is variation between humid and drier climate at the source of terrigenous sediment on neighboring land in the north. This was suggested by CECIL (1990). It is not yet known whether the principal marl-limestone successions no. 1–3 in the Oxfordian of northern Switzerland reflect variation in temperature of the pertinent climate. MILANKOVITCH (1941) calculated the periodicity of precession of the Earth's axis and periodicities in the eccentricity of the Earth's orbit around the Sun. He quantified the intensity of solar radiation arriving on the surface of the Earth which he called insolation, and he calculated the variation in insolation with time. MILANKOVITCH summarized the results of his calculations in the form of curves in Fig. 51 on p. 549, and he assigned certain insolation minima to particular glaciations of the Pleistocene. The calculations by MILANKOVITCH had subsequently to be revised. For instance, MILANKOVITCH (1941, p. 186) calculated the period of Earth's axial precession to be 25,735, or approximately 26,000 years. According to HINNOV in GRADSTEIN et al. (2004, p. 56 and Fig. 4.3 on p. 58), the average periodicity of precession of the Earth's axis is 21,000 years, and the eccentricity of the Earth's orbit around the Sun varied over the past 10 million years with the principal periodicities of 100,000, 400,000, and 2,360,000 years. Variation in the intensity of insolation of the Earth affects

the temperature of land, air, and surficial ocean water, and thereby causes the volume of ocean water to wax and wane slightly, as it was pointed out by SCHULTZ and SCHÄFER-NETH (1997).

The question arises from this whether minor sea level fluctuations and variation in the rate of climate-sensitive carbonate sedimentation in the area investigated in this study was tuned by variation in insolation of the Earth according to Earth's precessional and orbital periodicities. Orbital time calibration of sedimentation is at the present time possible, according to HINNOV in GRADSTEIN et al. (2004, p. 61), not further back than the Miocene-Oligocene time boundary at 23 million years before present. Data in Sect. 5.2 of this study are evidence that some of the rises of sea level documented in sediments from shallow water in northern Switzerland were indubitably rapid (Fig. 5.1). Observations described in Sect. 5.3 indicate that probably all of the sea level falls were slow. The curve of sea level fluctuation was therefore in these cases probably asymmetrical much like the curve of variation in temperature during the Quaternary which was derived from ice cores drilled in Antarctica (MCMANUS 2004). The principal marl-limestone successions no. 1–3 of the Oxfordian in Fig. 1.5 represent an unequal number of ammonite chrons. A list of Late Jurassic ammonite chrons is in Fig. 1.6 in this study. It was indicated above that the proportion of time represented by marl and by limestone sedimentation is different in every one of the three principal Oxfordian marl-limestone successions. The elementary marl-limestone cycles in units of micritic limestone from deeper water were averaged in every unit. The averaged elementary cycles represent widely different time spans: about 2,730 years in the lower Pichoux Formation and approximately 10,000 years in the Letzi Member. The average duration of an elementary cycle in the lagoonal Paulin Member was calculated to be at least 25,000 years.

Recognizing elementary cycles in the shallow-water realm which represent about as much time as elementary cycles in sediments from deeper water was impossible, especially in units of mostly massive calcareous oolite like for instance the Chestel or the Verena Member. It is stated below that it was necessary to revise several correlations between sections which GYGI (2000a) had measured in the shallow-water realm. This illustrates how difficult it is to correlate sections within carbonate platforms in detail. Time correlations are prone to error especially within the Günsberg Formation. It is therefore at least premature to calibrate shallow-water sedimentation of the Oxfordian in northern Switzerland with MILANKOVITCH-type cycles. Both OLSEN and KENT (1999) and WEEDON et al. (1999) claimed that they could recognize MILANKOVITCH cycles in Jurassic sediments. STRASSER (2007, p. 427) concluded from sections of Late Jurassic sediments in the Swiss Jura Mountains that "it can be shown that Oxfordian and

Kimmeridgian platform and basin carbonates hold a record of orbital (Milankovitch) cycles" of 20,000, 100,000, and 400,000 years. MILANKOVITCH (1941) clearly distinguished periods of precession of the Earth's *axis* from longer periods of eccentricity of the Earth's *orbit* around the Sun. Some of the sections shown in STRASSER (2007) were measured by the author of this study and are represented in plates by GYGI (2000a). The above conclusion of STRASSER (2007) is scrutinized in Sect. 11.6.2 below.

2.4   Rates of Sedimentation

Meaningful, net sedimentation rates can as a rule be calculated only in deposits from deeper water on the basin floor, where erosion, transportation, and re-deposition of fine-grained sediment by bottom currents were of a minor scale. The situation is different in the shallow-water realm. Much calcareous mud was chemically precipitated for instance in the peritidal area in which the Vorbourg and the Röschenz Member were sedimented in the lower Vellerat Formation landward of the Günsberg carbonate platform, or in the shallow lagoon where the Courgenay Formation was laid down landward of the Balsthal carbonate platform. Some of the calcareous mud which was produced in very shallow-marine water initially settled from suspension in the environment where it was precipitated. Currents driven by tides and by occasional storms then stirred the mud up from the bottom, and it was carried away by currents in suspension along with the mud which never settled, across the carbonate platform of the Günsberg Formation, or across the wide sand bank of calcareous ooids of the Balsthal Formation. Both terrigenous and calcareous mud were resedimented in deeper water of the Rhodano-Swabian Basin. This is indicated in Sect. 4.14.3. Varying rates of sedimentation on the carbonate platform of the St-Ursanne Formation are calculated in Sect. 8.1.

2.5   Provenance and Quantity of Calcareous Mud

Calcareous mud was quantitatively the most important primary constituent of the Late Jurassic sediments and bioconstructions investigated. GYGI (1969a) figured on Pl. 11, Fig. 41 part of a coccolith and on Pl. 5, Fig. 17 an unidentified organism of nannoplankton. Such nannoplankton is refigured in this study in Figs. 4.16 and 4.17. GYGI (1969a) showed in Fig. 1 on p. 24 that he found coccoliths only in micritic limestones which were sedimented in deeper water. The abundance of coccoliths increases in the Oxfordian sediments investigated with growing depth of deposition. 150,000 coccoliths were calculated to occur in

1 cm^3 of micritic limestone of the Letzi Member from a water depth of about 50 m in bed no. 88 of section RG 70 at Mellikon. Approximately 1,000,000 coccoliths per cm^3 occur in a limestone sample of the Wangental Member from around 100 m depth in bed no. 101 of section RG 91 at Immendingen in southern Germany. This is equivalent to about 0.004% of the rock volume. The sample from Immendingen was the richest of all out of the limestones searched for nanoplankton. GYGI (1969a, p. 24) therefore concluded that the contribution of nannoplankton to sedimentation of calcareous mud in the deeper part of the Rhodano-Swabian Basin was negligible *in the Oxfordian Age*. PITTET and STRASSER (1998, p. 161) arrived at a similar conclusion.

Bioerosion can produce fine-grained calcareous sediment. This was documented in Bermuda by NEUMANN (1966) and by GYGI (1969b, 1975, Fig. 11). A substantial amount of the fine-grained calcareous sediment produced by various organisms, possibly mainly by regular sea urchins, on coral reefs in the lagoon in the internal part of the calcareous platform of the upper St-Ursanne Formation is included in the Buix Member (Fig. 1.5). It is shown in Sect. 4.14 that the quantity of calcareous mud produced on carbonate platforms was less than what was at the same time laid down in the deeper part of the basin like for instance in the Effingen Member. The Effingen Member is a mixture of on average about 60% lime mud and of 40% of clay minerals and some silt of mainly detrital quartz. The Effingen Member includes much more $CaCO_3$ than the entire volume of the coeval carbonate platform of the Günsberg Formation, from where calcareous mud spilled into deeper water. This is evidence that most of the lime mud in the Effingen Member was produced in the peritidal environment landward of the Günsberg Formation, where the lower Vellerat Formation with the Vorbourg and the Röschenz Member was sedimented. Currents transported mud from there across the carbonate platform of the Günsberg Formation and into the deeper part of the basin, where it was incorporated into the Effingen Member. The carbonate platform of the Balsthal Formation is much wider than that of the Günsberg Formation below. Nevertheless, the great amount of calcareous mud in the Villigen Formation, which is adjacent to and coeval with the Balsthal Formation, cannot possibly have all been produced on the sand bank of calcareous ooids of the Balsthal Formation. The Villigen Formation extends far into adjacent southern Germany and grades southeastward into the coeval part of the lower, micritic Quinten Formation. Most of the pertinent calcareous mud was probably chemically precipitated in very shallow water landward of the Balsthal Formation, where the micritic limestone of the Courgenay Formation was sedimented (Fig. 1.5). All of this is approximately quantified and discussed in Sect. 4.14.

2.6   Water Depth and Sea Floor Topography

There is no evidence that the tidal range was more than a few decimeters during deposition of the sediments investigated. The position of mean sea level can therefore be inferred based on stromatolites from the upper intertidal zone. Such stromatolites are shown in Figs. 4.4–4.6 and 5.8. Water depth above Oxfordian sand banks on which calcareous ooids were accreted was probably as slight as that on the Recent ooid sand bank which is shown in the photograph of Fig. 4.7. The greatest water depth above the top of a carbonate platform was calculated to have been 10 m. This depth was reached after rapid sea level rise no. 5 (see Fig. 5.1) in the internal part of the carbonate platform of the St-Ursanne Formation. The initial depth of the lagoon, into which the Buix Member was sedimented (Fig. 1.5), is calculated in Sect. "Rise no. 5" in part 1 of Chap. 5. RANKEY (2004) concluded that no "deterministic link" exists between depositional facies and water depth in shallow water on carbonate platforms. Hermatypic corals could live in the environments investigated in a paleobathymetric range which probably nowhere exceeded approximately 20 m. This is calculated based on section RG 307 at Péry in Canton Bern, which is represented in Pl. 22 in GYGI (2000a). The pertinent calculation is in the middle of Chap. 6. The paleobathymetric interval between 20 and 30 m is characterized by a macrofauna with large bivalves of the genus *Pholadomya*, which can be very abundant in the *Crenularis* Member of Canton Aargau, in the Reuchenette Formation near Olten (GYGI 1986, Fig. 6B), and that are fairly abundant in the upper half of the Sornetan Member.

The thin marker bed with a rich macrofauna of mainly ammonites, which separates the lower half of the distal part of the Sornetan Member from the member's upper half, was sedimented at a water depth of between 30 and 35 m. This is calculated in Sect. 4.2. Water depth at the time boundary between the Middle and the Late Jurassic near Blumberg in southern Germany, 18 km north-northwest of Schaffhausen, was 100 m. The depth is calculated semiquantitatively in Sect. 4.1.1. This is an important mark of paleobathymetry which can be observed in both vertical and lateral facies boundaries. Water depth at the beginning of the Oxfordian in northwestern Switzerland and in Canton Aargau can only be estimated. The depth was probably about 80 m (see Figs. 1.5 and 4.15). The argumentation of how this estimate was arrived at is to be found in the last part of Sect. 4.2.

The grain size of sediments is undiagnostic of water depth. Mud-grade sediment occurred from tidal flats down to the greatest water depth of up to well over 100 m on the floor of the epicontinental basin (Fig. 4.9). On the other hand spherical, primarily hard oncoids of calcium carbonate with a diameter of up to 36 mm were periodically rolled at the sediment surface of calcareous mud by currents of bottom water driven by exceptionally violent storms on the floor of the epicontinental basin at the water depth of as much as approximately 120 m (see Sect. 4.3.2). Ellipsoidal nodules of limestone with a diameter of up to 30 cm occur in a matrix which was originally mud in sediments from a water depth approaching 100 m. Such nodules are the product of subsolution (or corrosion, see below) of initially continuous limestone beds. Nodules formed by subsolution at a water depth of close to 100 m are represented in Fig. 8.3. ALLENBACH (2002, p. 336) thought that large nodules at the base of the Oxfordian like those shown in Fig. 8.3 of this study were formed "beyond doubt . . . in an agitated environment close to storm-wave-base". The term wave base is avoided in this study, because fair weather wave base varies within days or even hours between sea level on an exceptionally calm summer day and a depth of at most a few meters. Storm wave base during exceptionally violent cyclones could reach down to the floor of the epicontinental basin at the greatest water depth of well over 100 m, which was concluded from the sediments investigated. This is discussed in Sect. 4.3. The vertical scale in Fig. 1.5 is exaggerated a hundred times. It can be read from this that the inclination of the slope of the Pichoux Formation (PIC in Fig. 1.5) and of the Villigen Formation between Villigen and Siblingen was very slight. Sea floor topography was subdued at any stage of sedimentation in the region investigated.

2.7   Subsidence

Endogenic subsidence is caused by processes deep in the interior of the Earth. *Exogenic subsidence* of the basement is caused by the weight of sediments or of additional water after a sea level rise. GYGI (1986, p. 472) calculated exogenic subsidence to have been approximately two-thirds of the partially compacted thickness of Oxfordian sediments in northwestern Switzerland. This is in agreement with ZIEGLER (1982). According to ZIEGLER (1982, p. 106), a eustatic rise of sea level causes the basement to subside by 40% of the rise. Differential exogenic subsidence of neighboring basement blocks caused by the lateral shift of belts with maximal sedimentation was imaged first by GYGI (1986, Fig. 3). A revision of that is shown in Fig. 4.15 in the present study.

2.8   Synsedimentary Tectonics

Synsedimentary tectonics were documented with photographs of normal faults in outcrops by GYGI in GYGI and PERSOZ (1986, Figs. 2 and 8). Such normal faults were active and caused a displacement of as much as 60 m in the St-Ursanne Formation between section RG 306 at Liesberg in Canton Basel-Landschaft (GYGI 2000a, Pl. 31) and the unpublished section RG 397 at Kleinlützel in Canton Solothurn (GYGI 1990c, Fig. 5), or of the same amount near Riniken in Canton Aargau (GYGI 1990c, Fig. 6). Synsedimentary faulting caused the substantial differences in thickness of coeval carbonate members in formations of the shallow-water realm. This becomes evident when loose foldout plates in GYGI (2000a) are laid out side by side. Synsedimentary faulting ceased almost entirely when the Hauptmumienbank Member and the coeval Steinibach Member were sedimented. Thicknesses within these members differ only slightly.

2.9   Sea Level Variations

Rises of sea level can be quantified, and their rate could be proved in many cases to have been geologically speaking rapid. Falls could be documented qualitatively, but quantification was impossible. This is discussed in Chap. 5. Rapid relative sea level rises were quantified with different methods. According to Sect. 5.2, they were driven by eustatic, global rises. Rapid relative rise no. 10 in Fig. 5.1 occurred at the end of the *Gredingensis* Chron. It is the smallest of all of the rapid rises which could be documented in sediments of Late Jurassic age in northern Switzerland. The eustatic component of rise no. 10 was quantified in the following way. The amount of relative rise no. 10 was so small that it had no visible effect on the vast shoal of calcareous ooid sand which is now the Holzflue Member in Fig. 1.5. Nevertheless, the rise left distinct traces in micritic limestones of the Villigen Formation from deeper water in the Rhodano-Swabian Basin. The rise can be quantified thanks to *lack* of traces of it in the oolitic Holzflue Member of the Balsthal Formation, and using the mineral-stratigraphic correlation J by PERSOZ in GYGI and PERSOZ (1986, Pl. 1A). Calcareous ooids are *accreted* in Recent marine environments in water no deeper than 6 m, according to the concurring results of several authors. This is stated in Sect. 4.1.3. Rise no. 10 left no traces on the sand bank of calcareous ooids of the Holzflue Member, because it amounted to less than 6 m and therefore did not interrupt ooid accretion.

The load of additional water brought into a basin by a eustatic sea level rise causes the basement to subside isostatically by 40% of the rise according to ZIEGLER (1982, p. 106). Isostatic equilibration of the basement is shown in Sect. 11.3 to be prompt as compared with sedimentation rates. Relative sea level rise no. 10 had consequently a eustatic component of 60% of at most 6 m, or of not more than 3.6 m. In spite of the small amount, the rise caused temporal omission (non-deposition) and thereby a distinct, corroded transgressive surface directly below the base of the slightly glauconitic Knollen Bed in the Villigen Formation at Villigen in Canton Aargau. This is well visible in Fig. 12 in GYGI et al. (1998).

The corroded surface is above sequence boundary O8 in GYGI et al. (1998, Fig. 2). The Knollen Bed is a thin marker bed which can be used in long-distance time correlation of micritic limestones in the Villigen Formation. The limestones of this formation were sedimented in the deeper part of the Rhodano-Swabian Epicontinental Basin (Fig. 1.5). Omission at the base of the Knollen Bed and formation of some glauconite within the bed (Fig. 4.1E) because of a very low sedimentation rate were the effect of rise no. 10, because the rise thoroughly reduced or even interrupted for a short time export of calcareous mud from the lagoon of the La May Member of the Courgenay Formation. According to the text at the end of Sect. 4.15.1, most of the calcareous mud sedimented in the Villigen Formation was produced in the very shallow lagoon of the Courgenay Formation landward of the carbonate platform of the Balsthal Formation. Tidal and occasional storm-driven currents transported most of the mud produced in the lagoon out of the environment and in the distal direction across the vast shoal of calcareous ooid sand of the Balsthal Formation, from where the mud spilled into deeper water of the basin (see Fig. 1.5).

2.10   Oxygen Content and Salinity of Bottom Water

Optimal oxygenation of bottom water in the deep part of the Rhodanian-Swabian Basin is indicated by the abundant and diverse assemblage of benthic organisms thriving while the Birmenstorf Member was sedimented at a low rate. The oxygen content of the pertinent bottom water is estimated in Sect. 4.4 by comparison with the Recent northwestern Gulf of Mexico. The clayey marl of the *Renggeri* Member was sedimented in a time span when the lower part of the water column above the sea floor was *permanently* dysaerobic. This is investigated in Sect. 4.4 judging from a comparison of the diameter of burrows in the lower part of the member (GYGI 2000a, Fig. 7 on p. 19) with trace fossils of the Recent. Bottom water was *intermittently* dysaerobic when the Effingen Member was laid down. This was documented by GYGI (1969a, p. 107) with less than millimeter-size casts of iron sulfide of bivalves and of gastropods in marl. Above-average salinity in shallow water is documented by dolosparite (GYGI 2000a, Fig. 16 on p. 26), dedolomite (Fig. 3.3 in this study), calcite pseudomorphs after sulfate minerals (Fig. 3.3), and occasionally by gypsum in calcareous oolite of the Verena Member (GYGI 2000a, p. 45, and Pl. 35, bed no. 113 in section RG 400). A mass occurrence of the worm *Cycloserpula socialis* (GOLDFUSS) in shallow-water limestone of the Laufen Member (GYGI 2000a, Fig. 19 on p. 29) is probably an indication of sedimentation in water with a salinity below the normal level.

2.11 Paleoclimate

An either tropical or subtropical climate can be concluded from coral reefs which occur mainly in the St-Ursanne Formation and in the boundary beds between the Effingen Member and the Günsberg Formation. Distinctly delimited accretion bands with a regular thickness in the skeletal part of hermatypic corals in the Liesberg Member and in the St-Ursanne Formation which were figured for the first time by GYGI in GYGI and PERSOZ (1987, Fig. 3), then by INSALACO (1996b), and which are shown in Fig. 3.1 in this study, are annual. The accretion bands in Fig. 3.1 are by comparison with Recent corals in subtropical Bermuda an indication that the temperature of shallow water varied substantially between summer and winter during growth of the reefs investigated in northern Switzerland. This is explained in detail in the following Chap. 3. Oxfordian climate in the region investigated was consequently *subtropical*. The growth rate of Oxfordian hermatypic coral colonies in northern Switzerland can be calculated from the thickness of their accretion bands. The lower limit to the paleobathymetric range of Oxfordian hermatypic corals is calculated in Chap. 6 to be at the water depth of approximately 20 m.

The principal marl-limestone succession nos. 1–3 of the Oxfordian Age (Fig. 1.5) are diagnostic of a long-term transtion from humid to drier climate. When rainfall and runoff on adjacent land in the northwest and north grew plentiful, weathering on land and supply of terrigenic sediment to the area investigated increased a great deal. Siliciclastic silt in succession no. 2 was concentrated by elutriation mainly in tidal currents in some beds in the peritidal Röschenz Member and in small, distal turbidites and possibly in tempestites of the Effingen Member in deeper water. Evidence of humid climate on land is gyrogonites of characean algae and shells of limnic ostracods from freshwater ponds (OERTLI and ZIEGLER 1958, Pl. 1) as well as lignite coal seams from freshwater swamps. Dolomite, calcite pseudomorphs after sulfate minerals, and some gypsum in sediments from very shallow-marine water are evidence of a drier climate. Supply of fine-grained terrigenic sediment of mainly clay minerals and of some siliciclastic silt diminished to a minimum in times of drier climate. This was the case when the extremely pure limestone of the Buix Member in the St-Ursanne Formation was sedimented into a lagoon in uppermost succession no. 1. No detrital quartz at all was found in the Verena Member, neither in thin section nor by x-ray diffractometry. The Verena Member is the youngest Oxfordian sediment on the carbonate platform of the Balsthal Formation (see Sect. 11.5.2).

Evidence of a *long-term* tendency of increasingly drier climate throughout the Oxfordian Age is absence of dolomite and of calcite pseudomorphs after sulfate minerals in the St-Ursanne Formation of succession no. 1. Some dolomite first appears in the Günsberg Formation (GYGI 1969a, Pl. 7, Fig. 30), and the earliest calcite pseudomorphs after sulfate minerals were found at some localities in the Hauptmumienbank Member in uppermost succession no. 2 (this study, Fig. 5.5C). A semiarid climate in the late Oxfordian during sedimentation of the Balsthal Formation of succession no. 3 (see above) is indicated by pervasive dolomitization (GYGI 2000a, Figs. 16 and 30), by abundant calcite pseudomorphs after sulfate minerals (this study, Fig. 3.3), and by some gypsum found in the Verena Member at one locality (GYGI 2000a, Pl. 35, thin section Gy 7706). The climate did not only become drier, but possibly also warmer during the Oxfordian Age. Measurements of paleotemperature listed by RAIS (2007, Chap. 4, Table 1) based mainly on belemnites from the author's collection at Basel, gave no unambiguous evidence of rising temperature of bottom water from middle Callovian to late Oxfordian time. A strong argument that bottom water *did* become warmer from early Oxfordian to early Kimmeridgian time is the paleobathymetric distribution of ammonites of the boreal genera *Cardioceras* and *Amoeboceras*. This is discussed in Sect. 11.5.5.

An island extended in the Late Jurassic like a broad strip from London in the west over the Ardennes at least to the Rhenish Massif and possibly to the Bohemian Massif (GYGI 2000a, Fig. 2, adapted and simplified after ZIEGLER 1988, Pl. 13, and refigured here as Fig. 1.1). Sediment was then shed from the island south- and southeastward into the epicontinental, Rhodano-Swabian Basin (Fig. 1.2). The epicontinental basin was filled in the process in northwestern Switzerland up to close below sea level for the first time in a *vast* area when the Vorbourg Member was laid down. Evidence of this is stromatolites with dewatering (prism) cracks and birdseye pores which were formed in the upper intertidal zone (Figs. 4.4–4.6 and 5.8). Very large, circular and shallow footprints of sauropod dinosaurs on an ancient tidal flat in the Reuchenette Formation of Kimmeridgian age were found by the author on the upper bedding plane of bed no.17 of section RG 434 that he measured in July of 1986 in the eastern quarry of Steingrueben on the territory of the township of Oberdorf, northwest of the city of Solothurn (GYGI 2000a, Pl. 42, and GYGI 2003, p. 156). The very distinct footprints had a diameter of 0.5 and 0.7 m, respectively. Many more dinosaur footprints were found thereafter in the Reuchenette Formation, mainly in the region of Ajoie in Canton Jura.

The paleolatitude of northern Switzerland was in Late Jurassic time, according to FIRSTBROOK et al. (1979), about 35°N. DERCOURT et al. (eds., 2000, maps 9 and 10) indicated 27°N for the middle Callovian and 28°N for the early Kimmeridgian. The paleoclimate is documented in the region investigated by hermatypic corals, which colonized first the biostromes of the Liesberg Member, and then built bioherms and true reefs, mainly in the St-Ursanne Formation and below the base of the successive Günsberg Formation. The high diversity of the coral assemblage in the St-Ursanne Formation indicates that the climate in middle Oxfordian time was, in spite of the comparatively high latitude, either tropical or subtropical. Well-delimited accretion bands with a regular thickness of between 4 and 6 mm were found in the

hard parts of some of the coral colonies from the St-Ursanne Formation, when the colonies were cut through the center (GYGI and PERSOZ 1987, Fig. 3, or GYGI 2003, Fig. 172, which is refigured in Fig. 3.1 of the present study). The width of the accretion bands in the cross-cut, thickly-branching coral skeleton shown in Fig. 3.1 is interpreted to document annual accretion of primarily aragonitic calcium carbonate to the hard part of the coral colony, according to the following argumentation.

The coral reefs on the Bermuda Atoll at 32°N in the North Atlantic are close to or at the northern boundary where coral reefs can grow in the Recent in this ocean. The temperature of sea water near the surface is in Bermuda around 28°C in July, and substantially cooler during the winter. Air temperature can drop to below 0°C in a particularly cold winter. The seasonal lows of water temperature at shallow depth are documented in the skeletal part of scleractinian coral colonies in the reefs of Bermuda by thin accretion bands with a lower porosity between thicker, more porous bands which are accreted in summer months. A patch reef in the lagoon of Bermuda, about 3 km south of North Rock, was visited in the summer of 1968 by the participants, including the author, in the Seminar on Organism-Sediment Interrelationships that was then held by R. N. GINSBURG at the Bermuda Biological Station. The location of the reef, which was informally named "Grid Reef" by the participants in the seminar, is indicated in the map of Bermuda which is represented in GYGI (1969b, Fig. 1) or in GYGI (1975, Fig. 1).

A living colony of the brain coral *Diploria strigosa* DANA with the diameter of 36 cm was collected by the author during the seminar on the top of "Grid Reef" at a water depth of about 1 m. The hemispheric skeleton of the colony, which is now specimen no. 72 of the Bermuda collection in the Basel Museum of Natural History, was then divided through the center with a diamond saw. It is uncertain whether the point of nucleation of the colony was hit in the procedure. Distinct accretion bands were thereby revealed on parts of the sawed surface. Where bands can be clearly

Fig. 3.1 Skeletal part of a colony of the thickly-branching, hermatypic coral *Cryptocoenia limbata* GOLDFUSS of middle Oxfordian age (*Transversarium* Chron) with distinctly delimitated bands of annual accretion. The thickness of the accretion bands is between 4.3 and 5.7 mm. Polished slab Gy 4323 in the Museum of Natural History, Basel. The slab is out of a bioherm in the uppermost St-Ursanne Formation which is intersected by the road leading from the village of Courtételle in Canton Jura to the farm Les Fouchies. The bioherm is 3 km south of the village in the bend of the road in the eastern part of Forêt de Vainé and is in unit no. 11 in the unpublished section RG 370. Scale bar is 2 cm. Refigured from GYGI (2003, Fig. 172) with permission of Schweizerische Paläontologische Abhandlungen

discerned, their width varies between 3 and 5 mm. The accretion bands are delimited by thin, darker rings in which the skeletal material is less porous than in the midst of the thick, light part of the bands. The thin bands resemble in this respect dark growth rings in the trunk of trees growing in temperate latitudes. The length of a radius that was measured between near the point of nucleation and the surface of the skeletal part of the colony is 178 mm. When the average thickness of the accretion bands is rated at 4 mm, the number of accretion bands in the skeleton of the colony can be calculated to be 45. IAMS (1969, Table 1) calculated the average thickness of accretion bands in *Diploria strigosa* DANA from three shipwrecks at the outer rim of the Bermuda Atoll to be 3.2 mm per year.

The passenger and cargo ship "Madiana" ran aground in water about 8 m deep and was wrecked in 1903 on the outer rim of the Bermuda Atoll about 5 km west-southwest of North Rock. The position of North Rock is indicated in the map in GYGI (1975, Fig. 1). The "Madiana" wreck was visited by the participants in GINSBURG'S 1968 seminar, and the author photographed it underwater (GYGI 1970, Fig. 12). Living colonies of the coral *Diploria strigosa* DANA are abundant on the wreck and are conspicuous in the photograph mentioned. The diameter of the largest of the *Diploria* colonies photographed was about 40 cm. IAMS (1969) estimated the diameter of the largest *Diploria labyrinthiformis* on this wreck in the caption to Fig. d in Pl. 1 at 60 cm and calculated the pertinent growth rate at 4.6 mm per year. The maximal age of these colonies was less than 1968 minus 1903 years or 65 years, because some time elapsed after the wreckage, until the first coral colonies nucleated at the steel surfaces. The maximal *possible* age of the largest colonies of *Diploria strigosa* DANA seen on the Madiana wreck in 1968 can be estimated at 60 years. Ocean water above the wreck at the depth of about 8 m outside the Bermuda Atoll is in the summer months possibly somewhat cooler than water in the lagoon on top of "Grid Reef". The accretion bands which could be discerned on parts of the cross-cut surface of the *Diploria strigosa* DANA with a diameter of 36 cm from the top of "Grid Reef" consequently document how much aragonitic calcium carbonate per year was added to the skeleton of the colony in a similar environment like on the "Madiana" wreck. Some of the accretion bands are well-delimited, because the rate of calcium carbonate accretion in scleractinian coral colonies of Bermuda diminishes during the winter, or else part of the accretion bands would not be clearly delimitated.

The latitude of Bermuda is with 32°N close to the paleolatitude of northern Switzerland of around 28°N in the Late Jurassic, but it should be borne in mind that the width of the tropical and of the subtropical belt was substantially greater in the Late Jurassic than it is in the Recent (see below). The well-delimited accretion bands in the skeletal part of the thickly-branching coral colony of *Cryptocoenia limbata* GOLDFUSS from the St-Ursanne Formation which is represented in Fig. 3.1 are interpreted to be bands of annual accretion, because their average thickness of somewhat less than 5 mm is *regular* and close to the average thickness of 4 mm of the accretion bands in the cross-cut, Recent specimen of the massive brain coral from "Grid Reef" in the lagoon of Bermuda. The coral reefs of Bermuda are growing in a climatic zone where the difference between air temperature in the summer and occassionally below freezing in the winter is pronounced. Seasonal variation in air temperature in Bermuda is more pronounced than variation in the temperature of shallow water in which hermatypic corals are growing. Seasonal variation in water temperature around Bermuda reefs is such that variation in the rate of accretion of aragonitic calcium carbonate to the skeleton of scleractinian corals between summer and winter can be measured. Less solar irradiation or insolation during the winter months reduces photosynthesis by symbiotic zooxanthellae in tissue of reef-building corals and primary production of plankton in sea water. Less primary production engenders less animal food supply to the corals (coral polyps feeding are imaged in Fig. 8 in GYGI 1970). The distinctly delimited accretion bands with a regular thickness in the coral colony from the small reef in the uppermost St-Ursanne Formation south of Courtételle in Canton Jura are consequently evidence that the paleoclimate was in northern Switzerland seasonal in Late Jurassic time, and that the temperature of shallow sea water varied substantially between summer and winter like in Bermuda today. The type of sediments associated with coral reefs in the St-Ursanne Formation, mainly calcareous oolite in the Tiergarten Member, document a subtropical climate as well.

INSALACO (1996b, Pl. 1, Figs. 1–2) figured polished slabs of *Isastraea explanata* (GOLDFUSS) with accretion bands from the Liesberg Member at Liesberg. He explained on p. 416 that the banding is caused by thinner septa and fewer dissepiments in a low density band and by thicker septa with more numerous dissepiments in a high density band. INSALACO added on pp. 427–428 that it was proved in modern (Recent) corals that the low density band is deposited during the summer and the high density band during the winter. It can be read from Text-Fig. 2 in INSALACO (1996b) that the annual growth rate of *Isastraea explanata* in reefs of the St-Ursanne Formation is slightly less than 4 mm near Courtételle, and slightly more than 4 mm per year near St-Ursanne. INSALACO (1996b) continued on p. 428 that the growth rate in an environment with a high rate of argillaceous sediment influx like in the Liesberg Member (from deeper water) is less than in an environment with pure carbonate sedimentation like in the St-Ursanne Formation (from shallower water).

Additional evidence of a subtropical climate in the region investigated is the fact that the diversity of hermatypic coral assemblages in northern Switzerland is less than that of such assemblages in a lower paleogeographic latitude. This was visualized in the compilation by ZIEGLER (1967, Fig. 14). Accretion bands which are as distinctly delimited as those shown in Fig. 3.1 of this study are exceptional in hermatypic corals from Oxfordian sedimentary rocks in northern Switzerland. No accretion bands could be distinguished on a surface broken with a hammer in the field in most of the coral colonies which were collected in the region investigated. This cannot be interpreted to be unequivocal evidence that seasonal variation in the temperature of shallow water was slight at that time. For instance, annual accretion rings in the trunk of the red fir *Picea abies* (LINNÉ) with low density wood being accreted during the summer and high density wood accreted in the fall are well-delimited in the temperate climate at low altitudes in Switzerland, whereas annual growth rings in the trunk of *Aesculus hippocastanum* (LINNÉ), the horse-chestnut tree growing in the same environment, can hardly be distinguished in a sample of wood which is attached to p. 351 in GUGGENBÜHL (1980). A similar variability in distinctiveness of delimitation of annual accretion bands in hermatypic corals of different species growing in the same environment may have existed in Oxfordian corals in biostromes and reefs in what is now the Swiss Jura Mountains.

Long time spans with a relatively humid climate alternated with times with less rainfall in the Late Jurassic of northern Switzerland. These long-term climatic changes were probably significant. They were documented for the first time by GYGI and PERSOZ (1987), and again by GYGI (2000a, p. 45, 2003, p. 155). A humid climate with enhanced rainfall caused an ample supply of mainly argillaceous sediment with a varying, minor content of siliciclastic silt by running water from land in the northwest, when the lower part of successions no. 1, 2, and 3 was laid down. Inorganic-chemical and biochemical deposition of carbonates prevailed during drier times, when the upper part of these

Fig. 3.2 Siliciclastic siltstone with pelletoids (dark) in the Röschenz Member. Angular particles of detrital siliciclastic silt are light. Thin section photograph of bed no. 50 in section RG 400 in the small gorge northeast of farm La Providence near Corban in Canton Jura. Refigured from GYGI (2000a, Fig. 15 on p. 25)

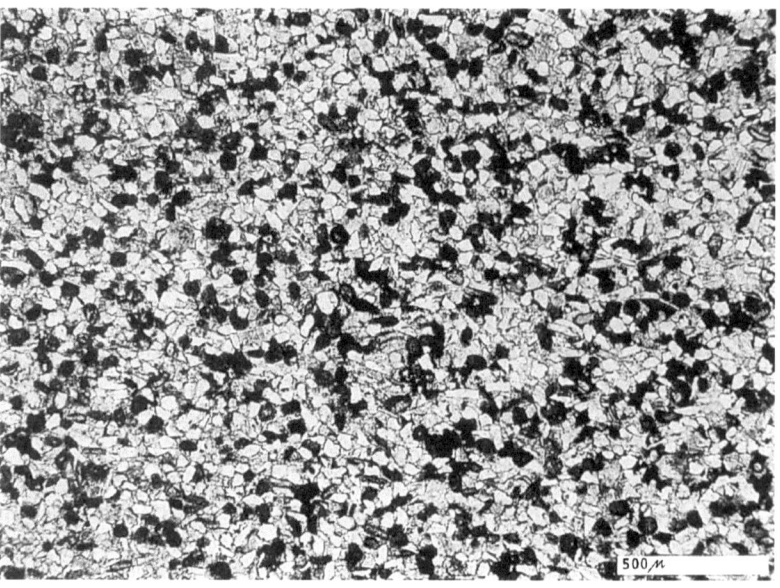

three principal successions was sedimented (see also CECIL 1990, p. 536). It was calculated above that the duration of major time spans with humid or drier climate was unequal. Alternation of the principal, unequal wetter and drier climatic terms in successions no. 1–3 consequently cannot have been caused by variation in the eccentricity of the Earth's orbit around the Sun ("astronomic tuning") in the mode of 100,000, 400,000, or 2,360,000 years, respectively (see above). The length of the principal climatic time spans was estimated using ammonite biochronology as well as time correlations based on mineral and sequence stratigraphy, the time scale by GRADSTEIN et al. (2004), and it was based on the simplifying assumption that ammonite chrons were of equal duration of on average approximately 370,000 years.

Each of the principal marl to limestone successions no. 1–3 shown in Figs. 1.5 and 1.6 represents a climatic cycle beginning with humid and ending with drier times. The early part of succession no. 2, which was sedimented in a humid climate, is a special case. The succession begins in the shallow-water realm with the Vorbourg Member. This is a pure limestone for instance in the Vorbourg section RG 366 near Delémont, bed nos. 13–24 in Pl. no. 23 in GYGI (2000a). The Röschenz Member above is a mixed succession of marl alternating with beds of marly limestone. Silt-grade siliciclastic minerals, mainly quartz and some feldspar, can be very common in some beds of the Röschenz Member (Fig. 3.2). Concentration of siliciclastic silt in the figured bed no. 50 from section RG 400 near Corban in Canton Jura must have occurred by elutriation of argillaceous and carbonate fines in flowing water. Both the Vorbourg and the Röschenz Member grade laterally into the coeval carbonate platform of the Günsberg Formation. This platform is much narrower than that of the St-Ursanne Formation below and that of the Balsthal Formation above. The Effingen Member in the

epicontinental basin is the time-equivalent of the Günsberg Formation (Fig. 1.5). A great amount of clay minerals and of some siliciclastic silt was supplied by water flowing from land in the northwest and north into the epicontinental sea (Fig. 1.2). The siliciclastic silt which is concentrated in some thin layers in the Effingen Member consequently bypassed the Röschenz Member and the Günsberg Formation in water currents during a time span with a humid climate (see discussion in Sect. 11.5.3 below).

A *long-term* climatic change throughout the Oxfordian Age, from on average rather humid climate in earliest Oxfordian time when the clayey marl of the *Renggeri* Member was laid down, to predominantly semiarid conditions during sedimentation of the calcareous oolite of the Verena Member in the upper Balsthal Formation before the end of the Oxfordian Age, can be traced through all of the successions no. 1, 2, and 3. There are no carbonate-rich intercalations in the homogenous clayey marl of the *Renggeri* Member in lower succession no. 1 (GYGI et al. 1998, Fig. 8, and GYGI 2000a, Pl. 30). On the other hand, carbonate-rich intercalations are common in the predominantly marly Effingen Member in the lower part of succession no. 2. The calcareous intercalations in the Effingen Member that are represented as black bands in section RG 37 in GYGI (1969a, Pl. 17), and are visible in Fig. 10 in GYGI et al. (1998), indicate a drier climate than that during sedimentation of the *Renggeri* Member. Thin layers of lignite coal with gyrogonites of characean freshwater algae and remains of land plants (see for instance GYGI 2000a, Pl. 22) were recorded in the Röschenz Member and in the coeval Günsberg Formation by several authors, beginning with HEER (1865, p. 125). This is evidence that climate continued to be humid during sedimentation of the greatest part of succession no. 2.

The Buix Member in the upper St-Ursanne Formation is the extremely pure carbonate end-member of succession no.

1 in its proximal, thick part (Fig. 1.5). Pümpin (1965, Fig. 21) figured part of a large, well-preserved leaf of a land plant which he had found in the Buix Member. The leaf is evidence of proximity of land with enough rainfall to support vegetation. The only traces of primarily ferroan dolomite (ankerite) found in the Buix Member are isolated millimeter-size, partially empty rhombohedra. Some powder of goethite was found in such otherwise empty rhombohedra. Such rhombohedra are common in bed no. 8 of the unpublished section RG 344 in a quarry 1 km west of the village of Courtemaîche, north of Porrentruy in Canton Jura. No traces of evaporite minerals were discovered to date in the Buix Member of succession no. 1. Calcite pseudomorphs after calcium sulfate minerals appear for the first time in the pure carbonate end-member of succession no. 2, in the Hauptmumienbank Member near Liesberg (Fig. 5.5C), but they are uncommon in this unit. Abundant calcite pseudomorphs after calcium sulfate that intersect both calcareous ooids and dedolomitized matrix were found in parts of the calcareous oolite of the Verena Member in the upper Balsthal Formation of succession no. 3 near Liesberg. Such pseudomorphs are conspicuous in a thin section photograph shown in Fig. 3.3. These pseudomorphs are evidence that dedolomitization was part of early diagenesis in the oolite of the Verena Member. Dolosparite is uncommon in that member. It was figured in a thin section from the Verena Member by Gygi (2000a, Fig. 16 on p. 26). Gypsum in the thin section Gy 7706 from the upper part of bed no. 113 in section RG 400 near Corban in Canton Jura that is mentioned in Pl. 35

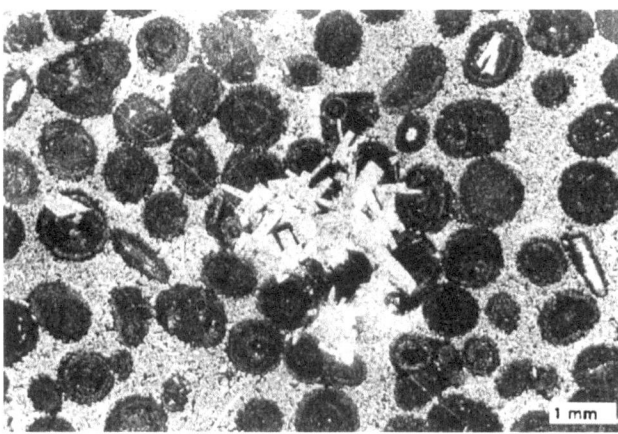

Fig. 3.3 Micritized calcareous oolite in Verena Member with secondary cement which was dolomitized and then dedolomitized. Pseudomorphs of calcite after sulfate minerals in the center replaced ooids and cement. Thin section photograph of bed no. 54 in section RG 398 in the limestone quarry of the former cement works at Liesberg, Canton Basel-Landschaft. Refigured from Gygi (2000a, Fig. 26 on p. 43)

and on p. 45 in Gygi (2000a) is more than 50% by volume of the calcareous oolite of the Verena Member. Calcrete nodules occur in a paleosol at the base of the Laufen Member in the unpublished section RG 15 in Horngraben Gorge south of Aedermannsdorf and at the base of the Reuchenette Formation at Balsthal (Fig. 5.9). Both villages are in Canton Solothurn.

4.1 Minerals, Fossils, and Sediments Diagnostic of Climate and of Water Temperature, Oxygenation, and Depth

4.1.1 Minerals

Goethite was accreted to crusts of iron ooids when the ooids were rolled at the surface of mud-grade marine sediments of Late Jurassic age in the region investigated from a minimal water paleodepth of about 10 m in a time equivalent of the uppermost Liesberg Member (GYGI 2003, p. 176), and down to a depth of approximately 100 m in the Schellenbrücke Bed in northeastern Canton Aargau (Figs. 4.10 and 4.11). The *Lamberti* Bed in Canton Schaffhausen and near Blumberg in adjacent southern Germany includes brown iron ooids, like in bed no. 20 of section RG 87 in the landslide of Bleiche on the western slope of Mt. Eichberg northwest of Blumberg (GYGI 1969a, Pl. 16). Very fine-grained glauconite was found in the Vorbourg Member, which is a sediment from the intertidal and shallow subtidal zone in the lowermost Vellerat Formation. The same type of glauconite occurs in a thin section from bed no. 23 in section RG 340 in a quarry at La Rasse south of Porrentruy in Canton Jura (GYGI 2000a, Pl. 17). Bed no. 23 of this section is a sediment of probably the *Evolutum* Chron from very shallow water in the lower Reuchenette Formation. Very fine-grained, light-green glauconite was found in a thin section of limestone bed no. 18 in section RG 434 in the eastern quarry of Steingrueben, which is 2 km west-northwest of Oberdorf in Canton Solothurn (GYGI 2000a, Pl. 42). This slightly glauconitic bed in the Reuchenette Formation is directly above an ancient tidal flat with sauro-pod footprints. The age of this bed is probably the *Caletanum* (or *Eudoxus*) Chron. EL ALBANI et al. (2005) found and identified glauconite in beds of lagoonal Purbeck facies (Early Cretaceous) in southwestern France. Glauco-nite occurs in the *Crenularis* Member of the Villigen For-mation in Canton Aargau, which was sedimented at a water

depth approaching 30 m. An aggregate of submicroscopic glauconite crystals with a blurred outline was found to have grown from the center outward in an isolated echinoderm ossicle in bed no. 32 of the Knollen Bed in section RG 79 in the quarry at Tenggibuck near Neunkirch in Canton Schaffhausen (Fig. 4.1E). The depth of deposition of the Knollen Bed near Neunkirch can be estimated at 80 m. Larger and mature, cauliflower pellets of pure glauconite with a clear-cut outline of the pellets were formed in thin beds which were sedimented at a particularly low rate on the basin floor from a minimal water paleodepth of about 100 m downward (Fig. 4.1A, B, C). Cauliflower pellets of pure glauconite are what ODIN and MATTER (1981, Fig. 4 C) called highly-evolved glaucony.

Glauconite pellets with a lustrous surface floating in a matrix of initially calcareous or argillaceous mud, which was sedimented at a low rate in thin beds on the level floor of the Rhodano-Swabian Epicontinental Basin, have the aspect of tiny cauliflowers with a diameter of typically between 200 and 400 µm (Fig. 4.1A). At least part of these glauconite pellets originated by early diagenetic transforma-tion of individual crystals of biotite to an aggregate of submi-croscopic glauconite crystals with a diameter of 1–3 µm (Fig. 4.1C). This is suggested by the presence of sparse crystals of fresh biotite with a diameter of 200–300 µm among abundant glauconite pellets and fine-grained detrital quartz in the Glaukonitsandmergel Bed of Canton Schaffhausen (GSM in the lower right of Fig. 1.5, or in Fig. 8.3). Biotite crystals of that size were also found in the stratigraphically condensed, lowermost bed of the Birmenstorf Member and in the glauconitic lower Baden Member. Biotite is a mineral with a slight resistance to abrasion. The isolated, fresh biotite crystals in the beds men-tioned were therefore probably not brought into the environ-ment by water currents. It is more likely that these biotites settled into the sediments out of a cloud of volcanic ash.

Two isolated crystals of biotite, one the size of 200 µm and the other of 250 µm, were found in a thin section from

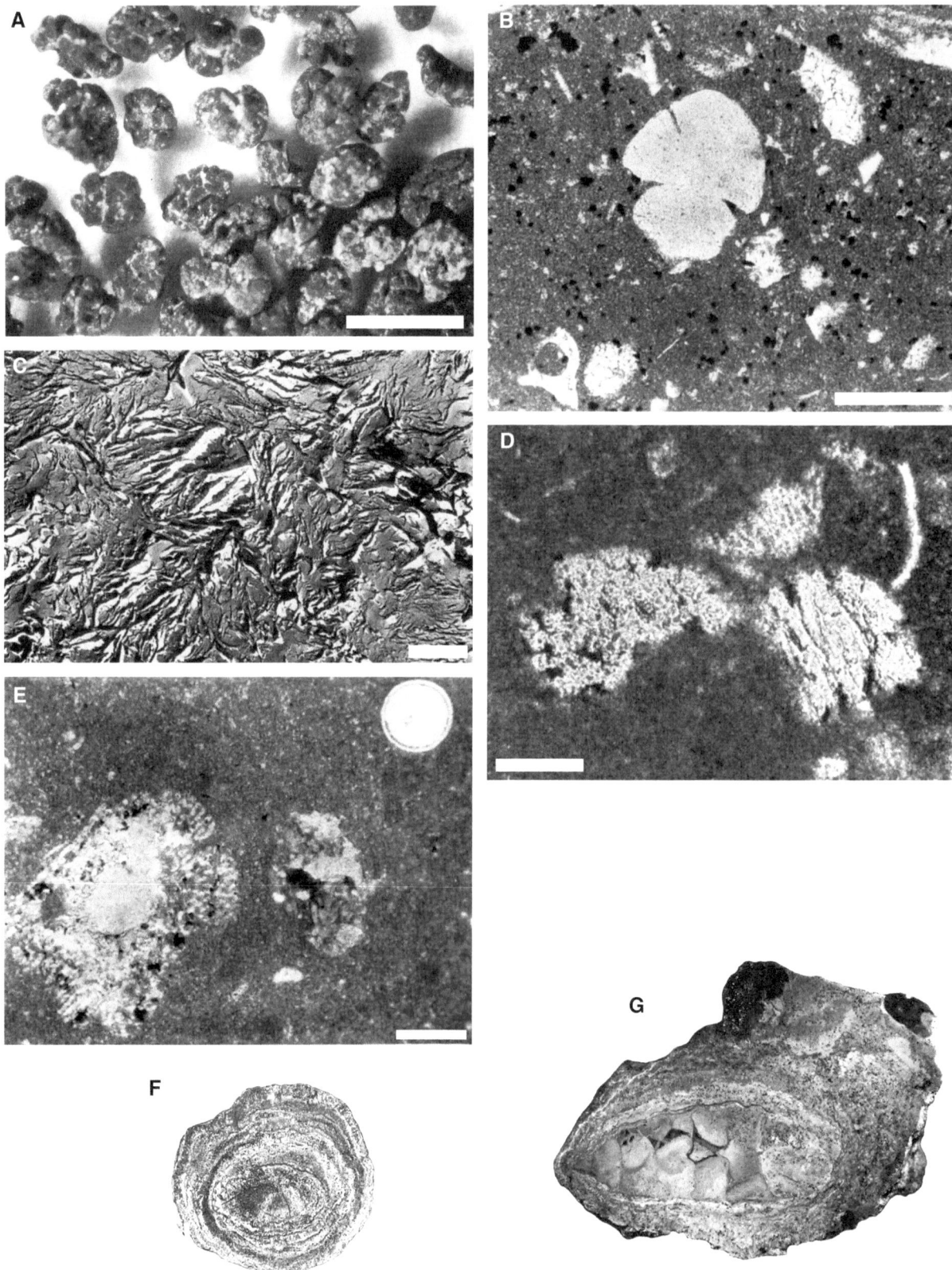

Fig. 4.1 (**A**) Cauliflower pellets of pure glauconite with surface cracked by swelling. Scale bar is 500 μm. Pellets isolated from

Glaukonitsandmergel Bed, no. 11 in section RG 81b, Gächlingen, Canton Schaffhausen (section shown in Fig. 8.3). Refigured with the lowermost bed of the Birmenstorf Member near Dangstetten in southern Germany (bed no. 5 of section RG 73 on the hill of Berchenwald). A photograph of this thin section is shown in Fig. 4.1D. The two biotite crystals are preserved at different stages of glauconitization. The grain on the right in the photograph is weakly glauconitized, whereas about 50% by volume of the biotite crystal on the left is replaced by dispersed, submicroscopic glauconite crystals. GALLIHER (1935, Fig. 12/14) showed that biotite crystals swell in the process of glauconitization, and that cracks open at the surface of evolving pellets of glauconite. The surface of the cracks grows convex when the pellets further swell during ongoing glauconitization. This is documented by a mature, pure glauconite pellet which is cross-cut in a thin-section from the Mumienkalk Bed (photograph of Fig. 4.1B). All of the investigated, mature glauconite pellets which were formed on the floor of the epicontinental basin are aggregates of minute, individual crystals of glauconite with a diameter of between 1 and 3 µm each (Fig. 4.1C).

GALLIHER (1935, Fig. 12/14) documented that Recent cauliflower glauconite pellets on the floor of Monterey Bay off the coast of California, USA, are growing by glauconitization of biotite at a low rate. He stated on p. 1582 that glauconite pellets become abundant from a minimal water depth of about 90 m downward. Monterey Bay is at latitude 37°N, this is to say 14° north of the tropic of cancer. The bay is therefore well outside the Recent subtropical belt. KUENEN in CLOUD (1955, p. 488) indicated that glauconite is now formed in Indonesia near the equator at a water depth greater than 55 m. PORRENGA (1967) found that glauconite formation now occurs in Africa at about 4°N off the Niger delta from a minimal water depth of between 60 and 70 m downward. He

showed in Fig. 2 that glauconite becomes abundant in that region from a water depth of 125 m downward.

Marine sand of calcareous ooids associated with coral reefs is forming in the Recent on the Bahama carbonate platform and off the Gulf shore of the United Arab Emirates not further north than approximately 25°N latitude. This is only slightly north of the tropic of cancer which is by definition the northern boundary of the northern subtropical belt. The northernmost coral reefs now living in the North Atlantic Ocean are those of Bermuda at 32°N. The northern boundary of the coral-oolite facies was found to be in the middle Oxfordian Age north of Yorkshire in England at a paleolatitude of more than 40°N (ARKELL 1933, p. 422 and Fig. 75 on p. 426). The belt of tropical and of subtropical climate was consequently in the Late Jurassic much wider than it is in the Recent. Bands of annual skeletal accretion with a regular thickness in the hard parts of Late Jurassic hermatypic coral colonies from northern Switzerland at the paleolatitude of about 28°N are not often as clearly visible as they were on a surface broken with a hammer in the colony shown in Fig. 3.1. Visibility of the delimitation of accretion bands in this specimen was enhanced by grinding and polishing the surface cut with a diamond saw. The fact that accretion bands in hermatypic corals from rocks in the region investigated often become distinctly visible only on polished surfaces like those figured by INSALACO (1996b), can be interpreted to be an indication that these corals grew, in spite of the relatively high latitude, not far north of the northern boundary of the tropical belt in the Late Jurassic, where seasonal variation in the temperature of shallow sea water was slight.

The minimal water depth from where downward prolific formation of mature cauliflower glauconite pellets occurs in

---

Fig. 4.1 (continued) permission by GSA (Geological Society of America) from FISCHER and GYGI (1989, Fig. 7). (**B**) Cross-section of a cracked cauliflower pellet of pure glauconite in micritic matrix of calcite with fossil fragments. The surfaces of the crack on the left side are convex because of swelling of the pellet in the course of glauconitization. Scale bar is 200 µm. Thin section photograph of Mumienkalk Bed, bed no. 24 of section RG 87 in the landslide of Bleiche on the western slope of Mt. Eichberg northwest of Blumberg in southern Germany (detailed section RG 87 shown in GYGI 1969a, Pl. 16). Refigured from GYGI (1969a, Pl. 4, Fig. 15). (**C**) Cross-section through a cauliflower pellet of pure glauconite, polished surface etched with HF. Scale bar is 1 µm. Photograph made with a transmission electron microscope of a pellet from Mumienkalk Bed, bed no. 3 of section RG 88 in the former open-cut iron mine at the southern foot of Stoberg Hill northeast of Blumberg, southern Germany (section RG 88 is shown as no. 9 in GYGI 1977, Pl. 11). Refigured from GYGI (1969a, Pl. 4, Fig. 14). (**D**) Two single crystals of biotite in micritic matrix of calcite. The biotites were identified with the petrographic microscope by the procedure described in GYGI (1969a, p. 44). The biotite grain on the right is weakly glauconitized with submicroscopic crystals of glauconite being interspersed in the biotite crystal. The biotite on the left is replaced to at least 50% of its volume by submicroscopic glauconite crystals. Scale bar is 100 µm. Thin-section photograph of bed no. 5 in the unpublished section RG 73, lowermost, stratigraphically condensed bed of the Birmenstorf Member,

Berchenwald hill near Dangstetten, southern Germany. Refigured from GYGI (1969a, Pl. 3, Fig. 9). (**E**) Bioclast of echinoderm ossicle on the left with aggregate of submicroscopic glauconite crystals in the core. Glauconitization of the ossicle nucleated in the core, and from there spread outward. To the right is a small algal nodule which is partially glauconitized. A calcisphere is in the upper right. Scale bar is 100 µm. Thin section photograph of Knollen Bed, bed no. 32 of the unpublished section RG 79 in quarry on Tenggibuck near Neunkirch, Canton Schaffhausen. Refigured from GYGI (1969a, Pl. 10, Fig. 38). Figures B-E refigured with permission of Beiträge zur Geologischen Karte der Schweiz, Neue Folge. (**F**) Spherical, deep-water oncoid in polished cross-section, natural size. Small black dots are cauliflower pellets of pure glauconite. Oncoid isolated out of Mumienkalk Bed with a rich macrofossil assemblage of mainly ammonites, bed no. 16a in section RG 207 near Siblingen in Canton Schaffhausen. Water depth approximately 120 m. Section RG 207 is shown in GYGI and MARCHAND (1982, Fig. 2). Refigured from GYGI (1992, Fig. 35). (**G**) Fragment broken off a cast of a large cardioceratid ammonite which was subsequently enveloped with a thick, oncolitic crust. Polished cross-section, natural size. Minute black dots are cauliflower pellets of glauconite. Oncoid isolated out of Mumienkalk Bed, bed no. 3 in section RG 88 in the former opencut iron mine at Blumberg, southern Germany (section RG 88 is shown as no. 9 in GYGI 1977, Pl. 11). Refigured from GYGI et al. (1979, Fig. 14b)

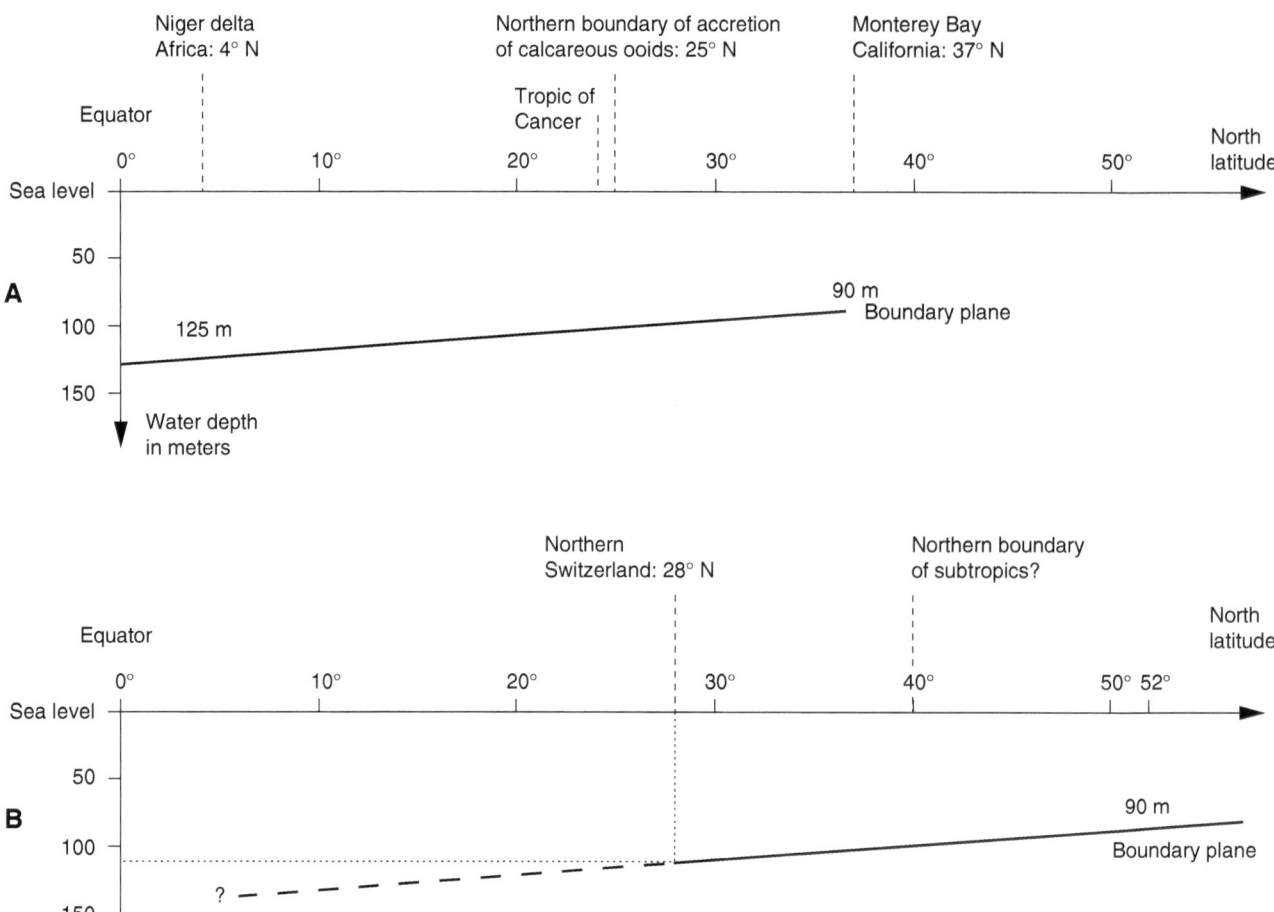

Fig. 4.2 Cross-sections of the hypothetical bathymetric boundary plane, below which formation of mature, cauliflower pellets of pure glauconite is prolific in the Recent (**A**) and was presumably prolific in the Late Jurassic (**B**). The boundary plane of the Late Jurassic is drawn with the same declivity as that of the boundary plane in the Recent. This declivity in cross-section B and the resulting, minimal depth of close to 110 m of prolific glauconite formation during the Late Jurassic in what is now northern Switzerland is rated in the text to be greater than it was in reality. The Late Jurassic boundary plane is concluded to have been in northern Switzerland at the water depth of approximately 100 m. The boundary planes are schematically drawn, without regard to warm or cold currents in the ocean

the Recent in Monterey Bay, California, at latitude 37°N is 90 m, and it is 125 m near the equator at 4°N off the Niger delta in Africa. The declivity of a hypothetical boundary plane of minimal water depth of prolific formation of mature glauconite pellets between 37°N and 4°N in the Recent could be concluded from this. A cross-section of such a hypothetical boundary plane is represented in Fig. 4.2A. The northern boundary of the area where *marine* calcareous ooids can be accreted in the Recent is at approximately 25°N. It must be mentioned in this context that the latitude where *limnic* calcareous ooids are accreted in the Recent in Great Salt Lake in Utah, USA, is 41°N (EARDLEY 1938). As mentioned above, the northern boundary of the area where marine calcareous ooids were accreted during the Late Jurassic was at a paleolatitude of more than 40°N. This is at least 15° latitude further north than where the boundary is now. It

was stated above that the minimal water depth of prolific formation of mature glauconite pellets at 37°N is now 90 m.

Provided that the corresponding water depth of 90 m was in the Late Jurassic at least 15° latitude further north like the northern boundary of calcareous ooid accretion at that time, then the minimal water depth of abundant glauconite pellet formation from 90 m depth downward would have been at a paleolatitude of at least 52°N in the Late Jurassic. When the additional assumption is made that the boundary plane of minimal water depth of prolific formation of mature glauconite pellets had the same declivity in the Late Jurassic as the corresponding boundary plane has in the Recent, then the minimal depth of prolific formation of cauliflower glauconite pellets is to be expected to have been in the Late Jurassic at a water depth of about 110 m in the Rhodano-Swabian Epicontinental Sea, at the paleolatitude of about 28°N where northern Switzerland is now (Fig. 4.2B). This assumption of minimal paleodepth of prolific glauconite formation is made without regard to warm or cold currents in the oceans of the Late Jurassic.

The figure of approximately 110 m of minimal water paleodepth for prolific, cauliflower glauconite pellet formation during the Late Jurassic concluded in northern Switzerland appears to be too great, when the following citations are taken into account. It is stated below in Sect. 11.6.2, that Arctic waters were in the early Oxfordian possibly almost as cold as they are now. The North Atlantic seaway between the Arctic sea and the epicontinental, Rhodano-Swabian Basin was as yet narrow in the Oxfordian, according to ZIEGLER (1988, Pl. 13). The quantity of cold water flowing from the Arctic towards the south was therefore probably modest. Temperatures of air and of shallow sea water in lower latitudes were in the Late Jurassic globally higher than they are in the Recent. ABBINK et al. (2001, p. 243) estimated an average annual air temperature of approximately 21°C for the early Oxfordian in the region of the southern North Sea. This may have engendered enhanced circulation of air and sea water between temperate latitudes and the tropics. The temperature gradient of both air and of near-surface sea water between higher latitudes and the equator was therefore in the Late Jurassic possibly less than it is now (compare with the citation of FRAKES, 1986, in MIALL 1997, p. 210). The declivity of the boundary plane as it is shown in Fig. 4.2B, which is the same as it is assumed to be in the Recent, is therefore possibly greater than it was in fact in the Late Jurassic. The assumption appears to be defendable, that the paleobathymetric boundary between iron ooid accretion in shallower water and formation of cauliflower glauconite pellets in deeper water, both above or within primarily argillaceous or carbonate mud, was in Late Jurassic time at a water depth of approximately 100 m on the floor of the epicontinental basin at the paleolatitude of 28°N where northern Switzerland is now.

Glauconite is an *authigenic* mineral in almost all of the sediments investigated. Authigenic glauconite was found in thin section in limestones from shallowest water like for instance in the peritidal Vorbourg Member and in peritidal parts of the Reuchenette Formation like bed no. 23 from the shallow subtidal zone of section RG 340 in the lower Reuchenette Formation in the quarry of La Rasse south of Porrentruy (GYGI 2000a, Pl. 17). Aggregates of authigenic glauconite crystals in limestones from very shallow water are tiny and of light green color. They have an indistinct, blurred outline. Authigenic aggregates of light-green glauconite crystals filled algal tubes in minute oncoids of the *Crenularis* Member (GYGI 1969a, Pl. 5, Fig. 21) from a water depth approaching 30 m, or algal tubes in larger oncoids in the lower Reuchenette Formation, in bed no. 47 of the unpublished section RG 28 at Schönenwerd in eastern Canton Solothurn.

Such a glauconitic oncoid from Schönenwerd was figured by GYGI (1992, Fig. 17). The Kimmeridgian oncoids near Schönenwerd have a maximal diameter of 14 mm, and they were formed at a water depth of about 40 m. This depth is concluded, because ammonites prevail in the macrofauna in bed no. 47 of section RG 28, and because section RG 28 is not far in the distal direction from section RG 21 near Olten, where bivalves of the genus *Pholadomya* prevail in the macrofauna of the coeval bed no. 57 from a depth of about 30 m (GYGI 1969a, Pl. 18, 1986, Fig. 6B).

Pure aggregates of glauconite crystals were found to have nucleated in the core of isolated ossicles of crinoidal echinoderms. Ossicles of echinoderms which are partially replaced by glauconite are dispersed in micritic limestone of the Knollen Bed in Canton Schaffhausen (Fig. 4.1E). The mode of formation of such unambiguously authigenic glauconite, which replaced skeletal calcite during diagenesis at a water depth that is estimated at about 80 m, is not understood. The cauliflower, authigenic pellets of glauconite which were formed, at least partially, by transformation of fresh, individual biotite crystals (Fig. 4.1D) at a water depth of more than 100 m, have always a clear-cut outline in thin section (Fig. 4.1B) and a lustrous surface in authigenic pellets after separation from marl-clay. Glauconite pellets which are doubtless *allochthonous* are very uncommon. Such pellets were found in thin sections of thin turbidites with a carbonate matrix in the Effingen Member. Between 40 and 50% by volume of such turbidites are grains of silt-size detrital quartz and of some feldspar. The rare pellets of glauconite in these turbidites have a clear-cut outline, but their surface is not smooth with cracks like in cauliflower pellets, probably because of abrasion during transport. The grain size of such allochthonous glauconite pellets in thin turbidites is the same as that of associated siliciclastic silt.

There are both vertical and lateral facies boundaries between the environment of iron ooid accretion in shallower water and the environment of authigenic growth of pure, cauliflower glauconite pellets in deeper water on the basin floor. A *vertical* facies change from iron ooid accretion mainly in the early Callovian Blumberg iron ore below to growth of glauconite pellets in the early Oxfordian Glaukonitsandmergel unit above (GSM in Fig. 1.5) was recorded for the first time by ZEISS (1955, Fig. 31) in the *Lamberti* Bed in the then opencut iron mine at the southern foot of Stoberg hill northeast of Blumberg in southern Germany. ZEISS (1955) found iron ooids and glauconite pellets to be dispersed side by side in the *Lamberti* Bed in the iron mine. The thickness of the bed is about 5 cm at that location. The *Lamberti* Bed is named after the ammonite *Quenstedtoceras lamberti* (SOWERBY). This is the index of the *Lamberti* Chron, which is the second-youngest ammonite chron of the Middle Jurassic. The Glaukonitsandmergel unit directly above the *Lamberti* Bed includes abundant cauliflower glauconite pellets, but no iron ooids (ZEISS 1955, Fig. 2).

GYGI (1966, Fig. 1) documented the same *vertical* facies change in three sections in adjacent Canton Schaffhausen and in northern Canton Aargau. GYGI (1986, Fig. 4) gave evidence that iron ooid accretion at a time of minimal sedimentation rate changed *vertically* in these sections to growth of glauconite pellets, because water was deepening in the course of eustatically rising sea level. GYGI and MARCHAND (1982) proved with figured ammonites that the thin, iron oolitic Schellenbrücke Bed from a water depth of less than 100 m in Canton Aargau grades *laterally* into the glauconitic, coeval Glaukonitsandmergel Bed in Canton Schaffhausen that was laid down in water deeper than 100 m (this study, Fig. 4.9). The vertical facies change occurred near Blumberg five ammonite chrons or approximately 1.85 million years earlier than in Canton Aargau, because water depth was in the *Lamberti* Chron near Blumberg, directly before the end of the Middle Jurassic, of the order of 100 m and at the same time approximately 80 m in Canton Aargau (Figs. 1.5 and 4.15). The vertical facies change near Blumberg was caused by the process of increasing water depth. According to Fig. 4 in GYGI (1986), water depth continued to grow after the time boundary between the Middle and the Late Jurassic. Figure 21 in NORRIS and HALLAM (1995) is at variance with this. Water depth at Blumberg became too great for iron ooid accretion (see below) just before the time boundary between the Middle and the Late Jurassic, and in Canton Aargau in the middle Oxfordian *Densiplicatum* Chron, when water depth above the iron-oolitic Schellenbrücke Bed of the early Oxfordian increased to more than approximately 100 m in that region at the time when sedimentation of the Birmenstorf Member began.

The youngest iron ooids which were found in some sections in Canton Aargau in the lowermost, thin and stratigraphically condensed bed of the Birmenstorf Member are uncommon and scattered. This bed was found to be on average 10 cm thick in excavation RG 208 near Ueken north of the Herznach iron mine in Canton Aargau, but it includes in that section both the *Densiplicatum* Zone and the *Antecedens* Zone. This was documented with ammonites which were figured by GYGI and MARCHAND (1982, Pl. 12, Fig. 3a–b), and by GYGI (2001, Figs. 32 and 62c). The condensed, lowermost layer of the Birmenstorf Member is a regional marker bed, and it rests in section RG 208, like in other sections, above the iron-oolitic Schellenbrücke Bed. The Schellenbrücke Bed was laid down during the *Cordatum* Chron in the latest part of the early Oxfordian, and it is the uppermost unit of the Herznach Formation (Figs. 4.10 and 8.2, beds no. 8 and 9). The bulk of the Herznach Formation is of Callovian age. Accretion of iron

ooids ended in Canton Aargau in most sections at the end of the *Cordatum* Chron when sedimentation of the Schellenbrücke Bed was completed, but it could apparently continue sporadically to the *Densiplicatum* Chron. This can be said, because iron ooids and glauconite pellets as well as ammonites of both the *Densiplicatum* and the *Antecedens* Chron occur side by side in the lowermost, condensed bed of the Birmenstorf Member, which is bed no. 10 in excavation RG 208 at Ueken (see section no. 2 in Pl. 11 in GYGI (1977)).

Accretion of iron ooids was replaced by growth of cauliflower glauconite pellets in Canton Aargau when water depth increased to more than approximately 100 m about 1.85 million years later than near Blumberg. Water then became too deep for iron ooid accretion because of relative sea level rise and because of some endogenic subsidence (see below). This is evidence that the sea floor at the beginning of the Oxfordian was in Canton Aargau at a lesser depth than near Blumberg. Sea floor topography is documented and quantified in Sect. 4.2 below. Only cauliflower pellets of glauconite occur in bed no. 22 of section RG 307 in the quarry of La Charuque at Péry in Canton Bern (GYGI 2000a, Pl. 22). The ammonite *Perisphinctes (Dichotomosphinctes) antecedens* SALFELD, the index taxon of the *Antecedens* Zone, was found in bed no. 22 of that section. This ammonite was figured by GYGI (1990a, Pl. 5, Fig. 4). The glauconitic bed no. 22 in section RG 307 is the lowermost bed of the Pichoux Formation. Bed no. 22 in section RG 307 is close to the toe of the sloping Pichoux Formation in the epicontinental basin.

Both the vertical and the lateral facies boundaries between environments in shallower water with iron ooid accretion and in deeper water with growth of cauliflower glauconite pellets, are consequently an important mark of water paleodepth at approximately 100 m *in the paleogeographic context investigated*. Some overlap of the paleobathymetric ranges of iron ooid and of cauliflower glauconite pellet formation cannot be ruled out to have existed. Provided that such an overlap did exist, it must have been insignificant. This can be concluded from the observation by ZEISS (1955) that the vertical facies change was going on near Blumberg during approximately one ammonite chron or less than 400,000 years. ALDINGER and FRANK (1943, p. 332) stated that the occurrence of abundant glauconite and of oolitic ironstone exclude each other in time and space in southwestern Germany. The vertical facies change from iron ooids below to cauliflower glauconite pellets above occurred near Blumberg during the *Lamberti* Chron in the latest Callovian. The same vertical facies change occurred in Canton Aargau and in Canton Solothurn five ammonite chrons later, at or just after the end of the early Oxfordian *Cordatum* Chron. This vertical facies change was almost contemporaneous in the whole region between Gansingen in northeastern Canton Aargau (sections

RG 60, 210, and 230, see Pl. 11 in GYGI 1977) and Günsberg in Canton Solothurn (section RG 14 in GYGI 1969a, Pl. 18). This is compelling evidence that the sea floor was in the pertinent area flat at that time (Fig. 1.5).

4.1.2   Fossils

A summary of variation in the composition of macrofossil assemblages depending on increasing *water depth* is Fig. 4.9 in this study. More data diagnostic of the close interrelation between water depth and fossil assemblages are included in Sect. 4.3. Of special interest in that text is that spherical, primarily hard calcareous oncoids with a diameter of up to several centimeters (Fig. 4.1F) could grow from very shallow lagoons down to a water depth of probably approximately 120 m (see main conclusion no. 21). Hermatypic corals of the *Oxfordian* are excellent indicators of water depth in the Rhodano-Swabian Epicontinental Sea, because they occurred from probably directly below low tide level to a maximal water depth of approximately 20 m. The lower limit to this rather narrow paleobathymetric range is quantified in Chap. 6. The bathymetric range of *Recent* hermatypic corals is much greater. Recent reef corals live only in warm water and are, therefore, diagnostic of a tropical or of a subtropical climate. The skeletal part of the coral colony with distinctly delimited accretion bands from the *Transversarium* Chron, which is refigured in Fig. 3.1 of this study, is evidence of a *subtropical* climate with a substantial variation in the temperature of shallow water between summer and winter in middle Oxfordian time. Recent hermatypic corals require adequate circulation of well-oxygenated water with a normal salinity in their habitat. There are no indications that the *same* ecological needs of the Oxfordian hermatypic corals investigated were substantially different. Therefore, the assumption can be made that water circulation in the lagoon of the Buix Member landward of the calcareous-oolitic sand bank of the Tiergarten Member was good, because lush growth of coral patch reefs occurred at least at an early stage of the lagoon's evolution (Fig. 4.19B). An additional indication that water exchange between the lagoon and the open epicontinental basin was adequate is that large perisphinctid ammonites were not uncommon in the lagoon, although water depth was never more than approximately 10 m at the initial stage of evolution of the lagoon. Ammonite casts from the very pure calcareous, chalk-like inter-reef sediment of the Buix Member were figured by GYGI (1995, Figs. 4, 6, 14, and 15).

The reason why the lower paleobathymetric limit to the Oxfordian hermatypic corals in northern Switzerland was no deeper than about 20 m could not be found. Turbidity of the water was probably not the critical parameter of this. This assumption can be made, because not a single hermatypic coral or coral colony was found among an estimated 3,000 ammonites in the very rich macrofossil assemblages with mostly ammonites in the Mumienmergel Bed (MUM) and Mumienkalk Bed (MUK in Fig. 1.5). Both of these beds include abundant cauliflower glauconite pellets. They are intercalated between the Glaukonitsandmergel Bed below and the unnamed glauconitic bed of marl or limestone above (Fig. 8.3). The two beds were laid down at a minimal sedimentation rate in probably particularly clear water at a depth of presumably 120 m (Fig. 4.9). In the Recent, hermatypic corals can live in very clear water of the Red Sea at a depth of up to 160 m (see below).

Fossils diagnostic of a *humid climate* are abundant gyrogonites of characean algae and shells of limnic ostracods in sediments from freshwater ponds in supratidal intercalations in the Günsberg Formation, which is mainly calcareous oolite. Thin intercalations of lignite coal were found as well in that formation (see below). Conversely, the Verena Member that was mostly a vast sand bank of calcareous oolite (Fig. 1.5) was sedimented toward the end of the Oxfordian Age in a relatively *dry climate*. Evidence of this is widespread early diagenetic dedolomite in the member. Calcite pseudomorphs after sulfate minerals grew in the Verena Member locally through both dedolomitized, originally sparitic calcareous cement and ooids (Fig. 3.3). Dedolomitization therefore occurred in the member during early diagenesis. Dolosparite was documented to occur by GYGI (2000a, Fig. 16). At least 50% by rock volume of gypsum replaced secondary cement of calcareous oolite in thin section Gy 7706 from 3 m above the base of bed no. 113 in the Verena Member of section RG 400 near Corban in Canton Jura (GYGI 2000a, p. 45 and Pl. 35). This indicates that water over the sand bank was locally hypersaline. Nevertheless, hermatypic corals were found at a few localities in the Verena Member, for instance in section RG 384 in Lochbach Gorge north of Selzach in Canton Solothurn (Pl. 29 in GYGI 2000a). Two small coral bioherms occur in the Verena Member at Seehof in Canton Bern in the unpublished section RG 419. They can be easily located because one of them projects out of bed no. 104 and the other out of bed no. 105. Local occurrence of more than 50% gypsum in calcareous oolite at one locality and of hermatypic corals as well as small coral bioherms in other parts of the member are an indication that there was regionally wide variation in the intensity of circulation and in salinity of water on the ooid sand shoal of the Verena Member.

4.1.3   Sediments

Humid climate on land in the northwest caused ample supply of terrigenous, predominantly argillaceous sediment to the Bärschwil Formation in the lower part of proximal, thick succession no. 1, and in succession no. 2 to the Effingen Member. Lignite seams and paleosols with fossilized rootlets

Fig. 4.3 Aggregates of radial, acicular calcite crystals which were isolated out of matrix of marl in the paleosol on top of Steinibach Member, bed no. 40B in the unpublished section RG 4 at cliff of Brocheni Flue 1.5 km west-southwest of Waldenburg in Canton Basel-Landschaft. Scale bar is 1 cm. Refigured from GYGI (2003, Fig. 176 on p. 162) with permission of Schweizerische Paläontologische Abhandlungen

(GYGI 2000a, Fig. 25) in the mainly oolitic Günsberg Formation are intercalations diagnostic of humid climate. Calcrete nodules in a paleosol are evidence of a drier climate. Radially arranged rays of acicular calcite below the surface of a calcrete nodule are shown in Fig. 5.9. Radial aggregates of acicular calcite crystals with an irregular orientation occur isolated within marl of a paleosol at the top of Steinibach Member (Fig. 4.3). Such supratidal sediments were formed on vast plains only slightly above high spring tide level as intercalations in parts of the carbonate platform of the Günsberg Formation. Widespread supratidal plains in the coeval, peritidal Röschenz Member landward from the Günsberg carbonate platform are documented by terrestrial intercalations with rootlets (GYGI 2000a, Fig. 25 on p. 42). Small and rapid relative sea level rises of a few meters like rises no. 6–8 (in Fig. 5.1) were therefore sufficient to flood such plains far into the interior of the shallow-water realm. Mud flats intersected by prism cracks (Figs. 4.4, 4.5, and 4.6), which evolved in laminated stromatolites with bird's-eye pores (fenestrae, Figs. 4.6 and 5.8), are indicative of the upper intertidal zone (KENDALL and ALSHARAN 2011, Fig. 16 (3) and 18). A laminated and crinkled stromatolite from the upper intertidal zone is in bed no. 65 of section RG 315 in Pichoux Gorge near Sornetan in Canton Bern. This stromatolite was figured by GYGI (1992, Fig. 12). Section RG 315 is shown in GYGI (2000a, Pl. 21). Intertidal stromatolites can be taken as indicators of mean sea level, because no evidence of a significant tidal range was found. The tidal range around the oceanic islands of the Bermuda Atoll or in the western Mediterranean Sea is approximately 30 cm. CHARLTON (1969, Fig. 2) measured in the protected waters of Harrington

Sound in Bermuda a tidal range of exactly one foot or 31 cm. This is possibly the order of magnitude of the tidal range in the Rhodano-Swabian Basin in the Late Jurassic.

Recent ooids of aragonite are accreted in the *marine* environment on sand banks in the bathymetric range between low tide level (Fig. 4.7) and, according to several authors, a water depth of at most 6 m (KENDALL and ALSHARAN 2011, Fig. 3). RANKEY and REEDER (2009) found ooids of Mg-calcite in the South Pacific to occur within the lagoon of Aitutaki Atoll, Cook Islands, in water depths of between 2.1 and 6.3 m. Pure sand of cleanly washed calcareous ooids is cemented during lithification by coarse-grained sparitic, secondary cement of low-magnesian calcite (GYGI 1969a, Pl. 8, Figs. 32–33, A and B cement: Fig. 36). Large wave ripples were conspicuous at the upper surface of bed no. 39 in the pure calcareous oolite of the Steinibach Member, when the unpublished section RG 448 was measured in the quarry of Mösliloch at Egerkingen in Canton Solothurn. A large sand wave of cleanly washed pelletoidal calcarenite from an environment similar to that of calcareous ooid accretion is shown in Fig. 5.5A.

Sand banks on which calcareous ooids are *accreted* in very shallow water like for instance on the *shoal* photographed in Fig. 4.7 differ from tidal *deltas* of calcareous ooid sand in which such sand is *shed* into water with a greater depth than the bathymetric range within which calcareous ooids can be accreted. A conspicuous, very large tidal delta of pure calcareous ooid sand was shed in Bajocian time from the basinward rim of the carbonate platform of the Hauptrogenstein Formation east of the villages of Auenstein and Veltheim in Canton Aargau. The ooid sand was shed in east-southeastern direction into water several tens of meters deep upon a surface of marly mud. The rim of the Hauptrogenstein carbonate platform runs north-south between Veltheim on the western bank of the Aare River and the village of Holderbank on the eastern river bank. The deltaic configuration of calcareous oolite in the quarries of Oberegg and Unteregg at Auenstein cannot be recognized when being viewed at close range. The foresets of the delta in the southern wall of the quarries only become apparent when viewed from a distance from the northern rim of the quarries. The top of the delta was at low tide level. The foresets dip from the top of the delta at an *apparent* angle of 10–15° from west to east. The exact, somewhat greater depositional declivity of the foresets could not be identified. The foresets resemble even planks reaching from the top of the formation through its entire thickness down to its base. The thickness of the Hauptrogenstein Formation was measured by the author to be 60.5 m in the western part of the quarry of Unteregg (section RG 226 in GYGI 1973, Fig. 3). It is concluded from this thickness that the toe of the tidal delta must have been at a water depth of several tens of meters. Calcareous oolite in itself is consequently undiagnostic of depth of deposition.

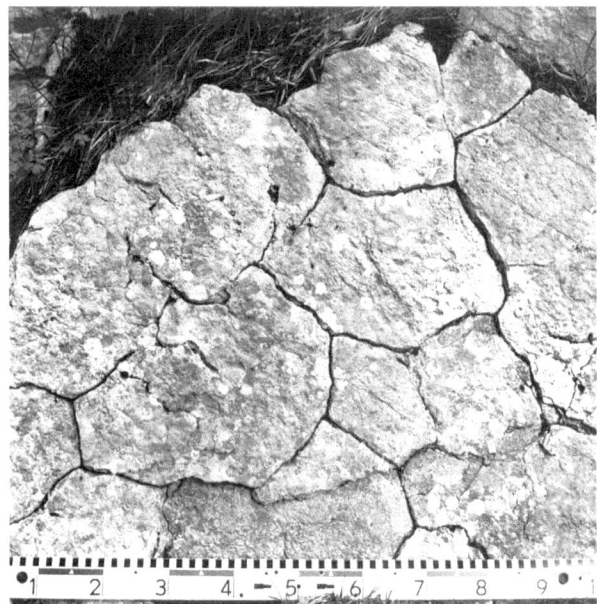

Fig. 4.4 Prism cracks at the surface of what was initially calcareous mud in the upper intertidal zone. Upper bedding plane of unit no. 42 in the unpublished section RG 417, upper Günsberg Formation, southern slope of Mt. Raimeux north of Crémines in Canton Bern. Numbers in scale bar indicate decimeters. Refigured from GYGI (1992, Fig. 5)

Fig. 4.5 Single prism with margins upwarped by dewatering of calcareous mud, taken out of the intertidal, mud-cracked stromatolite of bed no. 21 in the unpublished section RG 414, Günsberg Formation, southern slope of Mt. Raimeux north of Grandval in Canton Bern. Scale bar is 5 cm. Refigured from GYGI (1992, Fig. 6)

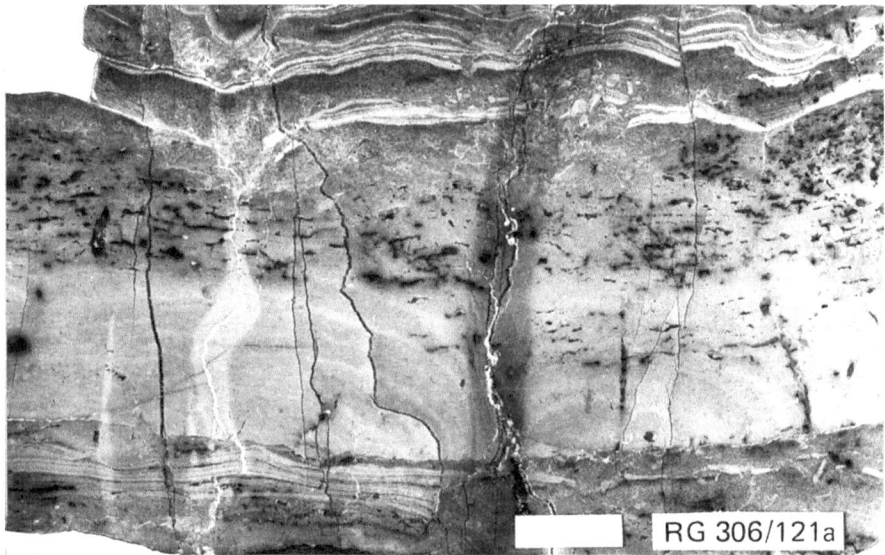

Fig. 4.6 Stromatolites from the upper intertidal zone, cross-cut in a polished slab. The flat-laminated stromatolite at the base is overgrown by a pincushion stromatolite. Above is a layer with abundant bird's-eye pores (fenestrae, black). A crinkled stromatolite is at the top. Thin, vertical dewatering cracks cut across the entire thickness of the bed. Unit no. 121a in Vorbourg Member on the crest of Chestel ridge, section RG 306 south of the village of Liesberg, Canton Basel-Landschaft. Scale bar is 1 cm. Refigured from GYGI (1992, Fig. 10)

The deltaic configuration of the entire Hauptrogenstein Formation in the two quarries at Auenstein and Veltheim is corroborated by the fact, that the formation pinches out completely east of the eastern end of the quarry of Unteregg.

The nearest outcrop on the eastern bank of Aare River is the quarry of Chalch at Holderbank, 1.3 km away. There is no trace of the Hauptrogenstein Formation in this quarry below the Spatkalk Member.

Fig. 4.7 Sand bank of calcareous ooids on which ooids are in the process of accretion. The photograph was made at low tide, when the sand bank is partially emergent and ripples at the surface are exposed. In the background is the tidal channel (dark gray) imaged in Fig. 4.8. The site where this photograph was made is marked in Fig. 4.8 with a circle. Refigured from GYGI (2003, Fig. 180 on p. 165) with permission of Schweizerische Paläontologische Abhandlungen

Fig. 4.8 Aerial view of some islands of the Joulters Cays north of Andros Island, Bahamas, looking toward the interior of the carbonate platform. To the left of the nearest island is a wide and deep tidal channel emptying into a delta of calcareous ooid sand in the foreground. The dark area in the lowermost part of the photograph is a coral biostrome growing in somewhat deeper water. The circle to the left of the tidal channel is around the spot from which the photograph of Fig. 4.7 was made. Refigured from GYGI (2003, Fig. 179 on p. 164) with permission of Schweizerische Paläontologische Abhandlungen

Undisturbed sediments settled out of suspension in water deeper than a few meters have a matrix of mud which gives no information of depth of deposition. Thin, laminated turbidites of carbonate mud mixed with siliciclastic silt (GYGI 1969a, Pl. 4, Fig. 12), or thick, pure mud turbidites which came to rest in scoured channels

(GYGI et al. 1998, Fig. 10), are evidence of deeper water. A clayey marl with dwarfed ammonite casts of iron sulfide is evidence of dysaerobic bottom water with an oxygen content of the order of 1 milliliter of dissolved oxygen per liter of water (see below) at a depth of several tens of meters.

4.2   Water Depth and Sea Floor Topography in the Deep Subtidal Zone

4.2.1   Water Depth

Sea level was assumed above to be indicated in sediments investigated approximately where there are stromatolites diagnostic of the upper intertidal zone (Figs. 4.4–4.6). The assumption could be made, because the tidal range in the area was very probably slight. Three clear-cut vertical facies boundaries separating well-defined paleobathymetric intervals were discerned in the epicontinental sediments investigated at the following, progressively greater depth of deposition and with uncertainty growing with depth. *First*, the most pronounced vertical change in the composition of macrofossil assemblages occurred when sedimentation in a shallowing-upward succession reduced water depth by aggradation to approximately 20 m. A mass occurrence of hermatypic corals in a biostrome appeared at this depth without a transition at the base of the Liesberg Member. This member is above the Sornetan Member with an assemblage with mainly bivalves and brachiopods in its upper half. The water depth of the same facies boundary at about 20 m in the uppermost Effingen Member is quantified in Chap. 6. Large bivalves of the genus *Pholadomya* are a conspicuous element of the macrofauna and can prevail in the macrofossil assemblage in the bathymetric interval between 20 and probably somewhat more than 30 m. *Pholadomya* are the main element in the macrofossil assemblage for instance in limestone of the Kimmeridgian *Lussasense* Chron in the large quarry on Mt. Born near Olten (GYGI 1986, Fig. 6B). *Second*, it can be read from Fig. 4.9 that ammonites prevail in the macrofossil assemblage in sediments laid down in water deeper than 30 m. Quantification of the bathymetric boundary where macrofossil assemblages with mostly bivalves are replaced in deeper water by assemblages with prevailing ammonites can only be made approximately. An argumentation going more deeply into the matter is necessary to explain this in the following section. *Third*, the deepest clear-cut vertical facies boundary between where iron ooids were accreted in shallower water and where cauliflower pellets of pure glauconite grew in deeper water was presumed above to have been at the water depth of approximately 100 m *in the region investigated*.

The upper half of the Sornetan Member in northwestern Switzerland with locally many large *Pholadomya* can be assumed to have been sedimented at a water depth of between 20 and 30 m according to the following argumentation. A thin marker bed with a rich assemblage of macrofossils separates the distal part of the Sornetan Member into a lower and into an upper half. This is the "Fossil Bed" of GYGI and PERSOZ (1986, Tab. 2 and Pl. 1A). GYGI and MARCHAND (1993, p. 1006) documented that the bed was sedimented during the *Cordatum* Chron at a low rate upon a transgressive surface after rapid sea level rise no. 4 (in Fig. 5.1). Sea level rise no. 4 caused a thorough change in the composition of the macrofossil assemblage in the fossil bed as compared with the assemblage below. Concluding from the bivalves *Pholadomya* above the fossil bed, the amount of rapid relative sea level rise no. 4 must have been modest. About 5 m is the best estimate that can be made. Ammonites prevail in the macrofauna of the fossil bed in section RG 399 at Bärschwil in Canton Solothurn, RG 373 at Vellerat in Canton Jura, and RG 314 at Sornetan in Canton Bern. Most of the ammonites are cardioceratids which were identified by the specialist D. MARCHAND from Dijon, France. Many of the ammonites from the fossil bed were figured by GYGI and MARCHAND (1993).

The fossil bed in the middle of Sornetan Member could also be discerned in the more proximal position of section RG 306 near Liesberg (GYGI 2000a, Pl. 31). The bed at that locality was sedimented below Chestel Member of the lower St-Ursanne Formation. The fossil bed at Liesberg includes much less macrofossils than further in the distal direction near Bärschwil, Vellerat, or Sornetan (see Fig. 1.5). Deposition of the sand bank of calcareous ooids in the upper part of Chestel Member ended near mean sea level. The fossil bed at Liesberg is no. 66 in section RG 306, and it is in that section 70 m below the top of Chestel Member. When the assumption is made that massive growth of hermatypic corals began at the base of the Liesberg Member at Liesberg at the same water depth of about 20 m as in the uppermost Effingen Member in section RG 307 near Péry (GYGI 2000a, Pl. 22), where the pertinent depth of 20 m could be quantified in Chap. 6, then the depth of deposition of the fossil bed at Liesberg can be approximately quantified in the following way. The thickness of Chestel Member in section RG 306 is 24 m, and that of the Liesberg Member directly below is 25 m. The fossil bed no. 66 is about 21 m below the base of the Liesberg Member. Water depth diminished from the base of the Liesberg Member at the depth of 20 m to 0 m at the top of Chestel Member. When a constant rate of sedimentation and even compaction of the marly Liesberg Member is assumed, then water depth diminished by 1 m per 2.45 m of measured, compacted sediment thickness of the aggregate thickness of 49 m of the Liesberg and of the Chestel Member. When water depth diminished at the same rate of 1 m per 2.45 m of measured sediment thickness in marl of the upper Sornetan Member with a thickness of 21 m below, then the fossil bed was deposited at Liesberg at the water depth of 29 m. This is

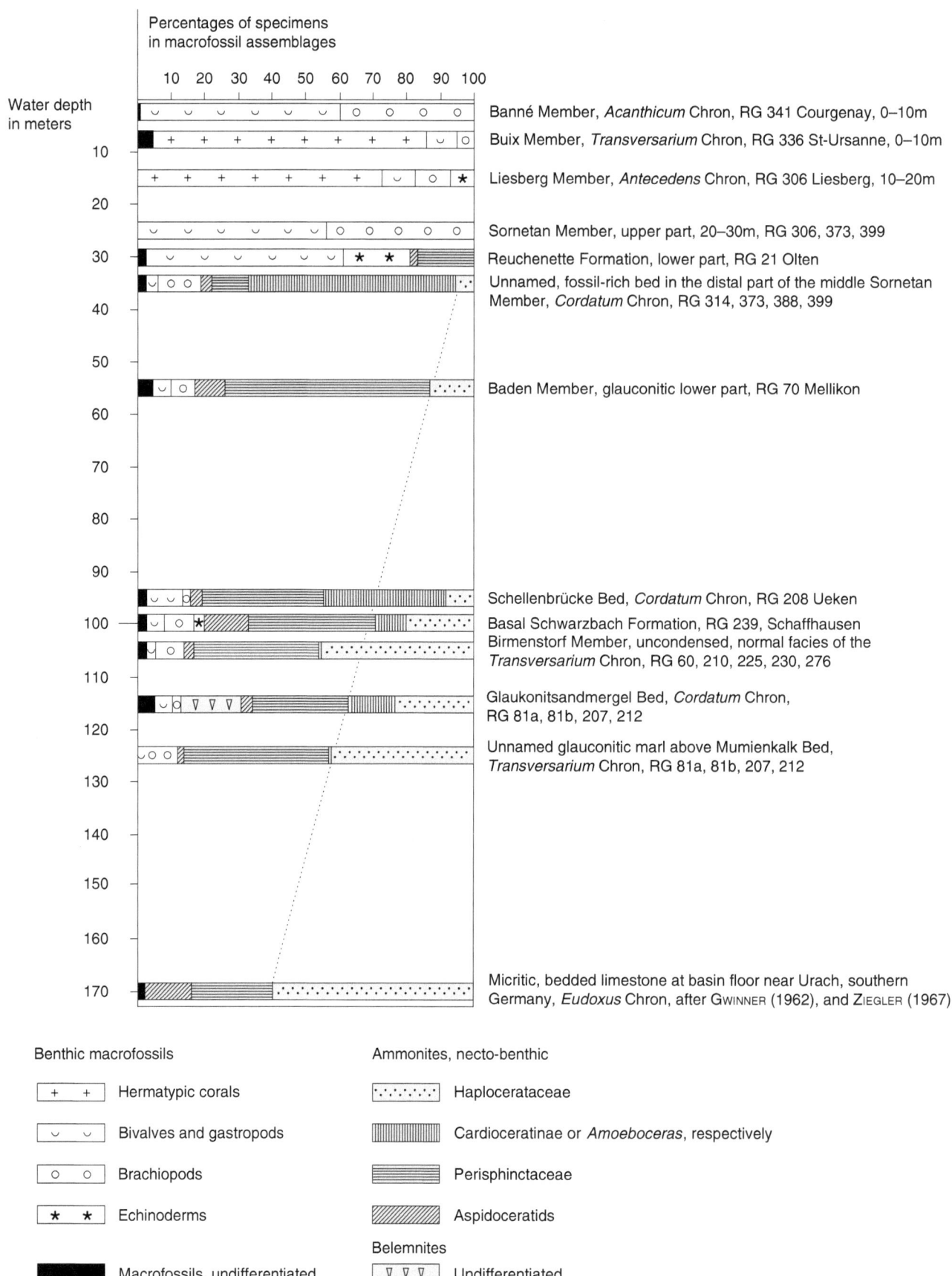

Fig. 4.9 Composition of macrofossil assemblages varying with increasing depth of deposition of the encasing sediment. Percentages in the Buix and the Liesberg Member are estimated

probably somewhat too much, because the muddy matrix of marl of the Sornetan and of the Liesberg Member was compacted, whereas the pure calcareous ooid sand in the upper part of Chestel Member was not compacted.

The depth of sedimentation of the fossil bed in the middle Sornetan Member below the cliff of Peute Roche southwest of the village of Vellerat in Canton Jura can be calculated based on sections RG 373, 389, and 451, which are located 800 m southwest of the village. The three sections were assembled into the composite that is represented in Pl. 24 in GYGI (2000a). Sedimentation of the coral biostrome of the Grellingen Member (Fig. 1.5) with a matrix of initially calcareous mud in the lower St-Ursanne Formation ceased at this locality, when the calcareous ooid sand of the Tiergarten Member above began to accumulate on the sediment surface of mud which was then at a water depth of probably at most 5 m (Fig. 1.5). The situation may have been similar to what is shown in Fig. 4.8 of the Recent carbonate platform of the Bahamas. The matrix of initially lime mud in the Grellingen Member was probably compacted about as much as the muddy matrix of the Liesberg Member and of the Sornetan Member below. The fossil bed in the middle Sornetan Member at Vellerat was excavated in the two trenches of section RG 373 and RG 388 (GYGI and MARCHAND 1993, Fig. 2) descending from the base of the Liesberg Member. The fossil bed at this location is no. 8 of section RG 373, and it is 85 m below the top of the Grellingen Member in contiguous section RG 451. The top of Grellingen Member in section RG 451 is at the top of bed no. 3. The vertical cliff of section RG 451 was measured and sampled on the rope.

The base of the Liesberg Member is at Vellerat at the base of bed no. 48 in section RG 373, or 60 m below the top of the Grellingen Member. On the assumption that hermatypic corals first appeared at the base of the Liesberg Member at the water depth of 20 m like at Péry in section RG 307, and that the rate of sedimentation was constant, then water depth diminished from the base of the Liesberg Member by 15 m to the top of Grellingen Member at the depth of about 5 m, or by 1 m every 3.93 m of measured rock thickness. The figure of 3.93 m is relatively great, because the thickness of the Grellingen Member is above average near Vellerat. The aggregate thickness of the Liesberg Member and of the Grellingen Member is at Vellerat 59 m. Fossil bed no. 8 in section RG 373 is 27 m below the base of the Liesberg Member. Depth of deposition of the fossil bed can then be calculated to be 27 m divided by 3.93, which is equal to an increment of water depth of about 7 m, plus the 20 m depth at the base of the Liesberg Member, plus 5 m water depth above the top of the Grellingen Member at the end of its sedimentation. The fossil bed at Vellerat was consequently sedimented at a calculated water depth of 32 m. The actual

depth was probably somewhat greater, because the unusually great thickness of the Grellingen Member at Vellerat biased the calculation. Ammonites prevail in the macrofossil assemblage in the fossil bed near Vellerat at a water depth which can be estimated to have been approximately 35 m. Ammonites were found to be uncommon in the macrofossil assemblage with mainly terebratulid brachiopods in the coeval fossil bed at Liesberg, where the bed was sedimented at a probable water depth of 25 m. This is evidence that the moderate increment in water depth from about 25 m near Liesberg across the threshold of 30 m to around 35 m at Vellerat was sufficient to cause the abundance of ammonites to augment abruptly and massively from slight to predominance in the macrofauna of the fossil bed not only near Vellerat, but also near Bärschwil and near Sornetan. Ammonites in the sediments investigated consequently prevailed in the macrofossil assemblage from a depth of sedimentation of approximately 35 m downward (see main conclusion no. 24 below).

The paleobathymetric interval of between 20 and 30 m in the upper Sornetan Member is characterized by a macrofossil assemblage with mainly bivalves. Large *Pholadomya* are bivalves typical of this facies which was called Pholadomyen by ÉTALLON (1862). Associated in the Sornetan Member are few, large or giant perisphinctacean ammonites like in the Geissberg Member in the large quarry on Mt. Geissberg west of Villigen, in the *Crenularis* Member in the former quarry of the unpublished section RG 65 at Würenlingen, as well as in the lower Reuchenette Formation in the quarry on Mt. Born west of Olten (section RG 21 on Pl. 18 in GYGI 1969a, 1986, Fig. 6B). Sedimentation of the Geissberg Member near Villigen (section RG 62, Pl. 17 in GYGI 1969a) ended at a water depth of more than 20 m, to judge from the absence of hermatypic corals. The macrofossil assemblage in the member includes on Mt. Geissberg very large ostreid bivalves with thick shells and *Pholadomya*. Rapid relative sea level rise no. 9 of several meters (Fig. 5.1) occurred at the beginning of sedimentation of the glauconitic *Crenularis* Member directly above the Geissberg Member and augmented water depth to probably close to 30 m. The mass occurrence of *Pholadomya* in the *Crenularis* Member of section RG 63 near Villigen is an indication that the optimal water depth for *large Pholadomya* was close to 30 m. The depth of deposition of the lower Reuchenette Formation near Olten with abundant *Pholadomya* and with some giant perisphinctacean ammonites, which were figured by GYGI (2003), was therefore probably approaching 30 m as well (GYGI 1986, Fig. 6B).

4.2.2   Sea Floor Topography

Water depth near Blumberg and in adjacent Canton Schaffhausen was at the beginning of the Oxfordian approximately 100 m. This was concluded from the occurrence of both iron ooids and cauliflower glauconite pellets in the thin *Lamberti* Bed near Blumberg, according to Zeiss (1955) and following the text on glauconite formation above. Water depth at the beginning of the Oxfordian Age in Canton Aargau is estimated below at 80 m. The net, relative sea level rise going on since that time increased water depth in this region to 100 m by *Densiplicatum* time. This is six ammonite chrons later than when water depth had increased near Blumberg to more than 100 m. The difference in water depth at the beginning of the Oxfordian was probably localized in a step which existed above the distal rim of the carbonate platform of the early Bathonian Spatkalk Member. The step was between Canton Schaffhausen and the level sea floor in Canton Aargau (Figs. 1.5 and 4.15A). This assumption can be made, because the lateral facies boundary between the Glaukonitsandmergel Bed in Canton Schaffhausen or in northwestern Canton Zürich and the coeval facies with iron ooids of the early Oxfordian Schellenbrücke Bed in northeastern Canton Aargau is approximately congruent with the basinward rim of the early Bathonian Spatkalk carbonate platform below. The course of this platform rim is shown in Fig. 1 in Gygi (1986). The lateral facies boundary mentioned is east of unit no. 3, the iron-oolitic Schellenbrücke Bed with a *Cardioceras* of the *Cordatum* Chron in section RG 73, in the forest on the hill of Berchenwald near Dangstetten north of the Rhine River 3 km northeast of Zurzach. The coeval bed in the exploration well which was drilled by Nagra 2 km east of Weiach in northwestern Canton Zürich (Nagra 1988) is the Glaukonitsandmergel Bed. The platform rim of the Spatkalk Member could act as a paleobathymetric template until mid-Oxfordian time, because the thickness of sediments of Callovian and of early Oxfordian age above the rim is of the order of decimeters or of a few meters at most, where these beds occur below the Birmenstorf Member between Canton Aargau and Canton Schaffhausen. Sediments of this age were not deposited at all in the region of Canton Aargau between Auenstein, Villigen, and Oberehrendingen (see pertinent sections in Mangold and Gygi 1997, Fig. 2). Gygi and Marchand (1982) documented with many figured cardioceratid ammonites that the Schellenbrücke Bed and the Glaukonitsandmergel Bed are coeval.

Gygi (1986, p. 488) estimated that the net, global sea level rise in the Oxfordian Age amounted to between 25 and 30 m. To judge from Fig. 4 in Gygi (1986), eustatic sea level rise from the beginning of the Oxfordian to the *Densiplicatum* Chron, plus basement subsidence under the load of the additional water, and slight endogenic subsidence during the pertinent time interval, resulted in a total of relative sea level rise that can be estimated at 20 m. This was enough to replace iron ooid accretion in the thin sediments of early Oxfordian age in Canton Aargau by formation of cauliflower pellets of glauconite in the *Densiplicatum* Chron of the middle Oxfordian, when water depth had increased to

100 m in the region. Water depth in Canton Aargau can therefore be assumed to have been about 80 m at the beginning of the Oxfordian. The step of the distal rim of the early Bathonian carbonate platform of the Spatkalk Member then had an elevation of probably 20 m (Fig. 4.15A).

The difference in depth of deposition between the Schellenbrücke Bed and the Glaukonitsandmergel Bed was comparatively slight, but it had a substantial influence on the composition of the ammonite assemblage in the beds. The iron oolitic Schellenbrücke Bed of *Cordatum* age in Canton Aargau directly below the Birmenstorf Member was sedimented at a water depth of slightly less than 100 m (Fig. 4.9). This depth is assumed to have been about 95 m, because few isolated pellets of glauconite were found in some sections in the Schellenbrücke Bed, for instance in bed no. 3 of section RG 73 near Dangstetten. Section RG 73 is located close to the lateral facies boundary between the Schellenbrücke Bed and the Glaukonitsandmergel Bed. Provided that endogenic subsidence proceeded at the same rate on both sides of the step of the basinward rim of the Spatkalk Member, then the Glaukonitsandmergel Bed must have been sedimented at a water depth which was greater than that of the Schellenbrücke Bed. The difference in paleodepth is indicated by the varying percentage of cardioceratid ammonites in the two beds which can be read from Fig. 4.9. Cardioceratinae in the iron oolitic Schellenbrücke Bed at a paleodepth of about 95 m are approximately three times more abundant than in the Glaukonitsandmergel Bed of Canton Schaffhausen with very abundant cauliflower pellets of glauconite. The tendency in haploceratid ammonites is the opposite. The abundance of Haplocerataceae more than doubles from the Schellenbrücke Bed to the Glaukonitsandmergel Bed, which was laid down in deeper water. This is an indication that the difference in depth of sedimentation between the two beds was considerable. Accordingly, the thin glauconitic marl or limestone (Fig. 8.3) above the Mumienkalk Bed in Canton Schaffhausen, which is time equivalent with the upper, uncondensed Birmenstorf Member in Canton Aargau, was sedimented at a greater water depth than the upper Birmenstorf Member. It is noteworthy, that haploceratid ammonites are somewhat less abundant in the thin bed above the Mumienkalk Bed than in the upper Birmenstorf Member from a lesser water depth. This is evidence that the percentage of Haplocerataceae in a sediment did not grow smoothly with increasing depth of deposition. But the dashed line in Fig. 4.9 outlines the general trend existing in Haplocerataceae to become progressively abundant in the ammonite assemblage with growing water depth.

Sedimentation of the Birmenstorf Member began at a water depth of around 100 m. It ended at a depth of approximately 105 m, according to the following calculation. Water depth had increased near Ueken in Canton Aargau (section RG 208) to about 100 m probably already during the *Densiplicatum* Chron, when sedimentation of the lowermost, stratigraphically condensed bed of the Birmenstorf Member was underway, and when accretion of iron ooids was replaced by growth of glauconite pellets in the thin bed. The rapid relative sea level rise which followed in the later part of the *Antecedens* Chron, plus isostatic basement subsidence under the additional water load amounted to a sum of approximately 10 m (see quantification of sea level rise no. 5 in Sect. 5.1 below). The measured, compacted thickness of the matrix of mainly calcareous and of some argillaceous mud in the uncondensed, normal facies of the Birmenstorf Member in the Rhodano-Swabian Basin is slightly more than 5 m both in section RG 14 near Günsberg in Canton Solothurn and in sections RG 210 and 230 near Gansingen in northeastern Canton Aargau. Sedimentation of the normal facies of the Birmenstorf Member in the epicontinental basin during the *Transversarium* Chron began after the relative sea level rise during the later part of the previous *Antecedens* Chron had augmented water depth in Canton Aargau to about 110 m. The initial, decompacted thickness of mud in the normal facies of the Birmenstorf Member was 7.5 m, and prompt isostatic basement subsidence under the load of this sediment of *Transversarium* age was about 3.5 m. A water depth of approximately 105 m resulted from this at the end of sedimentation of the Birmenstorf Member (Fig. 4.9) at the top of succession no. 1 in eastern Canton Solothurn and in Canton Aargau. Endogenic subsidence was insignificant at this time. Compaction and decompaction were quantified using the nomogram of Fig. 11 in PERRIER and QUIBLIER (1974). Compaction of lime mud was estimated after TERZAGHI (1940).

The earliest Oxfordian sediment in northwestern Switzerland is the uppermost bed of the Herznach Formation of the *Scarburgense* Chron. This is a regional marker bed with iron ooids in a matrix of marl. The iron ooids fade away completely from the base of the bed to its top. The thickness of the bed is everywhere approximately 25 cm. This is bed no. 20 in section RG 307 in the quarry of La Charuque near Péry (GYGI 2000a, Pl. 22). The bed was excavated in this quarry on a surface of more than 20 m^2. Macrofossils were found to be uncommon in the bed both in the excavation at Péry and in coeval bed no. 6 of section RG 280 at Liesberg (GYGI 1990a, Fig. 2). Ammonites prevail in the macrofossil assemblage everywhere in this marker bed. A macrofossil assemblage with mainly ammonites was concluded above to indicate a minimal water paleodepth of 35 m. Iron ooids in a matrix of what was primarily mud in a thin and widespread bed on the basin floor are evidence of a maximal water paleodepth of sedimentation of about 100 m. The depth of sedimentation of the thin iron oolitic marl at the base of the Oxfordian in northwestern Switzerland, for instance of the excavated bed no. 20 in section RG 307 at Péry, or of coeval bed no. 6 in section RG 280 near Liesberg (GYGI 1990a, Fig. 2), cannot be concluded directly from the macrofossil

assemblage, because well-preserved macrofossils in the marker bed are not numerous enough to be diagnostic. Therefore, the *assumption* must be made that water depth at the beginning of the Oxfordian in northwestern Switzerland was more than the 35 m which are the minimum depth from which ammonites prevail in the macrofauna. Water depth was less than 100 m because of the presence of iron ooids, and because cauliflower glauconite pellets are absent in the basal marker bed of the Oxfordian.

The lowermost bed of the Pichoux Formation in the quarry of La Charuque near Péry is bed no. 22 in section RG 307 (GYGI 2000a, Pl. 22). Abundant, cauliflower pellets of glauconite and the ammonite *Perisphinctes (Dichotomosphinctes) antecedens* SALFELD in this bed are evidence that a relative sea level rise augmented water depth from the beginning of the Oxfordian to more than 100 m in the *Antecedens* Chron. The lowermost bed of the Pichoux Formation near Péry grades laterally into the lowermost bed of the Birmenstorf Member. Cauliflower pellets of glauconite coexist in several sections in Canton Aargau with sporadic iron ooids in the basal bed of the Birmenstorf Member. The basal bed with an average thickness of less than 10 cm represents both the *Densiplicatum* and the *Antecedens* Chron. Glauconite pellets prevail by far over iron ooids in this bed. This is an indication that water depth increased during sedimentation of the lowermost bed of the Birmenstorf Member to more than 100 m in the *Densiplicatum* Chron rather than in the successive *Antecedens* Chron. The vertical change from iron oolitic facies of *Cordatum* age below to glauconitic facies above in the lowermost part of the Birmenstorf Member occurred at almost the same time between section RG 14 in the landslide of Gschlief, 1.5 km northwest of the village of Günsberg in Canton Solothurn (GYGI 1969a, Pl. 18), and section RG 210 in Eisengraben, 2 km south-southwest of the village of Gansingen in northeastern Canton Aargau (this study, Fig. 8.3).

This is evidence that sedimentation of the Birmenstorf Member between Günsberg and Gansingen began on a level sea floor, as it is shown in Fig. 1.5. It can be concluded from this that the sea floor in the whole region between northwestern Switzerland and northeastern Canton Aargau was already flat at the beginning of the Oxfordian Age, because the thickness of the early Oxfordian iron oolites below the Birmenstorf Member is nowhere more than a few decimeters where they occur in the pertinent region. These iron oolites pinch out entirely because of nondeposition between Auenstein, Villigen, and Ehrendingen in Canton Aargau. The iron-oolitic Schellenbrücke Bed of the Oxfordian reappears in bed no. 3 of the unpublished section RG 73 on the hill of Berchenwald 700 m north of the village of Dangstetten in southern Germany. Section RG 73 is 4.5 km north-northwest of section RG 70 near Mellikon that is shown in Fig. 1.5. Water depth near Blumberg 18 km north-northwest of Schaffhausen was at the beginning of the Oxfordian approximately 100 m. The step of the distal rim of the carbonate platform of the Spatkalk Member was concluded above to have been about 20 m. Water depth in Canton Aargau at the beginning of the Oxfordian Age was, therefore, assumed to have been about 80 m. Provided that the sea floor at the time boundary between the Middle and the Late Jurassic was indeed a level plain between Canton Aargau and Günsberg, then water depth at the beginning of the Oxfordian Age in northwestern Switzerland was approximately 80 m like in Canton Aargau.

4.3 How the Composition of the Macrofossil Assemblage in a Sediment Varied with Increasing Depth of Deposition

4.3.1 Variation in Main Macrofossil Assemblages

The upper limit of the paleobathymetric range of the hermatypic corals investigated, both at the top of biostromes and bioherms, was probably close below low spring tide level, as it is the case at the present time, for instance, in the Bermuda Atoll. Massive hermatypic coral colonies bored by bivalves of the genus *Lithophaga* are at the base of the Banné Member in the unpublished section RG 341 near Courgenay in Ajoie region of Canton Jura. These corals are associated with a mass occurrence of bivalves and brachiopods, and with some gastropods. The initial mixture of calcareous and argillaceous mud in the Banné Member has now a compacted thickness of 7.7 m in section RG 341. The mud was sedimented into a lagoon with a water depth of at most 10 m above a stromatolite with crinkled laminae from the upper intertidal zone. The stromatolite was indurated, covered with a crust of goethite, and then settled by ostreid bivalves, before sedimentation of the partially argillaceous Banné Member began. The stromatolite is bed no. 33 of section RG 341. The very fossiliferous Banné Member is, therefore, evidence of a rapid relative sea level rise which occurred during the *Acanthicum* Chron (see below, and rise no. 13 in Fig. 5.1). The lower limit to the habitat of hermatypic corals was in the Oxfordian of northern Switzerland possibly nowhere deeper than approximately 20 m below mean sea level, as it could be quantified in Chap. 6.

Bivalves mainly of the genus *Pholadomya* prevail in the macrofossil assemblage in the sediment on the depositional slope near Olten off the basinward rim of the carbonate platform of the lower Reuchenette Formation east of Balsthal (see this study, Fig. 1.5, Fig. 4.9, and GYGI 1986, Fig. 6B). There is a mass occurrence of *Pholadomya* in the *Crenularis* Member of the Villigen Formation in the cliff of section RG 63 located 300 m northwest of the ruin of Besserstein castle above the village of Villigen in Canton

Aargau. Both these sediments must have been laid down at a water depth approaching 30 m. The upper half of the Sornetan Member with locally numerous, large *Pholadomya* was sedimented in an interval of water depth between 20 and 30 m. This is below the paleobathymetric range of Late Jurassic hermatypic corals in the sediments investigated, and above the boundary of paleodepth from where downward ammonites begin to prevail in the macrofossil assemblage (Fig. 4.9).

A radical change in the composition of macrofossil assemblages was found to occur at the water paleodepth of between 30 and 35 m (Fig. 4.9). The percentage of well-preserved benthic animals in the macrofossil assemblage, with the exception of siliceous sponges, drastically drops when water paleodepth increases beyond approximately 30 m. Fossil benthos *except siliceous sponges* is less than 20% of the entire macrofossil assemblage from a water paleodepth of 30 m downward. Siliceous sponges were certainly more abundant than ammonites from a water depth beyond about 50 m. Siliceous sponges prevailed mainly in the sponge biostromes of the Birmenstorf Member and in those of the lowermost Effingen Member. Sponge biostromes occur in the lowermost Effingen Member between Veltheim, Holderbank, and Schinznach-Dorf in Canton Aargau, as well as near Oberbuchsiten in Canton Solothurn (GYGI 1969a, Pl. 19). Small bioherms of siliceous sponges with an elevation of 0.8 m above the flat sediment surface around them were found in the lowermost bed no. 22 of the Pichoux Formation in section RG 307 in the quarry of La Charuque at Péry in Canton Bern. These bioherms were overlooked in the author's field book no. 7, p. 6, when section RG 307 in Pl. 22 in GYGI (2000a) was drawn. Much larger bioherms of siliceous sponges with an elevation of several meters above the sediment surface around them occur in what GYGI (1969a, Pl. 17) improperly called *Crenularis* Member in the Villigen Formation at Mellikon in Canton Aargau. The largest bioherm in section RG 70 at Mellikon is drawn on p. 119 of the author's field book no. 2. A large sponge bioherm was intersected in the Hornbuck Member, when the cantonal road bypassing Merishausen in Canton Schaffhausen was built (GYGI 1969a, Pl. 19, upper assembled section).

Siliceous sponges were omitted from the benthic assemblages represented in Fig. 4.9, because their prospect of becoming fossilized was very variable. Few siliceous sponges, for instance in the normal facies of the Birmenstorf Member, are entirely preserved. Two near-complete, disk-like sponges of the genus *Discophyma* with a diameter of up to 0.5 m, one of them being preserved in living position and the larger one overturned at a water depth which was somewhat greater than 100 m, were photographed on a bedding plane of the Birmenstorf Member by GYGI (2000a, Fig. 17 on p. 27). Provided that siliceous sponges became fossilized at all, their remains varied as a rule between all intermediate stages of tuberoids and rare entire fossils of giant size. Tuberoid is a term used in southern Germany for particles between about the size of a sand grain and relatively large parts of incompletely fossilized siliceous sponges.

Ammonites prevail in the macrofossil assemblage in sediments from a paleodepth of approximately 35 m downward (Fig. 4.9). The lateral variation in the composition of the ammonite assemblage in coeval sediments from different water depth in the deep subtidal zone can best be read from beds of *Cordatum* age which are shown in Fig. 4.9. *Cardioceras* are more than 60% of the entire macrofossil assemblage at the water depth of about 35 m in the unnamed, very fossiliferous bed in the distal part of the middle Sornetan Member ("Fossil Bed" in Table 2 in GYGI and PERSOZ 1986). Numerous ammonite casts from this bed were figured by GYGI and MARCHAND (1993). Perisphinctids amount to 10% and Haploceratacea to 5% of the macrofossil assemblage in that bed (Fig. 4.9). The coeval Schellenbrücke Bed in Canton Aargau was laid down at a water depth of about 95 m. The percentage of *Cardioceras* and of perisphinctid ammonites is almost equal in the Schellenbrücke Bed (this study, Fig. 4.9, drawn after GYGI 1981, Table 1). The Glaukonitsandmergel Bed in Canton Schaffhausen was sedimented at a water depth of approximately 115 m. *Cardioceras* are less abundant than perisphinctids in the fossil assemblage at this paleodepth (this study, Fig. 4.9, drawn after GYGI 2003, Table 78 on p. 213). Conversely, the percentage of haploceratid ammonites grows with increasing depth of sedimentation from the fossil bed in the middle Sornetan Member in northwestern Switzerland through the Schellenbrücke Bed to the Glaukonitsandmergel Bed. This is outlined in Fig. 4.9 by a dashed line. Haploceratid ammonite casts are the most abundant group in the almost pure ammonite assemblage in Kimmeridgian beds of the *Eudoxus* Chron from the paleodepth of about 170 m near Urach in southern Germany (GWINNER 1962, Fig. 14, and Fig. 4.9 in this study).

The tendency in perisphinctacean ammonites is reciprocal to that of Haploceratacea. Casts of perisphinctaceans figured in GYGI (1995) are the only representatives of ammonites which were found in the lagoonal sediment from a water paleodepth of at most 10 m of the Buix Member. Perisphinctacean casts are 60% of the entire macrofossil assemblage in the lower, glauconitic part of the Baden Member in section RG 70 at Mellikon, which was sedimented at a water paleodepth of probably about 55 m (Fig. 4.9). The abundance of perisphinctacean ammonites and of representatives of the ammonite subfamily Cardioceratinae generally decreases in the macrofossil assemblage with growing depth of deposition of the encasing sediments. The tendency in *Amoeboceras*, a boreal ammonite genus which evolved from Cardioceratinae at the

end of the *Antecedens* Chron, is different from that in Cardioceratinae. No *Amoeboceras* were found in sediments of the *Galar* Zone in section RG 70 at Mellikon which were laid down at the water depth of around 50 m. Sediments of the *Galar* Zone in section RG 239 at locality Summerhalde near Schaffhausen (GYGI 2003, Fig. 160) were deposited at a paleodepth of approximately 100 m and, therefore, in probably cooler water than near Mellikon. It is stated in Sect. 11.5.5 that small specimens of *Amoeboceras* are quite numerous in some of the Oxfordian/Kimmeridgian boundary beds of section RG 239.

It is now generally agreed that phylloceratid and lytoceratid ammonites were prolific in the Late Jurassic only in waters more than about 200 m deep. The highest percentage of phylloceratid ammonite casts which was found in an epicontinental, macrofossil assemblage of Oxfordian age in northern Switzerland was calculated by GYGI (1981, Table 1) to be 0.9% in the Schellenbrücke Bed of the *Cordatum* Chron from a water depth of approximately 95 m. A single lytoceratid specimen out of the between 9,000 and 10,000 ammonite casts in the author's collection was found in the lower, glauconitic part of the Baden Member near Mellikon from a water depth of about 55 m (bed no. 124 in section RG 70, Pl. 17 in GYGI 1969a). ZIEGLER (1967) showed in his Fig. 11 the composition of a macrofossil assemblage of the Kimmeridgian *Eudoxus* Chron (or *Caletanum* Chron, respectively, in Fig. 1.6 of this study) from bedded limestone which was sedimented on the basin floor between very high sponge bioherms near Urach in southern Germany. As mentioned above, the depth of deposition of this limestone on the basin floor could be read from Fig. 14 in GWINNER (1962) to be approximately 170 m. There are almost exclusively casts of ammonites in the macrofossil assemblage at this water depth. Approximately 60% of the ammonite casts preserved are Haploceratacea ("Oppelien" of ZIEGLER 1967), and only about 25% are casts of perisphinctaceans (compare with Fig. 4.9 in this study). Neither phylloceratids nor lytoceratids were mentioned by ZIEGLER (1967) to occur at this locality.

Very large or giant perisphinctacean ammonite casts in section RG 70 at Mellikon are overrepresented in the lower Villigen Formation and in the lower, glauconitic Baden Member. The depth of deposition of both of these units is estimated to have been about 50 m. Casts of giant perisphinctacean ammonites are as well unusually abundant in Kimmeridgian sediments in section RG 21 in the quarry at Mt. Born west of Olten (Pl. 18 in GYGI 1969a). The pertinent limestones were laid down on the depositional slope off the basinward rim of the carbonate platform of the Reuchenette Formation east of Balsthal on the slope at a water depth approaching 30 m (Fig. 1.5). Some giant ammonite casts from sections RG 21 and RG 70 were figured by GYGI (2003). Giant *Lithacosphinctes* were found in the Geissberg

Member from a water depth of between 20 and 30 m in a quarry west of Aarau and in the quarry of Gabechopf (unpublished section RG 294) on Mt. Geissberg near Villigen. The percentage of giant perisphinctaceans is much less in the ammonite assemblage with a normal size distribution in the stratigraphically uncondensed facies of the Birmenstorf Member which was sedimented at a water paleodepth of somewhat more than 100 m. Large or giant perisphinctacean ammonites were consequently particularly abundant in sediments from the moderate paleobathymetric range of between approximately 20 and 50 m. A similar tendency is even more pronounced in nautilids. Giant *Paracenoceras ingens* TINTANT in TINTANT et al. (2002) were found only in sediments from very shallow water.

4.3.2 Spherical Calcareous Oncoids Growing at Widely Varying Water Depth

The most common fossils to be found floating in the micritic matrix of the sediment laid down in some very shallow lagoons are more or less spherical, calcareous algal balls or oncoids with a diameter of up to several centimeters. The oncoids were primarily hard. This is documented by holes that small bivalves of the genus *Lithodomus* bored into the oncoids. Such a bored oncoid was figured by GYGI (1969a, Pl. 10, Fig. 37). The oncoids were most often dispersed in what was initially soft, calcareous mud. One exception to this was found in a bed with calcareous oncoids in a matrix of what was initially mixed calcareous and argillaceous mud in the lower Röschenz Member, bed no. 2 of the unpublished section RG 393 in the quarry beside Kalkofen, 1 km southwest of the village of Fislis in Département Haut Rhin, France, or 18 km west-southwest of Basel. Calcareous oncoids within marl were also found by MARCHAND and TARKOWSKI (1992, p. 55 and Fig. 1) in the Jurassic overburden of the large quarry in Permian porphyry at Zalas near Krakow in southern Poland.

One of the Late Jurassic lagoons with large spherical oncoids floating in what was primarily calcareous mud extended from Péry in Canton Bern westward to the Chasseral range. The oncolite laid down in this lagoon was called Grüne Mumienbank (green oncolitic bed) by ZIEGLER (1956). This oncolite is at Péry about halfway up in the Günsberg Formation. It is bed no. 163 in section RG 307 which is represented on Pl. 22 in GYGI (2000a). Some of the oncoids in the Grüne Mumienbank Bed near Péry grew to a diameter of as much as over 5 cm. Minute, columnar stromatolites stemmed from the upper surface of the larger oncoids in this bed (see polished slab as figured by GYGI 1992, Fig. 14). The columnar stromatolites probably indicate that the pertinent oncoids were rolled or overturned only occasionally by currents driven by major storms. Such

storms probably occurred between long time intervals of quiescence. The time equivalent of the Grüne Mumienbank Bed near Péry is near Sornetan in section RG 315 in Pichoux Gorge probably bed no. 56 with oncoids and hermatypic corals in the upper part and a hummocky erosion surface at the top (GYGI 2000a, Pl. 21).

Another, younger and predominantly oncolitic carbonate unit with an average thickness of 5 m was sedimented into a very shallow lagoon, which extended from locality RG 395 (unpublished, 1 km southwest of the village of Wolschwiller in the southwestern part of Alsace, Département Haut Rhin, France, or 18 km southwest of Basel at Swiss coordinates 696,300/255,750) to Canton Neuchâtel in the southwest over a palinspastic distance of at least 70 km. The oncolitic member sedimented into this lagoon was called Hauptmumienbank (main oncolitic member) by ZIEGLER (1956) in order to distinguish this very widespread unit from local oncolites (akzessorische Mumienbänke of ZIEGLER 1956) in the younger Laufen Member directly above. The Hauptmumienbank Member is abbreviated HMB in Figs. 1.5 and 4.19A. PERSOZ in GYGI and PERSOZ (1986, Fig. 10) was right to conclude, based on his correlations with kaolinite, that this member also occurs in GYGI's section RG 307 near Péry, in bed nos. 198–199 (Fig. 5.2 in this study). This is not shown in GYGI (2000a, Pl. 22) because of an error. The largest oncoids in the Hauptmumienbank Member have a diameter of 4 cm (GYGI 2000a, photograph of polished slab shown in Fig. 14 on p. 24).

Net accumulation of lime mud in which oncoids floated occurred during time intervals when water in the shallow lagoons was quiet. The episodic currents driven by storms which rolled oncoids on the lagoon floor must have been essentially oscillating, not unidirectional, or else all of the mud which was churned up into suspension by turbulent water would have been winnowed out of the environment and settled out of suspension elsewhere. Currents of bottom water strong enough to roll oncoids with a diameter of up to 4 cm had a minimal velocity of approximately 1.5 m/s according to the diagram of Fig. 18 in HJULSTRÖM (1935). Spherical oncoids were rolled on the lagoon floor at the surface of calcareous mud, whereas the largest, ellipsoidal oncoids in the Grüne Mumienkalk Bed were perhaps only occasionally overturned during particularly severe storms. Both mud-grade calcareous particles in the matrix and centimeter-size, hard oncoids could be formed and could come to final rest in this way together in the same environment. Most or all of the calcareous mud in the primary matrix of these lagoons was probably chemically precipitated in whitings. This must have been a large-scale process. Evidence is given in Sect. 4.14.3 that probably most of the lime mud that was produced in the lagoons mentioned was occasionally churned up into suspension from the bottom during storms and was then carried out of the environment. This mud settled in deeper water of the epicontinental basin. Most of the calcareous mud produced in the lagoon of the Grüne Mumienbank Bed and landward from there in the Röschenz Member was finally sedimented in the basinal micritic limestone of the Gerstenhübel Beds. The major part of the lime mud precipitated in the very extensive lagoon of the Hauptmumienbank Member was finally sedimented in the Geissberg Member of Canton Aargau and in deeper water in the Hornbuck Member of the Rhodano-Swabian Basin. *Regular* and gentle water currents in a lagoon are indicated at the few localities, where branching hermatypic corals were found in oncolite of the Grüne Mumienbank Bed and the Hauptmumienbank Member with a matrix of calcilutite. Currents of bottom water strong enough to roll or overturn large oncoids were *episodic*.

There is no record that ammonites were ever found in any of the two pure calcareous, lagoonal oncolites of the Grüne Mumienbank Bed or the Hauptmumienbank Member. The probable reason for this is that water in the lagoons was too shallow to be a habitat for cephalopods and that water was at times hypersaline in parts of the lagoon of the Hauptmumienbank Member. Hypersaline water is documented by calcite pseudomorphs after calcium sulfate minerals which were found for instance within an oncoid in the Hauptmumienbank Member near Liesberg. The photograph of a polished slab with this oncoid is shown in Fig. 5.5C. Conversely, branching hermatypic corals living in other parts of the lagoon of the Hauptmumienbank Member are evidence of normal salinity. Branching corals occur, for instance, in the uppermost part of the Hauptmumienbank Member in the small quarry at the unpublished locality RG 420 south of Rüti, 500 m west of the village of Blauen. This village is 8 km northeast of Liesberg in Canton Basel-Landschaft. Branching hermatypic corals were also found in the Grüne Mumienbank Bed, bed no. 163 of section RG 307 near Péry (Pl. 22 in GYGI 2000a) and in its possible time-equivalent, bed no. 56 in section RG 315 in Pichoux Gorge near Sornetan (GYGI 2000a, Pl. 21).

Green, near-spherical oncoids with a diameter of up to 2 cm that are colored by dispersed, submicroscopic glauconite were stated above to occur in a bed with numerous ammonites from a water depth of approximately 40 m at Schönenwerd in eastern Canton Solothurn, west of Aarau. The oncoids in this bed are floating in micritic limestone. They probably grew *in situ* at the surface of calcilutite in the relatively deep water at the toe of the depositional slope off the basinward rim of the carbonate platform of the Reuchenette Formation. This is shown in Fig. 1.5 of this study and in Figs. 4, 16, and 17 in GYGI (1992). Figure 17 in that paper is the photograph of such a green oncoid cross-cut in a polished slab. Ammonites from this peculiar oncolitic unit were found in a quarry south of locality Löchli south of Schönenwerd and in bed no. 47 of the unpublished section RG 28 in the old quarry at locality Halden east of

Schönenwerd. The pertinent ammonites were figured by GYGI (2003, Figs. 70, 72, and 78).

Spherical, most probably primarily hard calcareous oncoids with a diameter of between 32 and 36 mm grew at a much greater water depth in the Mumienkalk Bed in what is now Canton Schaffhausen. The bed is abbreviated MUK in the lower right of Fig. 1.5. It is named after the abundant mummies it includes. Such a mummy is an ammonite cast (Fig. 8.4) or a fragment of a cast with a thick oncolitic crust all around (Fig. 4.1G). The Mumienkalk Bed with an average thickness of 20 cm has a matrix of calcilutite. Abundant cauliflower glauconite pellets are dispersed in the micritic matrix of the bed, in ammonite casts, in the oncolitic crusts around such casts, as well as in the thin and numerous concentric cortices of spherical oncoids. The spherical oncoids in the Mumienkalk Bed are at first glance similar to the oncoids in the Hauptmumienbank Member. The principal difference is that the oncoids in the Mumienkalk Bed include abundant, cauliflower pellets of glauconite (Fig. 4.1F), and that ammonites are most abundant in the macrofossil assemblage in the bed.

One of the spherical oncoids was taken from bed no. 14a of the Mumienkalk Bed in section RG 81b near Gächlingen (Fig. 8.3). This entire oncoid was figured by GYGI et al. (1979, Fig. 14a). Another such spherical oncoid with the diameter of 32 mm from the Mumienkalk Bed, bed no. 16a of section RG 207 near Siblingen, is photographed in polished cross-section in Fig. 4.1F in this study. Section RG 207 is shown in GYGI and MARCHAND (1982, Fig. 2). The cross-cut oncoid shown in Fig. 4.1F has ten concentric and well-delimited cortices with a thickness of up to 2 mm. Ammonites prevail by far in the rich macrofossil assemblage in the Mumienkalk Bed like in the unnamed glauconitic bed directly above which is shown in Fig. 4.9. It must be concluded from the composition of the macrofossil assemblage in the Mumienkalk Bed, from the presence of abundant cauliflower glauconite pellets, and mainly from a comparison with the depth of deposition of the Birmenstorf Member of approximately 100 m in Canton Aargau according to Figs. 4.9 and 4.15B, that the Mumienkalk Bed was sedimented at a water depth of around 120 m above the Mumienmergel Bed and the Glaukonitsandmergel Bed (Fig. 4.9). This depth is greater than the well-defined lower limit of approximately 100 m to the depth range in which iron ooids could be accreted when being rolled at intervals at the sediment surface of mud by gently oscillating currents of bottom water during storms. The step on the level sea floor at the beginning of the Oxfordian Age between Canton Aargau and Canton Schaffhausen was estimated in Sect. 4.2.2 to have had an elevation of about 20 m, as it is represented in Figs. 1.5 and 4.15A.

Deposition of the Mumienkalk Bed began before sedimentation of the stratigraphically condensed bed at the base of the Birmenstorf Member in Canton Aargau was completed. The Mumienkalk Bed with an average thickness of 20 cm was sedimented, according to Fig. 230 on p. 163 in GYGI (2001), during the *Antecedens* and part of the *Transversarium* Chron. Following the definition of stratigraphic condensation in Sect. 8.2, the Mumienkalk Bed is condensed in the strict sense. The thin bed probably represents a time span of the order of 400,000 years. The *average* rate of lime mud sedimentation in the matrix of the bed was consequently minimal. Water in the deeper part of the Rhodano-Swabian Basin was therefore probably very transparent at the pertinent time of sedimentation, and the lower limit of the euphotic zone was then at an unusually great depth. No hermatypic corals were found among hundreds of ammonites in the Mumienkalk Bed from a depth of deposition of approximately 120 m. Conversely, Recent hermatypic corals are living at a corresponding water depth in the particularly clear water of the Central Pacific or in the Red Sea. The micritic cortices of the spherical oncoids in the Mumienkalk Bed and mainly the oncolitic or rather stromatolitic crusts around very large perisphinctid ammonite casts in the Mumienmergel Bed directly below the Mumienkalk Bed were probably accreted principally by cyanobacteria at a low rate. This is to be expected at a water depth of approximately 120 m. Cyanobacteria are known to survive at low and even at very low levels of illumination. Mega-oncoids in the Mumienmergel and Mumienkalk Beds are shown in Figs. 8.3 and 8.4.

At least ten concentric cortices can be discerned in the near-spherical, cross-cut oncoid with a diameter of 32 mm from the Mumienkalk Bed at Siblingen, which is shown in Fig. 4.1F. This is evidence that every one of the distinctly delimited cortices of this oncoid must have been accreted following more than one event of rolling. Long time spans may have elapsed between events of rolling of this oncoid, but the quiet time intervals between rolling must have been limited. Growth of the oncoid would have become asymmetric, if more than a distinct, maximal amount of time of quiescence elapsed after an event of rolling, and if the oncoid stayed after an event of rolling for a long time at the sediment surface. This indicates that loose lime mud was churned up from the sediment surface in the event of turbulent bottom water during a storm. After the storm had died down, the clay-size calcareous particles settled from suspension and thereby smothered some of the spherical oncoids. Lime mud of the free sediment on the sea floor therefore remained to be unlithified for a long time. Conversely, calcareous mud *within* partially filled gas chambers in the phragmocone of the ammonite cast shown in Fig. 8.4 was lithified soon after fossilization of the pertinent ammonite shell. The numerous, well-delimited crusts with a similar thickness which are visible in Fig. 4.1F are evidence that the spherical oncoids in the Mumienkalk Bed were rolled

after more or less regular time intervals. An explanation should be found of how the few, almost perfectly spherical calcareous oncoids with concentric crusts and a diameter of as much as 36 mm in the Mumienkalk Bed could be rolled and accreted at a water depth of approximately 120 m at the surface of sediment which was originally calcareous mud. A bottom water current with the minimal velocity of approximately 1.5 m/s is required to roll such oncoids according to the diagram of Fig. 18 by HJULSTRÖM (1935).

It is concluded below following GYGI (1981) that iron ooids can be accreted on the level basin floor at the sediment surface of calcareous or argillaceous mud by gently oscillating bottom currents driven by storms in water no deeper than 100 m. Rolling of an ooid with the diameter of 2 mm on the sediment surface requires a minimal current velocity of the bottom water of 15 cm/s, according to the diagram by HJULSTRÖM (1935, Fig. 18). When water depth increased to beyond 100 m, iron ooid accretion was replaced without a transition by growth of cauliflower glauconite pellets. Such an abrupt *vertical* facies transition is exposed in the landslide of Bleiche at the western slope of Mt. Eichberg northwest of Blumberg in southern Germany. The boundary is between the iron oolitic marl no. 20 and the glauconitic marl no. 21 in section RG 87 that is shown on Pl. 16 in GYGI (1969a). The corresponding *lateral* facies boundary between the thin, iron oolitic Schellenbrücke Bed no. 3–4 in the unpublished section RG 73 near Dangstetten in southern Germany and the coeval, glauconitic Glaukonitsandmergel Bed which was recorded by the author at the depth of between 478.0 m and 478.3 m in the NAGRA exploration well 2 km east-northeast of the village of Weiach in northwestern Canton Zürich, 11.5 km south-southeast from Dangstetten, is well documented with cardioceratid ammonites of the *Cordatum* Chron that were figured by GYGI and MARCHAND (1982).

In view of what was said above about the environment where iron ooids are accreted, it appears to be questionable to presume that episodic bottom currents oscillating during particularly violent storms at the ten times greater maximal velocity of at least 1.5 m/s rolled the spherical oncoids in the Mumienkalk Bed at the sediment surface of loose lime mud at the water depth of approximately 120 m. Nevertheless, the only plausible mechanism which could lead to accretion of such spherical oncoids at the sediment surface on a level sea floor in water as deep as 120 m is rolling by probably oscillating currents of bottom water during rare, exceptionally violent storms or hurricanes. Intervals of quiet bottom water between events of oncoid rolling were during deposition of the Mumienkalk Bed mainly in the *Antecedens* Chron probably much longer than those in the Hauptmumienbank Member, which was sedimented during the younger *Berrense* Chron in water about 5 m deep at the bottom of an extensive lagoon.

The greatest wind speed in a hurricane is at the periphery directly outside the margin of the "eye" in the center of the vortex of the storm. The diameter of a hurricane eye is between 10 and 25 km. Wind speed adjacent to the margin of the eye is reported to be up to 200 knots or approximately 370 km/h. KUENEN (1960, p. 75) stated that waves 16 m high have been photographed, but that the greatest height of waves is reported to be well over 30 m. The height of waves depends on water depth and mainly on the length of fetch of the waves. According to KUENEN (1960, p. 75), a wind blowing for example with the velocity of 60 km/h over a distance of 5 km can generate waves with a maximal height of 0.9 m. When wind blows at the same speed over a distance (fetch) of 500 km, it can produce waves up to 6.2 m high. No method could be found in the literature of how to calculate, whether great waves could cause at the water depth of 120 m a minimal current velocity of bottom water of the order of 1.5 m/s strong enough to roll spherical oncoids with a diameter of as much as 36 mm at the sediment surface. According to KUENEN (1960, p. 228), rocks weighing half a kilogram were sometimes washed into lobster crawls at a depth of over 60 m off Land's End in Cornwall, England. MOORE et al. (1992, Fig. 25 on p. 146) indicated in their model of Late Jurassic climate, that the axis of mid-latitude storm tracks was at 30°E at latitude of 45° in the northern hemisphere. This is far north of the paleolatitude of the area investigated in this study.

WALKER and PLINT (1992, Fig. 1) indicated storm wave base in a near-shore, shallow marine environment to be at a water depth of about 25 m. This figure is probably typical of an average storm over shallow water. Storms strong enough to cause bottom water to oscillate sufficiently to roll and accrete iron ooids with a maximal diameter of 2 mm at a water depth of 100 m were therefore possibly of hurricane force. Currents oscillating at a peak velocity of at least 15 cm/s are required for this according to HJULSTRÖM (1935, Fig. 18). Bottom currents oscillating at the minimal peak velocity of the order of 1.5 m/s that could roll spherical oncoids with a diameter of close to 4 cm at the sediment surface approximately 120 m below sea level were probably driven by exceptionally violent storms which occurred at rare occasions. Probability that the periphery of the eye with maximal wind speed of an average hurricane passes over a given spot on the floor of an epicontinental sea is slight. The chance that a distinct locality be hit by the marginal part of the eye of an extremely severe hurricane is even less. Nevertheless, such a probability can be assumed to be sufficient that the few spherical oncoids in the Mumienkalk Bed could be rolled on the sea floor in the event of an extremely violent storm after long, more or less regular time intervals of quiet bottom water.

The depth of deposition of the earliest Oxfordian sediment of ferriferous marl with iron ooids in excavation RG 208 near Ueken was estimated above to have been approximately 80 m. It was at this depth that an episodic bottom current churned up unconsolidated mud of marl including iron ooids from the sediment surface and washed pockets up to 20 cm deep out of this sediment between calcareous concretions (Fig. 8.2B). The sediment stirred up into suspension was removed entirely by the current, which was consequently essentially unidirectional and episodic in a normally quiet environment. The event of erosion occurred during the *Scarburgense* Chron in the earliest Oxfordian Age. This is documented by ammonites figured in GYGI and MARCHAND (1982). Only an exceptionally violent storm could be the cause of such a process of erosion at a water depth of as much as 80 m. Another, similar event of erosion occurred at stage D in Fig. 8.2 at a water depth which had increased in the meantime to approximately 95 m. Ferruginous lime mud was then removed entirely from the same pockets. According to the diagram by HJULSTRÖM (1935, Fig. 18), a current velocity in bottom water of at least 1.5 m/s was required for these events of erosion. A current with the minimal velocity of 1.5 m/s can roll, following HJULSTRÖM (1935, Fig. 18), a particle with the diameter of 40 mm at the sediment surface. Therefore, the possibility cannot be ruled out that episodic bottom currents driven by exceptionally violent storms at rare occasions, but at more or less regular intervals, could roll spherical oncoids with a diameter of as much as 36 mm at the sediment surface in water as deep as approximately 120 m. It is stated below that episodic bottom currents washed large calcareous ammonite casts clear of lime mud in the Mumienkalk Bed, and clear of marly mud in the Mumienmergel Bed directly below the Mumienkalk Bed.

A polished cross-section of a fragment of a lithified cast of a cardioceratid ammonite with a thick oncolitic crust and with cauliflower glauconite pellets both within the cast and in the crust is represented in Fig. 4.1G. The specimen is from the Mumienkalk Bed, that is bed no. 3 of excavation RG 88 which was made in 1963 in the then open-cut iron mine near Blumberg (GYGI 1977, Pl. 11, section no. 9). The diameter of the broken ammonite cast including the crust around it is 6 cm. Overturning or rolling of oncolitically encrusted ammonite casts of this and of much greater size (see below) by bottom currents in water approximately 120 m deep is unlikely. BÖHLKE and CHAPLIN (1970, p. 444) wrote, that carnivorous hogfish were observed to overturn large nodules formed by red algae in search of small prey animals hiding below the nodules near Recent coral reefs in the tropical West Atlantic Ocean. It is therefore conceivable, that fish overturned large ammonite casts lying flat at the surface of the muddy sediment of marl that finally became the matrix of the Mumienmergel Bed. Such early lithified casts must have been winnowed entirely clear of embedding sediment of mud by strong bottom currents before they could be overturned. Some of the casts were subsequently broken, before the fragments were oncolitically encrusted (Fig. 4.1G). The fragments are normally close together and their edges show no traces of abrasion. The fragments were consequently not transported and abraded by the strong bottom currents which washed ammonite casts or fragments entirely clear of embedding mud.

The question about the *depth of storm wave base* arises because of these proved processes and because of the above assumption, that the spherical oncoids with a diameter of up to 36 mm in the Mumienkalk Bed *could* be rolled at the sediment surface at the water depth of about 120 m by oscillating bottom currents driven by exceptionally violent storms. In addition, a satisfactory explanation should be found for the fact that the lower limit to the paleobathymetric range in which iron ooids could be accreted is well defined at approximately 100 m below sea level. It is within the bathymetric range of between 10 and 100 m that the iron ooids investigated were accreted probably during storms when being rolled at the sediment surface by gently oscillating bottom currents (see below). Provided that the above argumentation is correct, a disparity would have existed between average hurricanes which could drive oscillating bottom currents with a minimal velocity of 15 cm/s strong enough to roll and accrete iron ooids with a maximal diameter of 2 mm at a water depth of as much as 100 m and rare, exceptionally violent storms that drove bottom currents with a minimal velocity ten times as much, of the order of at least 1.5 m/s that was sufficient to erode unconsolidated, mud-grade sediment including iron ooids in water between 80 and more than 90 m deep, or that could drive currents with the same velocity which rolled spherical calcareous oncoids with a diameter of several centimeters at the sediment surface 120 m below sea level. No explanation for the existence of such a disparity could be found. Nevertheless, the water depths concluded above were arrived at using different, independent methods, and the resulting figures are consistent. It follows from all of this evidence, that storm wave base cannot be at a fixed water depth. The base is in fact highly variable and can probably be much lower down than 100 m below the water surface (see main conclusion no. 22).

The Mumienmergel Bed is on average 15 cm thick and is directly below the Mumienkalk Bed in Canton Schaffhausen (Fig. 8.3). The bed is abbreviated MUM in the lower right of Fig. 1.5. The Mumienmergel Bed includes calcareous ammonite casts of both the *Densiplicatum* and *Antecedens* Chron and is consequently stratigraphically condensed in the strict sense. The matrix of the Mumienmergel Bed was, contrary to the Mumienkalk Bed above, initially argillaceous and calcareous mud mixed by bioturbation. Calcareous casts of perisphinctid ammonites with a diameter of as much as more than 30 cm were formed within shells which were

initially entirely embedded in calcareous mud. The casts of large perisphinctids are embedded with the plane of coiling parallel to the base of the bed. The diameter of these flat mega-oncoids can be as much as 40 cm. A minor such mega-oncoid is shown in Fig. 8.4. Ammonite casts are the great majority in the rich macrofossil assemblage in both the Mumienmergel and the Mumienkalk Bed, as well as in the unnamed glauconitic bed directly above the Mumienkalk Bed (Fig. 4.9). The age of the glauconitic bed above the Mumienkalk Bed is the *Transversarium* Chron. This is bed no. 15 in the large excavation RG 81b which was made in 1970 below Räckolterenbuck 3 km north-northeast of Gächlingen in Canton Schaffhausen. The bed was glauconitic limestone in the southern part of excavation RG 81b and passed laterally into marl with limestone nodules produced by subsolution, as it is shown in Fig. 8.3. The same bed was found to be glauconitic marl with small limestone nodules in the lower part of bed no. 10 in excavation RG 212 in Churz Tal near Siblingen.

The calcareous ammonite casts embedded in the marly matrix of the Mumienmergel Bed could only be formed within shells which were entirely embedded in calcareous mud for some time. The casts were early lithified and are therefore undeformed by compaction. Some of them were then broken by a process unknown. The edges of the fragments are unabraded, and the fragments stayed in most cases close together. The oncolitic crust of micritic calcium carbonate could only grow all around even the largest casts or fragments after the casts were washed clear of embedding sediment entirely, and if the casts were overturned from time to time. The possibility must be ruled out that the flat casts of very large perisphinctid ammonites with a thick oncolitic crust and with an overall diameter of as much as 40 cm could have been overturned by storm-driven bottom currents at a water depth of well over 100 m. A minimal depth of sedimentation of 100 m is indicated by the cauliflower pellets of glauconite which are abundant in the Mumienmergel Bed both within ammonite casts and in their oncolitic crusts.

Additional evidence in the Mumienmergel Bed that large ammonite casts with an oncolitic crust all around were sometimes overturned is early lithified, geopetal sediment of calcareous mud which partially filled empty chambers between septa in the phragmocone of a particular, stromatolitically encrusted perisphinctid ammonite cast. This cast was collected from bed no. 7 in excavation RG 212 near Siblingen in Canton Schaffhausen (Fig. 8.3). The photograph of a polished cross-section of this cast is Fig. 8.4. The cast in this figure is shown in the position, in which it was finally embedded in the Mumienmergel Bed. The diameter of the cast including the stromatolitic crust around it is 7.3 cm. The upper side of the cast in the position in which it was excavated was deeply corroded by subsolution (see Sect. 8.5). The corroded surface was subsequently encrusted with goethite. Thorough corrosion and the ferruginous crust covering the corroded surface are evidence that the upper side of the encrusted ammonite cast was in direct contact with bottom water for a long time. Many of the ammonite casts with an oncolitic crust in the Mumienmergel Bed were consequently overturned only infrequently after long time intervals. The ammonite cast shown in Fig. 8.4 was formed and early lithified at a time, when pure calcareous ooze was sedimented in the stratigraphically condensed Mumienmergel Bed. The geopetal lime mud in chamber no. 1 in Fig. 8.4 was possibly filled in and lithified at the same time as formation of the cast occurred. Early lithified lime mud in the two chambers no. 2 of the cast is proof that the cast was overturned after a time span which was at least sufficient for lithification of geopetal sediment in chamber no. 1. Only episodic and therefore *discontinuous,* strong currents of bottom water could be the cause of repeated winnowing of calcareous ammonite casts clear of sediment of mixed, clay-size carbonate and clay minerals.

4.4   Oxygen Content of Bottom Water

The maximal possible oxygen content of bottom water in the deeper part of the epicontinental, Rhodano-Swabian Basin of the Late Jurassic in northern Switzerland was estimated by GYGI (1999, p. 130) to have been approximately 5 ml of dissolved oxygen per liter of water. This estimate was made by comparison with water above the sea floor in the northwestern part of the Recent Gulf of Mexico, at latitude of approximately 27°N, according to RICHARDS (1957, Fig. 11). The highest oxygenation level of bottom water in the Rhodano-Swabian Basin existed probably in times of comparatively dry climate, when supply of siliciclastic sediment and nutrients from land was insignificant like for instance during the *Transversarium* Chron in the middle Oxfordian, when the uncondensed part of the Birmenstorf Member in Canton Aargau was sedimented at a water depth of slightly more than 100 m. GHASEMI et al. (1999, p. 40) found almost no marine palynomorphs in this part of the Birmenstorf Member. When rainfall and weathering on nearby land increased in a humid climate, and when runoff from land and tidal or occasionally storm-driven currents in the shallow-marine environment began to supply much fine-grained terrigenous material to the basin, nutrients from land supported plentiful plankton in the upper, normally oxygenated part of the water column. Dead, decaying plankton consumes oxygen dissolved in the water when settling to the bottom. The oxygen content of bottom water can thereby be substantially reduced when currents above the sea floor are particularly slack. This was the case in the proximal, open-marine part of the Rhodano-Swabian Epicontinental Sea when the homogenous, blue-gray clayey marl of the

Renggeri Member was sedimented in lowermost succession no. 1.

Ammonites in the *Renggeri* Member near Liesberg in Canton Basel-Landschaft are in section RG 280 most abundant between 2 and 3 m above the member's base. It is concluded in the following section that most of these ammonites are dwarfs. They are embedded in massive, burrowed clayey marl which is figured in GYGI (2000a, Fig. 7). The diameter of the burrows figured is seldom more than 1 mm. The burrows are filled with iron sulfide. This is evidence of a strong reduction of the oxygen content of bottom water according to SAVRDA et al. (1984, p. 1184). A minimally aerobic to dysaerobic environment must be concluded from this following BRETT and BAIRD (1986, p. 217 and Fig. 11). Oxygenation of the water directly above the sediment surface of the order of one milliliter of dissolved oxygen per liter of water can be concluded from the classification of trace fossils by EKDALE and MASON (1988, p. 720). According to GYGI (1999, Fig. 1), there is no evidence of short-term variation in oxygenation of bottom water during sedimentation of the *Renggeri* Member (see also SAVRDA and BOTTJER 1986).

The Bärschwil Formation is a shallowing-upward succession with a carbonate content that substantially increases from a low level at the base toward the top. Sedimentation of the formation began at a water depth of approximately 80 m. Water depth gradually diminished during aggradational sedimentation of the entire Bärschwil Formation, except late in the *Cordatum* Chron. The composition and the abundance of the macrofossil assemblage varied in the process of shallowing. Macrofossils are well represented in the lowermost part of the *Renggeri* Member. Ammonite casts prevail by far over few fossils of benthic animals in this part of the member. Most of the ammonites in the *Renggeri* Member are small. At least the inner whorls of them are casts of iron sulfide. Tiny ammonites in the member are preserved entirely as casts of iron sulfide. Ammonites that are at least partially preserved as casts of iron sulfide occur from the base to the top of the *Renggeri* Member. The abundance of ammonites substantially diminishes from near the base of the member upward. The oxygen content of bottom water was low when sedimentation of the *Renggeri* Member began. Oxygenation of bottom water probably augmented proportionally to the decrease in water depth in the course of aggradational sedimentation of the Bärschwil Formation. Indirect evidence of this is that upward from the upper boundary of the *Renggeri* Member, calcium carbonate is the substance of ammonite casts in the successive Sornetan Member.

Ammonite casts are few in the lower half of the Sornetan Member. They have a normal size distribution. This is diagnostic of fair oxygenation of water in the habitat of the pertinent ammonite community. Bivalves of the genus

Pholadomya were comparatively common in the macrofossil assemblage in the upper half of the Sornetan Member. Water depth diminished from 30 m to approximately 20 m at the time when sedimentation of the upper half of the Sornetan Member was completed. As soon as water depth diminished to less than about 20 m because of ongoing aggradational sedimentation of marl, an abrupt change to prolific growth of hermatypic, mainly dish-shaped corals in a biostrome occurred at the base of the Liesberg Member. These hermatypic corals thrived in a time span when the rate of supply of argillaceous mud from land had increased to a maximum (see Sect. 8.1). The color of fresh matrix in the Liesberg Member is blue-gray due to the oxygen-free interstitial water within the mixed calcareous and argillaceous mud just below the sediment surface. The mass occurrence of hermatypic corals growing on the substrate of this mud is evidence of well-oxygenated water directly above the sediment surface of the Liesberg Member.

4.5 The Influence of Bottom Water Oxygenation on the Composition and Preservation of Fossil Assemblages

The Birmenstorf Member was sedimented at a water depth of approximately 100 m. The oxygen content of bottom water was of the order of 5 ml of dissolved oxygen per liter of water. The rich and diverse macrofossil assemblage in the stratigraphically uncondensed, normal facies of the Birmenstorf Member is characterized by prolific, disk-shaped siliceous sponges of the genus *Discophyma,* which grew to a diameter of as much as 50 cm in densely populated biostromes. A bedding plane in the Birmenstorf Member at Holderbank in Canton Aargau with large *Discophyma* is shown in Fig. 17 on p. 27 in GYGI (2000a). The largest specimens among the abundant ammonites fossilized in the Birmenstorf Member are calcareous casts of giant perisphinctids. The diameter of full-grown, complete adults of these perisphinctids was greater than 400 mm. Two giant perisphinctid ammonites from the Birmenstorf Member were figured by GYGI (2001, Figs. 142 and 161). The smallest complete adults of ammonites which were found in the uncondensed part of the Birmenstorf Member are *Mirosphinctes* with a diameter of between 20 and 26 mm. Such dwarf *Mirosphinctes* were figured by BONNOT and GYGI (2001, Pl. 4, Figs. 3 and 5). *Glochiceras* are relatively common in the ammonite assemblage in the stratigraphically uncondensed facies of the Birmenstorf Member. Dwarf adult and complete specimens of several taxa of *Glochiceras* from the Birmenstorf Member were figured by GYGI (1991b). Dwarf size of *Glochiceras* in the Birmenstorf Member cannot have been caused by low oxygenation of bottom water. Insufficient food supply must be ruled out as well to be the

cause of dwarf size in such *Glochiceras* because of the abundance of ammonite casts of medium and large size and the occurrence of some giant ammonite taxa. The macrofossil assemblage including ammonite casts between dwarf and giant size derived from organisms that were living on and close above the sea floor in the Birmenstorf Member is rich in individuals and diverse, because oxygenation of bottom water and water depth were optimal for the assemblage. A size distribution in an ammonite assemblage like in the uncondensed part of the Birmenstorf Member is, therefore, classified here to be *normal*. Dwarf size of adult and complete ammonites fossilized in this member was consequently predetermined genetically.

The Herznach Formation in section RG 280 in the clay pit of Andil near Liesberg is represented in GYGI (1990a, Fig. 2). Sedimentation of the formation began at this locality at the water depth of approximately 5 m in the latest part of the early Callovian with bed no. 3, the Lumachellenbank of STÄUBLE (1959). The principal, calcareous bed no. 4 of the formation in this section has a matrix of gray micrite with at most 10% by volume of brown iron ooids of goethite. Bed no. 4 includes an abundant macrofossil assemblage, in which ammonite casts with a normal size distribution prevail. This assemblage indicates well-oxygenated bottom water and a depth of deposition of at least 35 m according to Fig. 4.9. Presence of authigenic iron ooids and the telescoping of ammonite zones in the Herznach Formation at Liesberg are evidence that the average sedimentation rate of these iron-oolitic deposits was very low. Sedimentation became increasingly argillaceous from above bed no. 4 in section RG 280. The sedimentation rate began slightly to increase during deposition of bed no. 6 and finally became too high for iron ooid accretion at the top, when low-rate sedimentation of the uppermost bed of the formation in section RG 280 was completed. Deposition of the Herznach Formation ended at Liesberg at the top of the marl-clay of bed no. 6 a very short time after the onset of the Oxfordian Age in the early *Scarburgense* Chron, when depth of deposition after a long time span with a minimal sedimentation rate had reached the maximum of approximately 80 m. Deposition of the Herznach Formation in the type area of the formation near Herznach ceased much later at the end of the *Cordatum* Chron, when sedimentation of the Schellenbrücke Bed ended at the water depth of close to 100 m (Fig. 4.9).

Iron ooids are about 5% by rock volume at the base of the brownish-gray marl-clay of the uppermost bed no. 6 of the Herznach Formation in section RG 280 at Liesberg. The iron ooids are brown to the top of the bed, where they fade away entirely. Goethite in the iron ooids indicates that good oxygenation of bottom water continued at the time, when such ooids were accreted in bed no. 6. The ammonite *Quenstedtoceras paucicostatum* LANGE excavated from the lower part of bed no. 6 at Liesberg is a calcareous cast of normal size. It is figured in GYGI (1990a, Pl. 3, Fig. 14). The taxon is the index of the last Callovian chron of the Middle Jurassic. A small *Cardioceras (Scarburgiceras)* cf. *scarburgense* (YOUNG and BIRD) was excavated from bed no. 20 in section RG 307 near Péry (GYGI 1990a, Fig. 4). The specimen is figured in the same paper on Pl. 3, Fig. 11. Bed no. 20 in section RG 307 is the uppermost bed of the Herznach Formation at Péry and is therefore the time equivalent of bed no. 6 of section RG 280 at Liesberg (GYGI 1990a, Fig. 2). Presence of goethitic iron ooids and of calcareous ammonite casts in the uppermost bed of the Herznach Formation both at Liesberg and at Péry is evidence of well-oxygenated bottom water.

The *vertical* facies change from marl-clay in the uppermost Herznach Formation with few brown iron ooids of goethite floating in the matrix and few ammonites preserved as calcareous casts of normal size, without a transition to blue-gray marl-clay at the base of the *Renggeri* Member with dwarf or stunted ammonite casts of iron sulfide was figured by GYGI (1990a, Fig. 2 at Liesberg, and Fig. 4 at Péry). The sudden vertical change in size and substance of ammonite casts, from calcareous casts of normal size in the Herznach Formation below to dwarf casts iron sulfide in the *Renggeri* Member above, indicates that oxygenation of bottom water significantly diminished at the onset of sedimentation of the *Renggeri* Member. Bottom water had become dysaerobic, before the first well preserved cast of iron sulfide of a *Cardioceras (Scarburgiceras) scarburgense* found at Liesberg 1.2 m above the base of the *Renggeri* Member was formed. The pertinent specimen is figured in GYGI (1990a, Pl. 3, Fig. 4). The lithologic boundary between the top of the mostly brown Herznach Formation with goethitic iron ooids and the base of the Bärschwil Formation above with blue-gray, homogenous clayey marl of the *Renggeri* Member with small ammonite casts of iron sulfide is clear-cut in all of the sections seen by the author.

When the sedimentation rate moderately augmented at Liesberg early in the *Scarburgense* Chron, iron ooid accretion ceased at the water depth of approximately 80 m and deposition of the blue-gray, homogenous marl-clay of the *Renggeri* Member began. Sediment supply of mostly clay minerals from land in the northwest (Fig. 1.2) then became ample. The sedimentation rate consequently further increased with time. Growing of the sedimentation rate during deposition of the lower and middle *Renggeri* Member is documented by the thicknesses of ammonite zones in the range chart which is represented in Fig. 3 in GYGI (1990a). The *Scarburgense* Zone is approximately 3 m thick in section RG 280 at Liesberg. Ammonites are most abundant in this zone, because the rate of sedimentation was low at the time. Ammonite abundance rapidly decreased upward in the succession concomitant with the increasing sedimentation rate. The *Praecordatum* Zone at Liesberg in Fig. 3 in GYGI

(1990a) is several times thicker than the *Scarburgense* Zone below. The thickness of the entire *Renggeri* Member in the lower Bärschwil Formation in the clay pit of Andil near Liesberg was measured to be close to 60 m (GYGI 2000a, Pl. 30). Depth of deposition diminished upward in the *Renggeri* Member because of aggradation (Fig. 1.5). Massive influx of siliciclastic material from land into the Rhodano-Swabian Basin beginning early in the *Scarburgense* Chron was probably combined with supply of a substantial amount of organic matter and nutrients. GRÜN and ZWEILI (1980), BERGER (1986), and GHASEMI in GHASEMI et al. (1999) figured a rich and diverse assemblage of plankton from the *Renggeri* Member at Liesberg. Dead plankton sinking out of the assemblage living in the well aerated, uppermost part of the water column and land-derived organic matter settled and consumed the greater part of oxygen dissolved in quasi-stagnant bottom water during sedimentation of the *Renggeri* Member in an unrestricted part of the Rhodano-Swabian Epicontinental Basin.

A vertical facies change similar to that at Liesberg previously occurred in the Callovian Age at Kandern in southern Germany, 18 km north-northeast of Basel. Iron oolite with goethitic iron ooids and calcareous ammonite cast of normal size is at Kandern below, and blue-gray, Callovian marl-clay of the lower Kandern Formation with dwarf ammonite casts of iron sulfide follows above. The lower part of the Kandern Formation of Callovian age is at Kandern more than 10 m thick and is lithologically like the *Renggeri* Member. This part of the formation thins in the distal direction and grades laterally into the coeval part of the upper Herznach Formation with iron ooids of goethite and calcareous ammonite casts that are diagnostic of well-oxygenated bottom water. A comparable *lateral* facies transition occurred in the early Oxfordian between the blue-gray marl with calcareous concretions of the lower Sornetan Member of the *Cordatum* Chron with a thickness of more than 25 m in section RG 306 at Liesberg (GYGI 2000a, Pl. 31, bed nos. 2–66) and the coeval, discontinuous Schellenbrücke Bed that is 25 cm thick or one hundred times thinner for instance at Ueken in Canton Aargau (Fig. 8.2, bed nos. 8–9 in this study). Autochthonous iron ooids of goethite are floating in this bed in what was primarily brown, ferriferous calcilutite laid down at a water depth that can be estimated at approximately 95 m (Fig. 4.9). Calcareous ammonite casts prevail in the rich macrofossil assemblage in the Schellenbrücke Bed. The bed also includes numerous benthic organisms (GYGI 1981, Table 1). GYGI (1981, Figs. 1–4) investigated the lateral facies change between the lower Sornetan Member (that was then called lower Terrain à Chailles) and the Schellenbrücke Bed. He showed how the lower half of the Sornetan Member thins toward the epicontinental basin between Liesberg and Ueken like a wedge. The

Sornetan Member pinches out entirely for instance in section RG 307 at Péry (GYGI 2000a, Pl. 22, see also GYGI 1995, Fig. 2).

Bottom water directly above the sediment surface of the *Renggeri* Member was probably permanently dysaerobic, because the lower part of the water column was quasi-stagnant except during a severe storm. This is documented by the diameter of burrows (see above) and by the fossil assemblage in which small or dwarf ammonites preserved partially or entirely as casts of iron sulfide are the great majority. Such ammonite casts are at Liesberg most abundant 2 m above the base of the marl-clay of the *Renggeri* Member. Many of these ammonites were figured in the photographic plates in GYGI (1990a). The small ammonites fossilized in the *Renggeri* Member could probably spend their entire life in a habitat where the content of dissolved oxygen was invariably of the order of 1 milliliter per liter of sea water, as it was stated above. The maximal diameter of casts of adult and complete ammonites in the *Renggeri* Member is around 40 mm, with the exception of the very few large specimens which are mentioned below. Casts of complete adults are quite abundant among the small specimens. At least the phragmocone of the small casts was replaced by iron sulfide of the minerals pyrite, marcasite, or a fine-grained intergrowth of the two minerals, according to GYGI (2003, p. 175). The ammonites living in the environment in which the *Renggeri* Member was sedimented reacted to the low oxygen content of the bottom water in their habitat with dwarf size of adults. The same was previously observed by SCOTT (1940, p. 311) in ammonites living in water with a low oxygen content in the Cretaceous of Texas, USA. SCOTT concluded from this that ammonites lived in close relation to the environment above the sea floor. He called their mode of life necto-benthonic. The very few, large ammonite casts that were found in the *Renggeri* Member must have originated from animals which died elsewhere, and whose empty shells then drifted into the environment of sedimentation of the member where they sank to the bottom and were fossilized. An example of this is given below.

Dwarfing, or stunting according to HALLAM (1965), of probably the great majority of the small ammonites which were fossilized in the *Renggeri* Member, is not easy to prove. A classification of size in ammonites living in normally oxygenated water was proposed by GYGI (2001, p. 12). A method must be found of how to discern genetically dwarf ammonite taxa living in well-aerated water from stunted, dwarf and small forms living in dysaerobic water. Casts of adult and complete ammonites with a diameter of less than 25 mm were rated as dwarfs by GYGI (2001). Casts of complete adults are much more common in small and mainly in dwarf ammonite taxa than in large ones. Tiny ammonites in the *Renggeri* Member like for instance *Scaphitodites scaphitoides* (COQUAND) with a diameter of complete adults

of not more than 16 mm (photographed in GYGI 1990a, Pl. 4, Fig. 4) are undiagnostic of stunting. Their size could be normal, since they were not found to date in sediments from another environment. Maturity can as a rule be recognized only in complete ammonite casts. Egression of the umbilical suture line on the last whorl of an adult and complete specimen is one morphologic feature which is unambiguous evidence of maturity in ammonites. Egression is the deviation of the umbilical suture line outward from the regular, logarithmic spiral. The umbilical suture line is on inner whorls a logarithmic spiral even in taxa with egression. By far not all of the ammonite taxa shape egression in mature specimens. When present, egression affects between 1/2 and the entire last whorl. Quantification of egression can be made according to Fig. 8 on p. 14 in GYGI (2001). Crowding of the last septa before the body chamber can be used as an indication of maturity in ammonite taxa which do not shape egression. But it must be noted that precocious crowding of septa was found in some ammonite casts on earlier ontogenetic stages before the beginning of the final body chamber. Morphologic differentiation of the aperture of the shell, for instance with lappets in microconch ammonites, is another criterion of maturity. Microconchs with lappets are shown in this study in Fig. 7.3E, F. Caution is mandatory with the criterion of a morphologically differentiated peristome, because formation of ontogenetically precocious peristomes was documented to occur in rare cases. An ammonite cast with a precocious peristome is shown in Fig. 7.3C. Peristomes are of limited use in taxonomic practice, because they are seldom preserved in ammonite casts, even in small taxa.

Most of the phragmocone of small or dwarf ammonites living in the environment of the *Renggeri* Member was after death of the animals filled with minerals of iron sulfide. Many casts of iron sulfide of particularly small specimens are preserved with part of or with the entire body chamber. Two small casts of *Cardioceras (Scarburgiceras) scarburgense* (YOUNG and BIRD) as figured by GYGI (1990a, Pl. 3, Fig. 1, and Pl. 4, Fig. 22) from the marl-clay of the *Renggeri* Member at Liesberg have both an undeformed phragmocone of iron sulfide with the diameter of 22 mm and a partially preserved body chamber of marl-clay that is compressed and flattened by compaction. A very well-preserved, wholly septate cast of iron sulfide of the same taxon with the diameter of 45 mm is figured on Pl. 6, Fig. 5 in GYGI (1990a). This undeformed cast of iron sulfide was probably shaped by the empty shell of an ammonite living in a time span of almost complete omission (nondeposition) in an environment with well-oxygenated bottom water basinward or southeast of Liesberg, where the *Renggeri* Member pinches out entirely (this study, Fig. 1.5, or GYGI 1990c, Fig. 2). The pertinent empty shell surfaced after the death of the ammonite, drifted to the northwest into the area where the *Renggeri* Member was sedimented into dysaerobic bottom water, sank to the sea floor, and was embedded there and partially fossilized, where the clay pit of Andil is now near Liesberg. The direction of the current of surficial water the shell drifted in was opposite to the direction of the bottom current of sediment-laden water flowing off land in the northwest into the area where the *Renggeri* Member was sedimented. The entire *Renggeri* Member extending in northwestern Switzerland from the upper *Scarburgense* Zone upward to the top of the *Costicardia* Zone (Fig. 1.6) pinches out completely toward the basin (GYGI 1995, Fig. 2). The index taxa of the pertinent zones are figured in GYGI (1990a) and in GYGI (1995). They are refigured in this study in Fig. 7.1.

The upper Herznach Formation near Herznach in Canton Aargau with *Cardioceras (Scarburgiceras) scarburgense* (YOUNG and BIRD), the index taxon of the *Scarburgense* Chron, is marl with calcareous concretions. Iron ooids are dispersed both in marl and in concretions (Fig. 8.2A). JEANNET (1951, Fig. 2) labeled these beds E2 in the Herznach iron mine. Two *Cardioceras (Scarburgiceras) scarburgense* that were collected probably from calcareous concretions in the pertinent beds in the Herznach iron mine were figured by MARCHAND et al. (2000, Pl. 1, Figs. 11–12). One of them has a phragmocone with the diameter of 38 mm. Part of the body chamber of the specimen is preserved. The diameter of the phragmocone of the second specimen is 55 mm. This phragmocone is probably complete with the beginning of the body chamber being preserved. The substance of the casts is calcareous micrite colored brown by ferric iron, and the casts include some brown iron ooids of goethite. It is documented in the following Sect. 4.8.1 that goethitic iron oolite was sedimented in well-oxygenated bottom water. The two casts figured are therefore probably incomplete fossils of large, adult specimens living in well-aerated water. They are much larger than all of the conspecific *Cardioceras* fossilized as casts of iron sulfide, some of them with part of the body chamber, in the *Renggeri* Member at Liesberg. The small specimens of the taxon in the *Renggeri* Member are consequently variants dwarfed or stunted by the environment.

The conclusion to be drawn from *Cardioceras (Scarburgiceras) scarburgense* found at Liesberg and Herznach is that specimens of the taxon living in the normally oxygenated water the Herznach Formation was sedimented in had a distinctly larger phragmocone (and adult size) than specimens fossilized in a sediment laid down in dysaerobic bottom water like the *Renggeri* Member. Small *Cardioceras (Scarburgiceras) scarburgense* with *crowded last septa* and at least part of the body chamber preserved were found in the *Renggeri* Member in section RG 280 at Liesberg. A photograph of one of them is shown in GYGI (1990a, Pl. 3, Fig. 1). Such specimens are most

probably adults in spite of their small size, because *precocious crowding of septa* is generally uncommon, and because precocious septal crowding in ammonite casts of comparably small size was never observed to the author's knowledge. A considerable disparity in size therefore existed between larger calcareous casts of *Cardioceras (Scarburgiceras) scarburgense* of ammonites living in well aerated water and small, probably adult representatives of the same taxon fossilized as casts of iron sulfide with part of the body chamber in sediment from dysaerobic water. Most of the ammonites fossilized in the *Renggeri* Member were small or dwarf, and the substance of their casts is partially or entirely iron sulfide. The conclusion by SCOTT (1940) and by ZIEGLER (1963, 1967), that *small* ammonite casts of iron sulfide are fossils of ammonites which were living in close relation to the sea floor, is thereby corroborated. A low oxygen content of water in the habitat of ammonites consequently reduced the *size* of adults, but it apparently did not noticeably reduce ammonite *abundance*.

The assemblage of macroorganisms with prevailing ammonites fossilized in the Schellenbrücke Bed of the early Oxfordian *Cordatum* Chron with autochthonous iron ooids in the uppermost Herznach Formation (Fig. 4.10) was living at a water depth of approximately 95 m. This can be presumed with a high degree of probability, because the Schellenbrücke Bed was laid down on the basin floor directly below the Birmenstorf Member. Sedimentation of the Birmenstorf Member with autochthonous pellets of cauliflower glauconite began at a water depth of about 100 m. It was concluded above that the boundaries between environments with iron ooid accretion in shallower water and environments with growth of cauliflower pellets of pure glauconite in deeper water, both vertical and lateral, were in the region investigated at the depth of deposition of approximately 100 m. It is stated in main conclusion no. 18 that this facies boundary is *independent* of the oxygenation of bottom water. Both the Schellenbrücke Bed with brown iron ooids of goethite floating in a matrix of initially calcilutite colored light-brown by some ferric iron in the type section of the bed in Canton Aargau (GYGI 2000a, Fig. 9 on p. 20) and the stratigraphically condensed, lowermost bed of the Birmenstorf Member with cauliflower glauconite pellets directly above include a normal macrofossil assemblage. The two beds were consequently sedimented in well aerated water.

The diameter of some of the large casts of full-grown and complete perisphinctid ammonites from the early Oxfordian Schellenbrücke Bed as figured by GYGI (1998) is around 250 mm. The largest ammonites living in the environment of the Schellenbrücke Bed were giant *Euaspidoceras*. One giant cast of *Euaspidoceras* was figured by BONNOT and GYGI (1998, Pl. 3, Fig. 1). The composition of the abundant and diverse macrofossil assemblage in the iron-oolitic

Schellenbrücke Bed with 13% of benthic animals was represented in detail by GYGI (1981, Table. 1). The well-represented benthos in the macrofossil assemblage in the bed documents that this iron oolite with iron ooids that were accreted at a water depth approaching 100 m was sedimented in well aerated bottom water (see below). Additional evidence of this is the normal size distribution in the ammonite assemblage which includes all sizes of casts between some giants and normal, genetic dwarfs. Such a genetically dwarf, adult and complete specimen with peristome and lappets is the *Glochiceras denticanaliculatum* GYGI which was figured by GYGI (1991b, Pl. 1, Fig. 1). A large number of ammonite casts from the stratigraphically condensed Schellenbrücke Bed in Canton Aargau (Fig. 8.2E) was figured by GYGI and MARCHAND (1982), GYGI (1991b), and GYGI (1998). It is evident from Figs. 2 and 4 in GYGI (1981) that the *Renggeri* Member sedimented in dysaerobic and quasi-stagnant bottom water in an unrestricted environment pinches out completely toward the epicontinental basin. The hiatus between iron-oolitic marl-clay with calcareous concretions of the earliest Oxfordian and the Schellenbrücke Bed of the *Cordatu*m Chron at Ueken in Canton Aargau represents, according to Fig. 2 in GYGI (1981), a time span from the latest *Scarburgense* to the end of the *Costicardia* Chron. This hiatus evolved because of nondeposition in well-oxygenated water at a depth of between 80 and 95 m. The hiatus below the thin Glaukonit-sandmergel Bed of the *Cordatum* Chron in Canton Schaffhausen encompasses all of the earlier ammonite chrons of the Oxfordian because of nondeposition in normally aerated water deeper than 100 m.

Bottom water above the sediment surface of the Effingen Member was intermittently dysaerobic (GYGI 1969a, p. 61). Bottom water with a low oxygen content in this member is documented by a dwarf ammonite cast of iron sulfide of a *Glochiceras* found by J. OGG on August 1 in 1987 in marl about 25 m below the base of the limestone succession of the Gerstenhübel Beds in section RG 37 east of Auenstein in Canton Aargau (GYGI 1969a, Pl. 17). The younger bed no. 54 in the upper Effingen Member of the same section RG 37 in the quarry of Jakobsberg includes minute bivalves the size of 160–320 μm and gastropods with a maximal size of 640 μm. Both groups of mollusks are preserved as casts of iron sulfide. Transformation from larvae to the tiny mollusks must have occurred in well-oxygenated water in the upper part of the water column. When the very young animals descended from surficial water in order to begin their benthic life, they suffocated in oxygen-deficient bottom water. The upper part of the water column, under which the Effingen Member was sedimented, was probably permanently well-aerated, to conclude from the rich assemblage of dinoflagellate cysts which GHASEMI in GHASEMI et al. (1999) collected and figured from the Effingen Member in section RG 37.

4.6 Hermatypic Corals and Their Greater Tolerance of Clay Mineral Sedimentation Compared to that of Ammonites

The *Renggeri* Member in the lowermost Bärschwil Formation at the base of succession no. 1 is homogenous clayey marl. Sedimentation began in dysaerobic bottom water at the depth of around 80 m. The initially low rate of sedimentation increased with time. This can be read from the upward growing thickness of ammonite zones in the member, as it is shown in Fig. 3 in GYGI (1990a). Ammonites prevail in the macrofossil assemblage in the entire member. They are most abundant and fairly numerous in the *Scarburgense* Zone, 2 m above the base of the member in the clay pit of Andil near Liesberg (section RG 280, Fig. 3 in GYGI 1990a). The thickness of the *Scarburgense* Zone near Liesberg is approximately 3 m. The abundance of ammonites and of macrofossils in general diminished upward in the succession concomitant with the growing sedimentation rate of predominantly clay minerals. The percentage of clay minerals in the *Renggeri* Member is more than 60% according to the single carbonate analysis that was published by PFRUNDER and WICKERT (1970). On the assumption that the *Scarburgense* Chron lasted 370,000 years, the rate of sedimentation of compacted, pure clay minerals was in the *Renggeri* Member at Liesberg of the order of 6.5 cm per 10,000 years during this chron and at a water depth of approximately 80 m.

Sedimentation of the Liesberg Member in the uppermost, aggradational and shallowing-upward Bärschwil Formation began at the water depth of approximately 20 m. This depth diminished because of ongoing aggradational deposition to about 10 m at the top of the member. The sediment of the Liesberg Member was probably primarily a homogenous mixture of argillaceous and of calcareous mud. No calcimetric analyses were made. Instead, the average carbonate content of the member was estimated visually to be at least 50% in the clay pit of Hinter Chestel at Liesberg in the fresh outcrop of section RG 306 when the pit was worked. The carbonate content of the sediment increases upward in the member in all of the sections measured. The carbonate content of the initial matrix of mud was partially unmixed during diagenesis and was concentrated in calcareous concretions with an irregular outline and with a slight content of white chert. The nodules in the Liesberg Member can thereby be distinguished from the pure calcareous, ellipsoidal concretions with a smooth surface in the Sornetan Member below. The average sedimentation rate of the Liesberg Member is calculated in Sect. 8.1 to have been at least 2.7 m of compacted sediment per 10,000 years. This is one of the highest rates of deposition recorded in Oxfordian sediments investigated, and it consequently implies a massive influx of clay minerals from land into the environment of deposition. Provided that the above estimate of the carbonate content of the Liesberg Member is of the right order of magnitude, the sedimentation rate of pure argillaceous mud may have been as much as 1.35 m of compacted deposit per 10,000 years, or approximately 20 times the pertinent rate in the lowermost *Renggeri* Member with fairly abundant ammonites, as it was calculated above. Hermatypic corals in the Liesberg Member are mostly plate-like colonies and are by far most abundant in the macrofossil assemblage (Fig. 4.9). They can be as much as 30% of the rock volume (Fig. 32 on p. 54 in GYGI 2000a). No ammonite was found by the author anywhere in the Liesberg Member. A. L. COE spotted a single ill-preserved, very large perisphinctid in the member in section RG 306 in the clay pit of Hinter Chestel at Liesberg which could not be identified. Living ammonites were probably primarily rare in the environment of the Liesberg Member mainly because of shallow water, and not so much because of the high rate of clay mineral sedimentation.

The Effingen Member in Canton Aargau is the bulk of succession no. 2. The member was sedimented in that region mainly by aggradation interacting with some progradation in the upper part (Fig. 1.5). The Effingen Member is a thick succession of marl with intercalations of marly limestone and relatively pure, micritic limestone like in the Gerstenhübel Beds. Sedimentation of the member began at the water depth of approximately 105 m, according to the calculation in Sect. 4.2. This depth was favorable to ammonites. Ammonites are not uncommon in the lowermost part of the Effingen Member, because sedimentation of mixed calcareous and argillaceous mud was initially going on at a moderate rate. The sedimentation rate increased upward in the succession. Accordingly, ammonite zones became progressively thicker with time like in the *Renggeri* Member. The Effingen Member includes the five ammonite zones from the *Schilli* to the *Hypselum* Zone. Ammonite abundance diminished upward in the succession with growing sedimentation rate. An exception are the 106 ammonite casts, most of them of micritic limestone, from the Effingen Member in the quarry of Hinterstein at Oberehrendingen in Canton Aargau, which were collected in the years of 1922–1924. The specimens are kept in the ROLLIER collection at the ETH Zürich. Mainly marl cropped out in the quarry of Hinterstein, which was worked for a cement plant. The quarry is now completely overgrown with vegetation. The author of this study searched ROLLIER's pertinent field book at the ETH Zürich in vain for information on the position and of the thickness of the limestone intercalation in the quarry of Hinterstein, in which the ammonites were found. ÉNAY in ÉNAY and GYGI (2001) identified and figured the ammonites from Oberehrendingen and assigned them to the *Grossouvrei* Subzone. This is equivalent to the *Bifurcatus* Zone in this study (Fig. 1.6). To judge from the micritic,

relatively pure limestone of the ammonite casts from Oberehrendingen, all of these specimens were probably found in the Gerstenhübel Beds.

It was calculated in Sect. 2.3.2 above that the averaged, compacted thickness of the Effingen Member in Canton Aargau is approximately 200 m. The member was laid down during approximately 1.8 million years. The average carbonate content of marl in the member is around 60%. When the assumption is made that the carbonate content of marl below and above the Gerstenhübel Beds is 60%, and that this marl was laid down at the average rate of the Effingen Member, a rate of sedimentation of pure, compacted argillaceous mud of 44 cm per 10,000 years results. The small ammonite cast of iron sulfide mentioned above which was found by J. G. OGG in marl below the Gerstenhübel Beds in section RG 37 in the quarry of Jakobsberg at Auenstein, was formed in marl with a carbonate content of the order of 55%. This can be estimated based on the detailed chemical analyses of cores from an exploration well, which was located by R. GYGI and drilled in 1962 by Jura Cement Company, Wildegg, in order to plan the deep road cut which was subsequently excavated between the quarries of Jakobsberg and Unteregg. The cores from the borehole were analyzed in the laboratory of the cement plant at Wildegg. The calcimetric results of the analyses are represented in GYGI (1969a, Pl. 17, section RG 37).

The depth of deposition of marl below the Gerstenhübel Beds in section RG 37 at Auenstein in Canton Aargau can be calculated to have been close to 80 m. This is approximately the same depth of deposition as that of the lowermost *Renggeri* Member at Liesberg. The oxygen content of bottom water, in which the pertinent marl of the Effingen Member was sedimented, was probably about as low as that of bottom water when the lowermost *Renggeri* Member was laid down at Liesberg. This assumption can be made based on the small haploceratid ammonite cast of iron sulfide from the Effingen Member mentioned above. Low oxygen content of bottom water when marl in this part of the Effingen Member was laid down is also likely judging from the minute casts of iron sulfide of mollusks in marl of the younger bed no. 54 of section RG 37, which were mentioned in Sect. 4.5 above. Ammonites are rare in the Effingen Member at Auenstein in marl with a sedimentation rate of pure clay minerals of 44 cm per 10,000 years, both below and above the Gerstenhübel Beds. Conversely, ammonite casts of iron sulfide are fairly abundant in the lowermost *Renggeri* Member at Liesberg, which was sedimented in water about 80 m deep and with an oxygen content of approximately 1 ml of dissolved O_2 per liter of sea water. It was calculated above that the sedimentation rate of pure argillaceous mud in the *Scarburgense* Chron was at that location 6.5 cm per 10,000 years.

It can be summarized that ammonites are fairly abundant in the *Scarburgense* Zone in the lowermost *Renggeri*

Member in section RG 280 at Liesberg in Canton Basel-Landschaft, where depth of sedimentation was approximately 80 m, bottom water was dysaerobic, and where the rate of pure clay mineral sedimentation was of the order of 6.5 cm per 10,000 years. When marl of the *Bifurcatus* Zone was sedimented in the Effingen Member in section RG 37 east of Auenstein in Canton Aargau at the rate equivalent to approximately 44 cm of pure, compacted clay per 10,000 years, bottom water at the depth of around 80 m had probably a similar low oxygen content like when the lowermost *Renggeri* Member was laid down in the *Scarburgense* Chron near Liesberg in water about 80 m deep. The rate of 44 cm of compacted, pure clay mineral deposition per 10,000 years in marl of the pertinent part of the Effingen Member with on average 40% clay minerals is approximately 7 times the corresponding rate of pure clay sedimentation in marl-clay with at least 60% clay minerals of the lowermost *Renggeri* Member at Liesberg. Ammonites were probably primarily rare near Auenstein because of the high rate of clay mineral sedimentation when marl of the Effingen Member was laid down below the Gerstenhübel Beds. The total sedimentation rate of calcareous and argillaceous mud combined was in marl of the Effingen Member near Auenstein much higher than total sedimentation rate in the predominantly argillaceous, lowermost *Renggeri* Member near Liesberg.

This is evidence of ammonite *intolerance* of a high rate of clay mineral sedimentation, as it was concluded before by GYGI (1999, p.136). Living ammonites were probably primarily uncommon in an environment with a high sedimentation rate of mixed argillaceous and calcareous mud (where sedimentation of clay minerals was diluted by calcareous mud), even at a water depth which was favorable to them. On the other hand, there was considerable ammonite *tolerance* of low oxygenation in bottom water. Ammonites reacted to this condition by dwarfing. Dysaerobic water reduced the size, but apparently did not noticeably reduce the abundance of ammonites. Conversely, the mass occurrence of hermatypic corals evolved in the Liesberg Member at Liesberg, when the rate of pure clay mineral sedimentation was as much as 135 cm per 10,000 years. This is 20 times the pertinent rate in the lowermost *Renggeri* Member. It is stated in main conclusion no. 27 that hermatypic corals could tolerate a much higher rate of clay mineral sedimentation than necto-benthic ammonites.

4.7   Sediment Transport from Land to Sea

Weathering on land in a subtropical climate was most effective at times with much rainfall. Water runoff from land where the Ardennes are now in the northwest and from the north (Rheinisches Schiefergebirge) supplied in times of humid climate most or all of the weathering products to the epicontinental sea of the Rhodano-Swabian Basin in the south. Supply of terrigenous sediment was ample when the Bärschwil Formation and the Effingen Member of the Wildegg Formation with their high content of clay minerals were laid down. Sediment transport in very shallow-marine environments probably occurred mainly by tidal currents and occasionally by currents driven by storms. Severe storms can displace a great amount of shallow water. Evidence of this is storm floods on the shore of the southeastern North Sea. Sediment supply from land diminished to a minimum in times of drier climate during deposition of the carbonate St-Ursanne, the Courgenay, and the Balsthal Formation. The interrelation of paleoclimate and sediment transport is discussed in Sect. 11.5.2.

4.8   The Mode of Formation of Thin and Widespread Iron Oolites in Relatively Deep Water on the Level Basin Floor

ALDINGER (1957b, p. 7) noted that the thin and very widespread iron oolites in the Jurassic sediments of southern Germany were sedimented far from land in relatively deep water on the floor of an epicontinental basin. TRÜMPY (1959, p. 447) wrote that there are rich macrofossil assemblages in the thin and widespread iron-oolitic boundary beds between the Middle and the Upper Jurassic Series in northern Switzerland, and that these assemblages are indicative of a normal marine habitat. TRÜMPY added that the pertinent macrofaunal communities possibly lived at a considerable water depth. He thereby confirmed in Switzerland what several German authors had concluded from oolitic ironstones of mainly Callovian age in southern Germany. These conclusions were summarized by ALDINGER (1957a, p. 5). The iron-oolitic beds which were sedimented in northern Switzerland mainly in the latest part of the Middle Jurassic are thin and widespread. Their matrix was initially marly or calcareous mud with a varying content of ferric iron, which gave the matrix a brownish color. The mud-grade matrix is evidence that bottom water was normally quiet. Ferric iron in the matrix and in the majority of the iron ooids indicates well-oxygenated bottom water. The iron ooids in such sediments are dispersed and float in the matrix which was initially soft mud.

4.8.1   Macrofossil Assemblages in Iron Oolites

The iron oolite of middle Bathonian age between Auenstein, Villigen, and Oberehrendingen in Canton Aargau is about 30 cm thick. The macrofossil assemblage in it is mostly bivalves with large ostreids near Auenstein (GYGI 1992,

Fig. 28) and abundant brachiopods. Ammonites are subordinate in the assemblage. Some of the ammonites from the middle Bathonian iron oolite in Canton Aargau were figured by MANGOLD and GYGI (1997). No hermatypic corals were found in the unit. This is evidence that the Bathonian iron oolite in Canton Aargau was sedimented in water deeper than 20 m. On the other hand, the minor percentage of ammonites in the macrofossil assemblage indicates, according to Fig. 4.9, that the depth of sedimentation of this iron oolite was nowhere more than approximately 30 m.

The iron-oolitic Herznach Formation was sedimented mainly during the Callovian. The formation is normally less than 1 m thick, but it can be followed over the long distance from eastern France over northern Switzerland to southern Germany. According to current knowledge, the formation pinches out only in the area between Auenstein, Villigen, and Oberehrendingen in Canton Aargau (MANGOLD and GYGI 1997, Fig. 2). Ammonites prevail in the macrofossil assemblage of the formation everywhere in Switzerland. Benthic animals are well represented (GYGI 1981, Table 1). The thin, but very widespread formation was, therefore, sedimented in well-aerated water at an average depth of more than 35 m. The size distribution in the ammonite assemblage varies between dwarfs (GYGI 1991b, Pl. 1, Fig. 1) and giants (BONNOT and GYGI 1998, Pl. 3, Fig. 1) and is, therefore, normal. This is additional evidence of well-oxygenated bottom water. The uniform macrofossil assemblage with mainly ammonites everywhere in the Herznach Formation is an indication that the depth of sedimentation did not vary significantly in the entire area where the formation was investigated.

The uppermost unit in the Herznach Formation is the Schellenbrücke Bed in the area where this bed occurs. The age of the bed is the *Cordatum* Chron. The pure iron-oolitic Schellenbrücke Bed was laid down at a water depth of close to 100 m (Fig. 4.9), because cauliflower pellets of glauconite appear for the first time in the lowermost, stratigraphically condensed bed of the Birmenstorf Member directly above. This thin marker bed is no. 10 of excavation RG 208 at Ueken north of the Herznach iron mine (GYGI 1977, Pl. 11, section no. 2). Bed no. 10 at Ueken was excavated on a surface of 119 m^2. The environmental change from iron ooid accretion at a water depth of less than 100 m to formation of glauconite pellets in water deeper than 100 m occurred during sedimentation of marker bed no. 10 in the *Densiplicatum* Chron. The vertical facies change was caused mainly by rising sea level. This change occurred simultaneously in the whole area between section RG 14 near Günsberg in Canton Solothurn in the west and section RG 73 near Dangstetten in southern Germany. Günsberg and Dangstetten are non-palinspastically 68 km apart. Precondition that the vertical facies change could be simultaneous in this vast area was that the sea floor was level

in *Densiplicatum* time in the entire pertinent area. The widespread, lowermost bed of the Birmenstorf Member with uniform facies is, therefore, unequivocal evidence that the entire Herznach Formation directly below was sedimented on the level floor of the Rhodano-Swabian Basin as well, as it is represented in Fig. 1.5.

4.8.2 How Iron Ooids did *not* Become Embedded into Thin, Widespread Beds of Argillaceous or of Calcareous Mud on the Basin Floor

Iron ooids in the thin and widespread beds on the level basin floor are dispersed in a matrix of what was initially soft *mud*, either argillaceous or calcareous. The macrofossil assemblage in such iron oolites is evidence of well-oxygenated bottom water. The grain size of the iron ooids which were figured by GYGI (1969a, Pl. 3, Fig. 8) and GYGI (2000a, Fig. 9) is of the same order as that of calcareous ooids. Ooids of calcium carbonate which were accreted in a *marine* environment are in most cases preserved in grainstone texture within a secondary cement of coarse-grained sparitic, low-magnesian calcite, as it was figured by GYGI (1969a, Pl. 8, Figs. 32 and 33). It is generally agreed that marine calcareous ooids are accreted when being rolled by currents in water which is not deeper than 6 m. Most authors, therefore, concluded that iron ooids could be accreted as well only in water as shallow as where calcareous ooids are formed in the Recent. As a consequence, these authors thought that iron ooids which are now embedded and dispersed in a matrix of mud must have been transported by a unidirectional current from the shallow-water realm, where they were presumed to have been accreted, into the environment of deeper water where they were finally embedded (see, for instance, Fig. 5 in BITTERLI 1979). There is no evidence that such unidirectional currents ever occurred over a long distance, like for instance on the depositional slope of the distal Sornetan Member, or above the muddy floor of the epicontinental basin of northern Switzerland during the early Oxfordian. Provided that such unidirectional currents *did* occur, they would have churned the mud at the sea floor up into suspension and would then have transported it out of the environment of initial sedimentation. A special situation exists once the mud has settled at the sediment surface (see below).

The Schellenbrücke Bed of *Cordatum* age in the uppermost part of the Herznach Formation was laid down on the level basin floor. Under the assumption that the iron ooids which are now in the Schellenbrücke Bed were accreted in a proximal shallow-water realm in *Cordatum* time, the iron ooids would have had to be transported through the coeval, highly fossiliferous bed in the middle of the aggradational, distal part of the Sornetan Member. This fossil bed was dated by D. MARCHAND to be of *Cordatum* age with many *Cardioceras* which are figured in GYGI and MARCHAND (1993). No iron ooids were found in the fossil bed in section RG 314, bed no. 1 near Sornetan; RG 373, bed no. 2, and RG 388, bed no. 2 near Vellerat; or RG 399, bed no. 46 near Bärschwil. The pertinent sections are shown in detail in Fig. 2 by GYGI and MARCHAND (1993). Moreover, the Sornetan Member pinches out in the distal direction (GYGI 1995, Fig. 2) and was not found below the Pichoux Formation, for instance, in section RG 307 near Péry (GYGI 2000a, Pl. 22). Neither can the problem of the occurrence of abundant iron ooids in a matrix of what was initially mud in thin and widespread sediments, which were laid down in relatively deep and well-oxygenated water, be solved by invoking the existence of hypothetical "swells" as the environment where iron ooids could have been accreted and concentrated within the epicontinental Rhodano-Swabian Basin, as ALLENBACH (2002) has done following NORRIS and HALLAM (1995). A swell or a sill is an areally restricted topographic high, whereas the thin, iron-oolitic sediments of the Herznach Formation of mainly Callovian age are very widespread and continuous over long distances in northern Switzerland. ALDINGER (1957a, p. 5) stated that oolitic ironstones of Callovian age can be followed without interruption from southern France over northern Switzerland to Swabia and even as far as to Franconia in southern Germany.

The question must be carefully addressed whether the iron ooids, which are dispersed in the matrix of what was primarily mud in thin and widespread beds with a macrofossil assemblage of mainly ammonites from a water depth of up to 100 m on a level basin floor, could indeed have been supplied from a shallow-water realm, and if yes, where the iron ooids came from. Iron ooids of goethite were found to be as much as 10% of the rock volume in the uppermost part of the Sornetan Member in section RG 456 near the village of Bure in Canton Jura. The Sornetan Member was initially a mixture of calcareous and argillaceous mud. The iron ooids in the Sornetan Member near Bure were 4.5 m below the top of the member in exploration well FN 2 south-southwest of locality Le Maira near Bure. The well log measured by the author is the unpublished section RG 456. Presence of iron ooids in the Sornetan Member near Bure is at the present state of knowledge unique and must be local, because no iron ooids were found in the corresponding level in exploration well N. E. B. 10, which was drilled 500 m to the north of well FN 2. The iron ooids in the uppermost Sornetan Member near Bure were embedded into a matrix of mud at a water depth of somewhat more than 20 m, because sedimentation of the Liesberg Member directly above the Sornetan Member began at the water depth of approximately 20 m. This is calculated in Chap. 6.

A low-grade oolitic iron ore was mined until after 1900 near Chamesol, a village in France that is located 20 km south of the city of Montbéliard and is adjacent to the Swiss border west of Porrentruy. This iron ore is mentioned on p. 7 of the "Note explicative" to the CARTE GÉOLOGIQUE DE LA FRANCE à 1/50,000, sheet no. 474 Montbéliard (1973). According to the text in the note and the assembled section on the left side of the map, the orebody is in the uppermost part of the time equivalent of the Swiss Liesberg Member. The age of the iron ooids both in the orebody near Chamesol in France and in the somewhat older, upper Sornetan Member near Bure in adjacent Switzerland is the *Antecedens* Chron. The paleogeographic position of these iron ooids is proximal and far in the interior of the shallow-water realm of the upper Bärschwil Formation. The distal time equivalent of the uppermost Sornetan Member and the Liesberg Member from shallow water in northwestern Switzerland is on the basin floor further east the lowermost, thin, and stratigraphically condensed bed of the Birmenstorf Member with sporadic iron ooids of goethite in some sections. The few iron ooids in this bed cannot have been transported over the long distance from the shallow-water realm to the floor of the Rhodano-Swabian Basin for two reasons. First, there are no iron ooids in the more distal part of both the uppermost Sornetan Member and the Liesberg Member in the thick, aggradational part of the upper Bärschwil Formation. Second, both the Sornetan Member and mainly the Liesberg Member pinch out in the distal direction before they reach the toe of the wedge-like part of the distal Bärschwil Formation (Fig. 1.5). The Liesberg Member, which was sedimented directly above the Sornetan Member on the flat top of the proximal, thick part of the aggradational Bärschwil Formation, pinches out at the outer rim of the thick part of the Bärschwil Formation. The Liesberg Member consequently pinches out in an even more proximal position than does the Sornetan Member below (Fig. 1.5).

The pinching out of the Sornetan Member and of the Liesberg Member in the distal direction was probably the effect of ongoing, net eustatic sea level rise since the beginning of the Oxfordian Age. A net relative sea level rise is the cause of the progressive evolution of a hiatus in the distal, wedge-like part of the Bärschwil Formation. The hiatus in section RG 315 in Pichoux Gorge near Sornetan in Canton Bern begins above the Sornetan Member in *Antecedens* time (GYGI 2000a, Pl. 21). The hiatus in this section represents a moderate time span. Further in the distal direction, in section RG 307 in the quarry of La Charuque near Péry, the hiatus begins much earlier above a very much thinned *Renggeri* Member (GYGI 2000a, Pl. 22). The hiatus at Péry encompasses four ammonite chrons or a time span of nearly 1.5 million years. The hiatus below the Pichoux Formation eventually became *regional* in a vast area which separated the Sornetan Member and the Liesberg Member in the proximal, shallow-water realm in the northwest *laterally* from time equivalents in the lowermost, stratigraphically condensed bed of the Birmenstorf Member in the distal, basinal realm in the southeast. This was schematically visualized in Fig. 2 in GYGI (1995). Weak contour currents moving above the depositional slope of the thinning, wedge-like part of the distal Bärschwil Formation were possibly active concomitantly with the relative rise of sea level. The two processes combined caused laterally progressive omission (nondeposition) on the depositional slope. Accordingly, contour currents are the most likely cause for the very pronounced thinning and partial stratigraphic condensation of the lower, glauconitic part of the Baden Member on a depositional slope, for instance, near Möriken, Villigen, and near Mellikon in Canton Aargau (Fig. 1.5).

When sedimentation of the locally dolomitic Oolithe rousse Member (abbreviated OOR in Fig. 1.5) was completed, the surface of calcareous, aragonitic ooids in the member was encrusted with a film of goethite. Ferruginization of the ooid surfaces occurred following emersion of the member, in a process which was similar to the one proposed by KIMBERLEY (1979). Emersion of the Oolithe rousse Member was caused by a minor sea level fall which is indicated in Fig. 5.1 to have occurred after sea level rise no. 9. The member was primarily a calcareous oolite with grainstone to wackestone texture. The best outcrop of the member is in section RG 315 in Pichoux Gorge near Sornetan, in bed nos. 108–111 on Pl. 21 in GYGI (2000a). A minor, rapid relative sea level rise occurred after the fall which caused emersion of the top of the Oolithe rousse Member. Evidence of the minor rise is large druses in bed no. 112 of section RG 315, which are partially filled with coarse crystals of calcite. The druses are probably remains of massive coral colonies. The same kind of empty druses partially filled with coarse crystals of calcite occur in bed no. 56 directly above the Oolithe rousse Member in the unpublished section RG 372 at Roches north of Moutier. The small rise is not shown in Fig. 5.1. According to GYGI (2000b, p. 137), the Oolithe rousse Member is in most sections a wackestone. Varying amounts of individual, euhedral dolomite crystals or of dedolomitized pseudomorphs after these crystals including a mosaic of anhedral calcite crystals are dispersed in the micritic matrix of many of the wackestones in the member. The crusts of goethite both around calcareous ooids and rhombohedra of dolomite or of dedolomite, respectively, where they occur in the Oolithe rousse Member, give the rock its characteristic red-brown color (rousse in French, or russet in English).

The time equivalent of the Oolithe rousse Member in northwestern Switzerland is the *Crenularis* Member in the Villigen Formation in Canton Aargau. Sedimentation of the *Crenularis* Member began after sea level rise no. 9 in Fig. 5.1. The thin, micritic limestone of the *Crenularis* Member with some very fine-grained glauconite includes mainly large bivalves of the genus *Pholadomya,* which are indicative of a water depth of close to 30 m (see above).

Time correlation between the base of the *Crenularis* Member in Canton Aargau and the base of the Oolithe rousse Member in the shallow-water realm was made according to the mineral-stratigraphic correlation I by PERSOZ in Pl. 1A by GYGI and PERSOZ (1986), because there are no ammonites in the Oolithe rousse Member. This correlation was calibrated with the ammonite figured in GYGI (2000a, Pl. 10, Fig. 5), which was then identified as *Epipeltoceras* cf. *bimammatum.* The pertinent ammonite was found by C. MOESCH in the glauconitic *Crenularis* Member in a quarry which is now the unpublished section RG 36 at locality Fahr east of Auenstein in Canton Aargau. The ammonite is a complete adult and is close to the taxon *berrense* which was named and described by FAVRE (1876, p. 59, and Pl. 3, Fig. 11). The *Epipeltoceras* found by C. MOESCH at Auenstein is a form transitional between the taxa *berrense* FAVRE and *bimammatum* QUENSTEDT. This is evidence that sedimentation of succession no. 3 in Canton Aargau began with the glauconitic *Crenularis* Member late in the *Berrense* Chron (Fig. 1.6).

Neither pure calcareous ooids nor such ooids coated with goethite occur in the slightly glauconitic *Crenularis* Member. Instead, there are minute oncoids including very thin algal tubes filled with glauconite that can be discerned in thin section and that give the oncoids a light-green color when being viewed under a hand lens. These oncoids are about the same size as calcareous ooids. Such oncoids in a polished slab from the *Crenularis* Member were figured by GYGI (1969a, Pl. 5, Fig. 21). Conversely, pure calcareous ooids were found by MOESCH (1867, p. 151) and by the author of this study in the region of Bözberg west of the city of Brugg in Canton Aargau in a stratigraphic position that the author of this study now assigns to the uppermost Geissberg Member. The maximal thickness of the pertinent thin, calcareous-oolitic unit in Bözberg area is 3.3 m. This thickness was measured in the quarry of the unpublished section RG 57 at locality Laufacher east of the village of Zeihen. Calcareous ooids in this unit occur in the sections RG 57 (Zeihen), RG 58 (Linn), and RG 59 (Effingen) which are listed with coordinates in Table 1 in GYGI (2000a). The three sections are schematically represented in the upper synthetic section on Pl. 19 in GYGI (1969a). No photograph of one of his thin sections with abundant calcareous ooids from the pertinent oolitic unit was figured by GYGI (1969a). Instead, the thin section photograph of a bioclastic calcarenite from the pertinent unit is shown as Fig. 22 on Pl. 5 in GYGI (1969a). It is now presumed that the calcareous-oolitic unit in the uppermost part of the Geissberg Member in Bözberg area was sedimented at the foot of a tidal delta of calcareous ooid sand that was shed from the most distal extension of the oosparitic Steinibach Member in the lower Balsthal Formation which must have prograded to close to the Bözberg area. The delta itself is not preserved because of erosion. The tongue at the foot of the delta with a variable

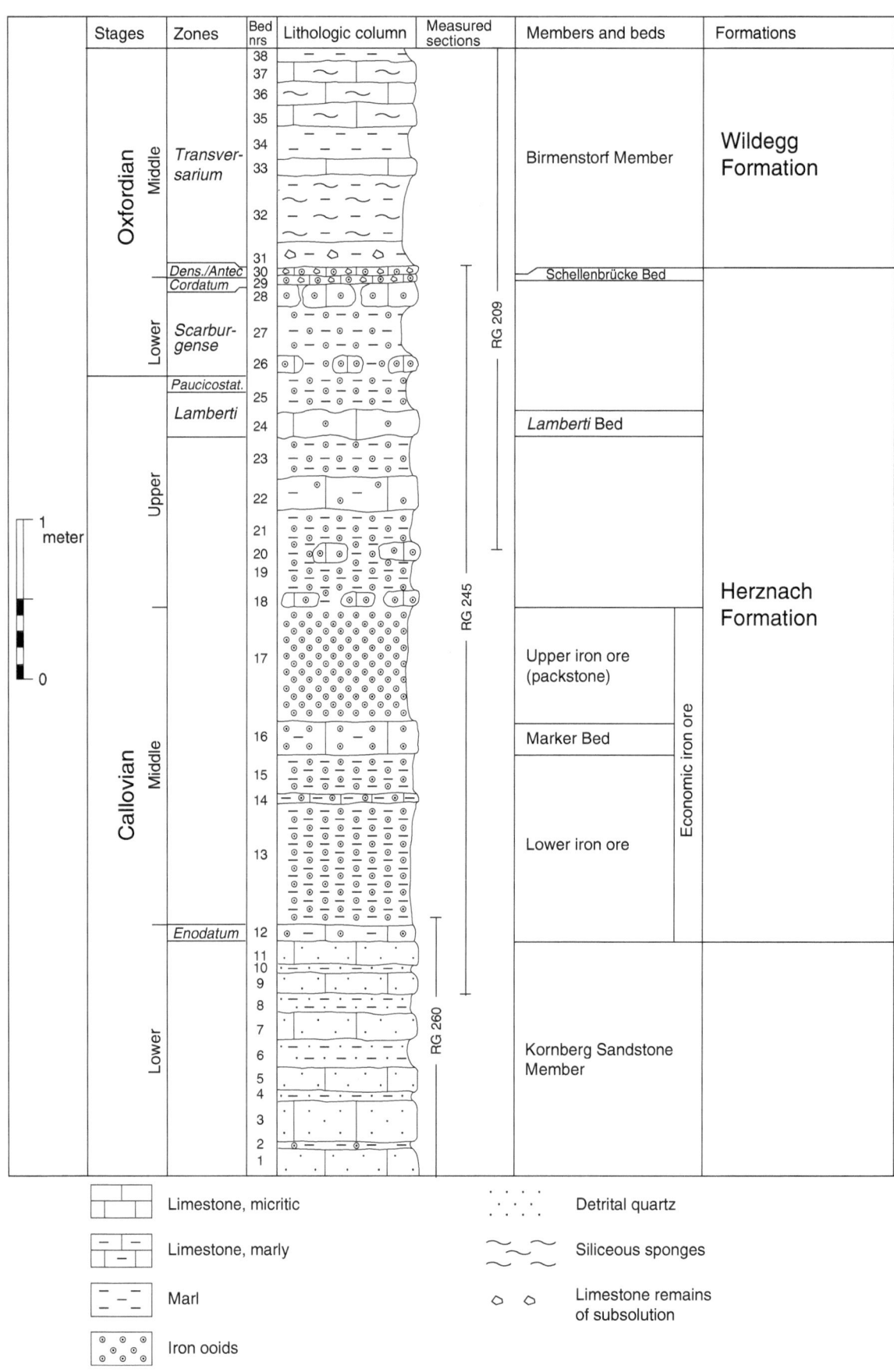

Fig. 4.10 Herznach Formation. Type section assembled from the following three partial sections, from below to above: RG 260, RG 245, and RG 209, which were measured by the author at different localities in the closed iron mine west of the village of Herznach in the Tabular Jura in Canton Aargau in 1971 and in 1976

content of calcareous ooids was probably sedimented at a water depth of more than 20 m, because it rests directly on micritic limestone with mainly bivalves of the Geissberg Member. No hermatypic corals of the Olten Member are intercalated in between.

The conclusion to be drawn from the sum of all of the data given above is that calcareous ooids with the characteristic crust of goethite which is diagnostic of emersion (see below) were certainly not transported from the Oolithe rousse Member in the central Jura Mountains to the Crenularis Member in Canton Aargau. It is important to note in this context the conclusion by BURKHALTER (1995, p. 68) that ferruginous ooids in the Passwang Formation of the Aalenian and Bajocian in the Swiss Jura Mountains originated in a fully marine environment. BURKHALTER continued on p. 69 that the iron ooids investigated by him were accreted in or near their environment of deposition. The iron ooids in middle Bathonian iron oolites and in the Callovian/Oxfordian iron-oolitic Herznach Formation were consequently neither derived from land nor from a lagoonal environment. They were not re-deposited in deeper parts of the epicontinental basin.

The *Herznach Formation* is all iron oolite with a total thickness of less than one meter in most of the area where the formation occurs. Nevertheless, the unit was given the rank of a formation by GYGI (2000a, p. 51) because of its wide geographic and temporal extent. The rank of a formation for these geographically widespread iron oolites is in agreement with the International Stratigraphic Guide, second edition, edited by SALVADOR (1994, p. 34). The geographic range of the Herznach Formation, following ALDINGER (1957a, p. 5), is with some interruptions between eastern France in the west, northern Switzerland, and southern Germany in the east. A section which was measured in the Herznach iron mine was published by JEANNET (1951, Fig. 2). JEANNET (1951, p. 157) mentioned *Kosmoceras enodatum* NIKITIN to occur in bed A5 of this section. A5 is the lowermost bed of the iron orebody. It is equivalent to bed no. 12 in Fig. 4.10 of this study. The base of the Herznach Formation *in the type region around Herznach* is in the *Enodatum* Zone (Fig. 4.10). This is the uppermost zone of the lower Callovian Stage. According to CALLOMON (1964, Table 1 on p. 288), the ammonite taxon which was called in a recent publication *Sigaloceras (Catasigaloceras) enodatum* (NIKITIN) is the index of the youngest subzone of the early Callovian. The top of the Herznach Formation is *near Herznach* at the top of the *Cordatum* Zone, which is the uppermost zone of the lower Oxfordian Stage. The *type section* of the Herznach Formation is shown in Fig. 4.10. The type section was assembled from sections RG 209, 245, 253, and 260, which were measured by the author of this study in 1971 and mainly in 1976 in the closed iron mine

west of Herznach. The mine was closed in 1967. JEANNET (1951) published the major part of the rich ammonite assemblage which was collected in the Herznach iron mine over the years when the mine was worked mainly during World War II. More ammonites from near Herznach and Ueken, mainly cardioceratids and perisphinctids which were not published by JEANNET, were described and figured by GYGI (1977), GYGI and MARCHAND (1982), GYGI (1998), MARCHAND et al. (2000), and GYGI (2001).

The problem of how iron ooids could become concentrated enough to form a local, economic orebody with a thickness of 2 m like in the Herznach Formation between Herznach and Wölflinswil remains to be unsolved. The same opinion was published before by ALDINGER (1957a, p. 6) on behalf of the oolitic iron orebodies of the Blumberg type in southern Germany. No explanation could be found, why the thickness of the Herznach Formation increased in a restricted area at Herznach and some distance to the west to several meters, and by which process iron ooids could become concentrated to packstone texture in the upper part of the Herznach iron ore. Iron ooids are more than 50% of the rock volume in the upper part of this orebody (in bed no. 17 in Fig. 4.10). The iron ore near Herznach was certainly not formed on a swell. The assertion that an oolitic iron orebody like that between Herznach and Wölflinswil, or the iron oolite in section RG 210 in Eisengraben near Gansingen (Fig. 8.3) could be formed on a submarine swell (NORRIS and HALLAM 1995) is incompatible with the regional facies relations and has to be rejected. The siliciclastic mud which is part of the matrix mainly in the lower part of the oolitic iron ore near Herznach was most probably not transported from deeper parts of the epicontinental basin *uphill* to a swell.

Ammonite casts prevail everywhere in the macrofossil assemblage of the Herznach Formation in northern Switzerland. The oldest part of the formation between northeastern Canton Aargau and Canton Schaffhausen is of early Callovian age like the iron oolite which was mined during World War II near Blumberg in adjacent southern Germany. The early Callovian iron oolite in northeastern Switzerland, for instance in Canton Schaffhausen and in the unpublished section RG 67 near Zurzach, has an approximate time equivalent in the lowermost part of the Callovienton Member in the former clay pit of Hinter Chestel south of the village of Liesberg in northwestern Switzerland. There are abundant iron ooids in the uppermost part of bed no. 11 in excavation RG 427, which was made on September 5 in 1985 in the clay pit of Hinter Chestel. The section of the excavation is represented as Fig. 1 in DIETL and GYGI (1998). One of the ammonites which were collected from the iron-oolitic bed no. 11 in excavation RG 427 is *Cadoceras tolype* S. BUCKMAN of the early Callovian *Koenigi* Chron. The specimen of this characteristic taxon was figured by Dietl and Gygi (1998, Pl. 5b and c). *Cadoceras tolype* is not uncommon in the coeval iron oolite with about 20% of iron ooids and with a rich, particularly well-preserved macrofossil assemblage of mainly ammonites in bed no. 7 of the unpublished excavation RG 265 near Anwil in Canton Basel-Landschaft. Excavation RG 265 was made on May 2–4, 1977, by R. and S. Gygi and D. Oppliger at locality Ächtelmatt northeast of the village of Anwil.

The text above can be summarized in the following way. Iron ooids in the uppermost Sornetan Member and in the nearby time equivalent of the uppermost Liesberg Member in adjacent France in the proximal, shallow-water realm were separated by a regional, *lateral* hiatus from the coeval, lowermost, and stratigraphically condensed marker bed of the Birmenstorf Member on the level floor of the Rhodano-Swabian Basin. The calcareous ooids in the Oolithe rousse Member of the shallow-water realm in the central Jura Mountains are characteristically encrusted with goethite. Such ooids do not occur in the coeval *Crenularis* Member from deeper water in the Villigen Formation in Canton Aargau. The hypothesis that iron ooids could only be accreted in water as shallow as where Recent calcareous ooids are formed and that the iron ooids were then transported by a unidirectional current into deeper water of an adjacent epicontinental basin is therefore untenable and must be dismissed. Aldinger (1957b, p. 7) concluded that such a transport of iron ooids, and mainly of primarily soft, plastic iron ooids (of berthierine which were figured by Gygi 1969a, Pl. 3, Fig. 9) is out of the question, and that accretion of iron ooids must consequently have been possible far from land in an epicontinental basin.

4.8.3 How Iron Ooids Dispersed at the Surface of Mud-Grade Sediment Could Be *Accreted* and Become Embedded at a Water Depth of up to 100 m

Ferric iron oxide is produced on land in soils by weathering in a tropical or subtropical climate with adequate rainfall. According to Aldinger (1957b, p. 8), the iron oxide was transported by rivers in solid state along with clay minerals from land to sea. Carroll (1958, p. 2) was more specific and stated that films of ferric iron oxide can coat particles of different minerals from clay to sand size in the process of weathering in soils. Such coatings like in the Oolithe rousse Member are therefore diagnostic of a terrestrial environment. Aldinger (1957a, p. 5, 1957b, p. 8) supposed that terrigenous, ferric iron oxide in mud-grade marine, nearshore sediment could be remobilized by solution in the interstitial water, in which the Eh is lowered by bacterial degradation of organic matter (compare with Carroll 1958,

p. 13). Ferric iron would be reduced in the process to the ferrous, soluble state. The ferriferous solution between minerals and particles in the sediment would then be driven by compaction of the deposit under its own weight upward to the sediment surface (Aldinger 1957b, p. 9) where the iron flocculated. From there, the floccules could be transported to the basin floor. Aldinger's views are compatible with all of the careful observations which were made on iron oolites of Aalenian to Oxfordian age in northern Switzerland.

At first glance it appears to be a contradiction that net accumulation of mud and accretion of grains the size of iron ooids could occur in one and the same environment and in well-oxygenated water down to a maximal depth of deposition of 100 m, as it was envisaged by Gygi (1981). Accretion of iron ooids when being rolled at the surface of muddy sediment *could* take place in the deep subtidal zone, when an occasional, particularly strong storm caused weak oscillating currents in the water above the sediment surface which was normally quiet. It can be read from the diagram by Hjulström (1935, Fig. 18) that mud-grade particles, once they are settled at the sediment surface, require (probably mainly because of their cohesion) a higher current velocity in order to be eroded and churned up into suspension and be transported, than particles the size of iron ooids require to be rolled on the sediment surface of mud. According to Hjulström (1935, p. 296), particles with a diameter of between 0.1 and 0.5 mm are those which are the easiest to be *reworked* (eroded) from the surface of a sediment. Figure 8 in Pl. 3 in Gygi (1969a) and Fig. 9 in Gygi (2000a) document that the average diameter of Jurassic iron ooids falls into this category of grain size. Particles with a greater as well as with a *lesser* size than the figured iron ooids require, once they are sedimented, a higher current velocity to be reworked. The study of Hjulström (1935) was made based on experiments with fluvial sediments, but it is probable that the basic results which were published by Hjulström (1935) can be applied to marine sediments as well.

Small particles rolled during a storm at the surface of argillaceous or calcareous mud could become selectively encrusted with iron hydroxide in the process. The cause of selectivity for iron hydroxide in iron ooid accretion has as yet to be found. A small bioclast with a thin ooidal crust of goethite and an iron ooid with a fragment of another iron ooid in the core were figured by Gygi (2000a, Fig. 9 on p. 20) from the early Oxfordian Schellenbrücke Bed in the type section RG 32 of the bed near Küttigen in Canton Aargau. An iron ooid of middle Bathonian age with a large bioclast as a core was figured by Gygi (1969a, Pl. 2, Fig. 5). This bioclast is bored by fungal hyphae with sporangia, and it is enveloped with a thin oolitic cortex of goethite. The source of iron hydroxide in the form of a sol was explained by Gygi (1981, p. 247) at that time without knowledge of, but in agreement with Aldinger (1957b, p. 9). The time intervals

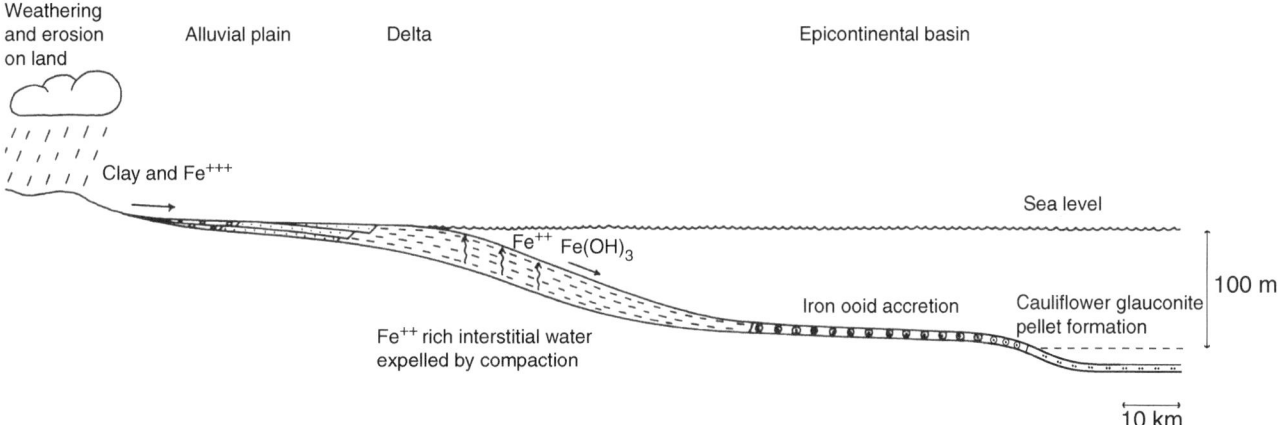

Fig. 4.11 Model of iron ooid accretion in the marine environment at the surface of loose, slightly ferriferous calcareous mud which was the initial matrix of the thin and widespread Schellenbrücke Bed. This bed was sedimented in well-oxygenated bottom water at a depth of approximately 95 m on the level floor of the Rhodano-Swabian Epicontinental Basin. The model was initially proposed by Gygi (1981). It is in agreement with Aldinger (1957b). The great exaggeration of the vertical scale should be noted. (1) Ferric iron oxide derived from weathering on land in a humid, subtropical climate, coats mineral particles from clay to sand size in a soil, according to Carroll (1958, p. 2). (2) Heavy rainfall washes the iron-clad mineral particles out of soil and running water transports the particles from land into the sea. (3) The particles are embedded in shallow water into the sediment of mixed argillaceous and calcareous mud in the lower half of the Sornetan Member. (4) After deposition of the mud, the organic matter in the sediment is decomposed by aerobic bacteriae until all of the oxygen dissolved in the interstitial water is used up. When anaerobic bacteriae continue to decompose organic matter, the Eh value of the interstitial water trapped in the mud becomes negative. Proof of a finally negative Eh is the blue-gray color of the Sornetan Member. At this stage, ferric iron in the coating of particles is reduced to the soluble ferrous state and is thereby dissolved and mobilized in the interstitial water. (5) The accumulating sediment of mixed argillaceous and calcareous mud is partially compacted under its own, growing weight. The process of compaction

drives part of the interstitial water with dissolved ferrous iron upward to the sediment surface. (6) As soon as ferrous iron arrives at the sediment surface, it is exposed to oxygen dissolved in the free bottom water. Ferrous iron is thereby instantly re-oxidized to the ferric state in the form of $Fe(OH)_3$. Iron hydroxide is insoluble in oxygenated sea water and coagulates in the form of a sol or as floccules. (7) This substance drifts in a multiphase process down the gentle depositional slope at the sediment surface (compare with Fig. 1.5), along with clay minerals and with calcareous mud. Elutriation of the coagulated $Fe(OH)_3$ occurs in the process, and the coagulate travels the greatest distance to and over the level basin floor, where it is concentrated. (8) Waves caused by infrequent, severe storms were strong enough to drive weakly oscillating currents in bottom water above the surface of mostly mud-grade sediment. Floccules of iron hydroxide or fine particles out of the sediment are rolled by the oscillating currents at the surface of calcareous mud. In times of nondeposition, episodic rolling of particles at the surface of loose mud can cause selective accretion of iron hydroxide to the surface of floccules or particles in well-oxygenated water at a depth of up to 100 m. Slightly ferriferous mud settles on the basin floor at a very low average rate during long, quiet time intervals between cyclones and smothers some of the iron ooids. When the process is repeated many times, the particles eventually evolve to iron ooids. Refigured with permission of SEPM (Society for Sedimentary Geology) from Gygi et al. (1998, Fig. 9)

between events of iron ooid accretion during a storm became longer, the deeper the water above the sediment surface was. The lower limit to the bathymetric range within which iron ooids could be accreted by gently oscillating currents of bottom water driven by a storm, was at the water depth from where downward an *average* hurricane could not generate oscillating currents in the water above the sediment surface strong enough to roll particles the size of iron ooids.

It must be noted in this context that Fig. 8.2A, B are evidence that *exceptionally violent storms* occurring at rare occasions could erode and churn up into suspension, and entirely remove loose, fine-grained sediment including iron ooids in it at a water depth of as much as 80 m. This observation was made in excavation RG 208 at Ueken in Canton Aargau. It corroborates what was stated in Sect. 4.3, that storm wave base is not at a fixed water depth. This is stressed below in main conclusion no. 22. The term *storm wave base* is, therefore, avoided in this study. It was concluded in

Sect. 4.3.1 that oscillating water currents driven by an extremely severe storm could roll spherical oncoids with a diameter greater than 3 cm at the surface of mud-grade sediment at a water depth of approximately 120 m.

The model of iron ooid accretion at the surface of muddy sediment in comparatively deep and well-oxygenated sea water which was conceived by Gygi (1981) and that was critically reviewed by Odin et al. (1988) is visualized once again in Fig. 4.11 of the present study. The model is explained in the caption to Fig. 4.11. The well-documented fact that the thin and widespread iron oolites which were sedimented at a very low average rate on the floor of the epicontinental, Rhodano-Swabian Basin as the distal time equivalent of a thick, partially argillaceous succession of mud, can be explained by this model. Burkhalter (1995, p. 60) concluded that all of the Aalenian and Bajocian ooidal ironstones investigated by him were sedimented at a low (average) rate and that iron ooids were accreted stepwise during time intervals of nondeposition (omission). He added on p. 57 that ferruginous ooids of the Passwang Formation grew at the sediment surface in an oxygenated, marine environment, and that these ooids were accreted in a relatively *broad bathymetric range*. This conclusion is in agreement with what was observed by the author of this study.

COLLIN et al. (2005, Fig. 5) figured brown, goethitic iron ooids from stratigraphically condensed, Middle/Late Jurassic boundary beds in Burgundy in eastern France. Their Fig. 5A shows an iron ooid with a nucleus of submicroscopic, extremely fine-grained goethite. The powdered mineral was identified in an x-ray diffractometer (op. cit., Fig. 7). This is evidence that growth of a goethitic iron ooid can begin without a small hard particle as a nucleus, like for instance a bioclast (GYGI 1969a, Pl. 2, Fig. 5), or silt-size particles of detrital quartz (SAEMANN 1921, Fig. 6). It is, therefore, conceivable that iron hydroxide can coagulate in the form of floccules *at the site of iron ooid formation* out of the sol which is supplied from argillaceous sediment in the process of compaction in a distant, proximal position. The soft floccules of iron hydroxide may serve as nuclei in iron ooid accretion. Subsequent accretion of cortices to the nuclei, when the nuclei were rolled on the sediment surface, could occur by direct adhesion of more coagulated iron hydroxide from sol in suspension in the bottom water. This mechanism would explain selective accretion of iron hydroxide to iron ooids independently of the substrate of either argillaceous or carbonate mud. If this be the case, then fresh cortices of goethitic iron ooids would be soft and would crystallize into solid cortices of submicroscopic goethite only some time after accretion of a cortex.

Provided that this assumption could be confirmed by experimental work, it would solve the problem posed by the observation that the surface of goethitic iron ooids is sometimes slightly blurred, and that such ooids can be surrounded by a diffuse brown halo of iron oxide in the adjacent matrix of calcareous mud the ooids are embedded in (GYGI 2000a, Fig. 9). Iron ooids of goethite must be solid from an early stage in their formation, because broken goethitic ooids are not uncommon. Broken brown iron ooids of goethite were figured by BURKHALTER (1995, Fig. 5) and by GYGI (2000a, Fig. 9). Conversely, green iron ooids of berthierine (chamosite in earlier publications by the author) are initially soft and can remain to be so until after their formation is completed. In this case the green ooids can be plastically deformed when mud with the embedded iron ooids is burrowed. Plastically deformed iron ooids with predominantly green cortices from an iron oolite of middle Bathonian age in section RG 51a near Oberehrendingen in Canton Aargau were figured by GYGI (1969a, Pl. 3, Fig. 8). Three authors are cited by GYGI (1969a, p. 19) who stated that chamosite is a paragenesis, not a mineral. Uncertainty persists relating to chamosite. BRINDLEY et al. (1968) were of the opinion that chamosite and berthierine are synonyms of one and the same mineral, and that berthierine has priority over chamosite. YOUNG (1989, p. xvi) stated that berthierine is transformed to chamosite during deep burial diagenesis at a temperature of about 150–160°C at a burial depth of 3 km. It is certain that green iron ooids and microbialites in northern Switzerland were never buried to the depth of 3 km. The pertinent green substance was not identified by x-ray diffraction by the author of this study.

4.8.4 Accretion of Brown Iron Ooids of Goethite and Green Iron Ooids of Berthierine Side by Side in Well-Oxygenated Bottom Water of the Deep Subtidal Zone

Pure, green iron ooids of berthierine were found to be intermingled with brown, goethitic ooids in bed nos. 18 and 20 of the unpublished section RG 253, which was measured in 1976 in the then closed Herznach iron mine. Iron ooids with alternating brown and green cortices were figured by GYGI (1969a, Pl. 3, Fig. 8) and by several earlier authors as cited by GYGI (1981, p. 248), and by BURKHALTER (1995, Fig. 12). Iron oolites of the early Oxfordian with such ooids include a macrofossil assemblage of mostly ammonites. The ammonite casts are evidence of a normal size distribution in the living community. It was shown above that such a macrofossil assemblage is diagnostic of well-oxygenated water directly above the sediment surface at a water depth of more than 35 m (Fig. 4.9). Accretion of iron ooids when being rolled at the surface of mud by weak, oscillating currents of well-aerated water in the deep subtidal zone could only occur during the short spells of severe storms between long time intervals when bottom water was quiet enough that some mud could be sedimented. When in the event of a storm small mineral grains, bioclasts, or floccules of iron hydroxide were rolled over the sediment surface on the basin floor, they selectively accreted cortices of iron hydroxide with ferric iron. This substance is presumed to have been present in the form of a sol in bottom water. The sol of $Fe(OH)_3$ originated at the surface of a thick unit of partially argillaceous mud being in the process of compaction under its own weight in a more proximal position, as for instance at the surface of what is now the lower half of the Sornetan Member (see caption to Fig. 4.11). Direct evidence of such a process is the thick crust of iron oxide which was found at the upper boundary of the Sornetan Member below the Pichoux Formation in the unpublished section RG 318 near Soulce in Canton Jura.

Iron hydroxide in the form of a sol or of floccules then drifted down the gentle depositional slope to the basin floor in the multiphase process that was described by GYGI (1981, p. 247). The sol could be accreted to iron ooids on the basin floor during a storm. In case of a severe storm, loose mud was churned up into suspension from the sediment surface on the basin floor. When the oscillating current of bottom water died down at the end of the storm, the mud in suspension settled down again and buried some of the iron ooids at the sediment surface. Mud, mainly argillaceous mud, has a very low permeability. Omnipresent organic matter trapped in the interstices between minerals and particles within the mud then decayed by bacterial activity and thereby used up the oxygen which was initially dissolved in the pore water. The initially positive oxidation potential (Eh) of the interstitial water in the fresh, muddy sediment turned to be negative in the process. Provided that enough silica, probably in the form of a sol, was present within the mud, then the outermost cortex with ferric iron of the buried iron ooids could be transformed to berthierine, which is colored green by ferrous iron. The initially plastic iron ooids of middle Bathonian age with alternating brown and green cortices, which were figured by GYGI (1969a, Pl. 3, Fig. 8), were embedded into mud together with a macrofossil assemblage which is evidence of a community living in well-oxygenated bottom water at a depth of between 20 and 30 m. The ooids retained a brown outer cortex, in case they remained at the sediment surface after a storm in the free, well-oxygenated bottom water. When an iron ooid was buried deeply enough by mud settling out of suspension after the storm-induced, oscillating currents had ceased in bottom water, then the outermost, goethitic cortex could be replaced by berthierine in the reducing interstitial pore fluid within the mud, after the oxygen initially dissolved in the interstitial water was used up by decaying organic matter. Repetition of these processes could produce iron ooids with alternating brown and green cortices. Buried green iron ooids plastically deformed by burrowing organisms, as were figured by GYGI (1969a, Pl. 3, Fig. 8), are additional evidence of oxygenated bottom water.

4.9 Aggradation, Progradation, and Backstepping

The term *aggradation* is used in this study for a pattern of sedimentation like that in the proximal, thick part of succession no. 1 (Fig. 1.5). Fine-grained sediments settling out of suspension in an aggradational unit accumulate vertically. The basinward rim of the proximal, aggradational and thick sediment stack of succession no. 1 neither prograded basinward nor receded landward to a significant extent during sedimentation in the time span of 8 ammonite chrons. This is equivalent to approximately 3 million years.

Progradation of shallow-water sediments, mainly of small coral reefs and calcareous oolite over advancing clinoforms of mixed calcareous and argillaceous mud in the proximal part of the Effingen Member typically occurred in the carbonate platform of the Günsberg Formation (GÜN in Fig. 1.5) and in the calcareous-oolitic Steinibach Member (STE in Fig. 1.5). This is shown in Fig. 1.5. Individual sigmoid clinoforms in the pertinent part of the Effingen Member are primarily (or depositionally) inclined. The clinoforms in this unit succeed each other in time and in the distal direction perpendicularly to depositional strike similar to foresets in a delta, but the clinoforms in the proximal part of the Effingen Member have a by far less depositional inclination than foresets in a delta. This can be read from Fig. 1.5. The vertical scale in the lower left of both Figs. 1.5 and 4.15 is exaggerated 100 times compared with the horizontal scale in order to visualize the very slight primary inclination of the clinoforms in the proximal Effingen Member. The depositional inclination of these clinoforms was in fact so slight that it would be imperceptible in Fig. 1.5 without great vertical exaggeration. The declivity of the depositional slope at the top of Pichoux Formation as it was at the end of sedimentation of succession no. 1 in the distal, thinning part of the succession can be read from Fig. 1.5 or 4.15B when the hundredfold exaggeration of the vertical scale of the two figures is taken into account. The depositional declivity of the top of the Pichoux Formation, as it was at the end of the formation's sedimentation, was secondarily enhanced during subsequent deposition of the thickest part of succession no. 2 in the adjacent epicontinental basin. The load of thick succession no. 2 in eastern Canton Solothurn and Canton Aargau caused differential, exogenic basement subsidence by the substantial amount which can be read from a comparison of Fig. 4.15B with 4.15C.

A special case of progradation is the lateral shift of a string of individual coral reef knolls in the boundary beds between the Effingen Member below and the Günsberg Formation above (Fig. 1.5). The string of coral reefs was aligned along depositional strike. This is shown in the paleogeographic map of Fig. 6 in GYGI (1990c). Hermatypic corals first appeared in biostromes in the uppermost Effingen Member. Patch reef knobs nucleated above the biostromes in the uppermost Effingen Member and continued to grow taller and wider into the lowermost part of the calcareous oolite of the Günsberg Formation above. This can be seen in the quarry of La Charuque near Péry. The pertinent section RG 307 is represented in Pl. 22 in GYGI (2000a). The row of closely spaced reef knolls grew in somewhat deeper water before the prograding front of a sand bank of calcareous ooids. The individual reefs at the base of the Günsberg Formation were nowhere seen to merge into a barrier reef. When the part of the sand bank, on which calcareous ooids were accreted, advanced to near the string of reefs, ooid sand was shed into the interspace between the reefs at such a rate that corals were eventually choked. The advancing bank of ooid sand smothered reefs in the course of further progradation. Calcareous oolite is the bulk of the Günsberg Formation.

Mixed terrigenous and calcareous mud as well as some siliciclastic silt were supplied from the peritidal realm of the Röschenz Member which is the time equivalent landward of the sand bank of calcareous ooids of the Günsberg Formation. Currents that were driven probably mainly by tides and by occasional storms transported the fine-grained sediment across the oolitic sand bank of the evolving carbonate platform of the Günsberg Formation and beyond into deeper water, where the sediment was laid down on the clinoforms of the proximal part of the Effingen Member. Meanwhile, new coral biostromes and another string of individual reef knolls above them began to grow further basinward on fresh mud in the uppermost part of a subsequent clinoform of the Effingen Member. The string of reef knolls thereby prograded perpendicularly to depositional strike over a subhorizontal, but *heterochronous* surface on top of the uppermost Effingen Member. Because of the juxtaposition of the prograding, temporally successive strings of reef knolls below the more or less horizontal base of the Günsberg Formation, the reefs look at first glance like being coeval. In fact, the reefs become progressively younger in the distal direction. This is shown in the schematic cross-section in Fig. 1.5.

The tidal delta of calcareous ooid sand near the Joulters Cays north of Andros Island in the Bahamas, which is shown in the aerial photograph of Fig. 4.8, progrades over a coral biostrome on level substrate in slightly deeper water. This situation in the Recent resembles the smothering of reefs in the boundary beds of the uppermost Effingen Member by sand of calcareous ooids of the lowermost Günsberg Formation. The internal geometry of individual sedimentary units within the thickest part of the Effingen Member in eastern Canton Solothurn and in the adjacent part of Canton Aargau which is shown in Fig. 1.5 is the result of a complicated interplay of progradation and aggradation. This could be visualized thanks to a combination of mineral stratigraphy with ammonite biostratigraphy by GYGI and PERSOZ (1986, Pl. 1A). The cause of the distinct variation between progradation and aggradation in the Effingen Member in Canton Solothurn and in Canton Aargau was variation in the rate of sediment supply and in the rate of eustatic sea level rise. Ample sediment supply in shallow water at a time of essentially stable sea level caused progradation.

Backstepping occurs when sea level eustatically rises and when sediment supply to a given locality is so slight that sedimentation cannot keep up with the growing accommodation space. The history of aggradation, progradation, and backstepping during the Oxfordian Age in the area investigated can be summarized as follows. The basinward rim of proximal, aggradational and shallowing-upward succession no. 1 stayed in the same position during the entire time of accumulation of the thick sediment stack. Rapid sea level rises no. 4 in the *Cordatum* Chron and no. 5 in *Antecedens* time caused no perceptible landward shift of the distal rim of the thick part of succession no. 1. Succession no. 2 was laid down in a time span when the total amount of eustatic sea level rise was slight (GYGI 1986, Fig. 4). The individual, small coral patch reefs in the boundary beds between the uppermost Effingen Member below and the lowermost Günsberg Formation above prograded rapidly over advancing sigmoid clinoforms of the Effingen Member. Progradation of the Günsberg Formation was halted when the early, gradual part of relative sea level rise no. 6 began. From that time on, the basinward margin of the carbonate platform of the Günsberg Formation remained in the same position for some time and then stepped back slightly in the *Hypselum* Chron because of rapid sea level rise no. 7. This is documented in section RG 14 above the landslide of Gschlief north of the village of Günsberg (GYGI 1969a, Pl. 18). Rapid progradation of shallow-water sedimentation resumed in the *Berrense* Chron with deposition of the uppermost part of the Effingen Member and mainly of the wide sand bank of calcareous ooids of the Steinibach Member above. Progradation rate slowed down when growth of the coral biostrome of the Olten Member began. Progradation then came to a stillstand which lasted during the *Bimammatum* and the *Gredingensis* Chron. The basinward boundary of the coral biostrome of the Olten Member became nearly vertical during these two chrons (Fig. 1.5).

Demise of the apparently slowly growing biostrome of the Olten Member was brought about when the rate of gradual sea level rise accelerated in the *Planula* Chron and became greater than the vertical growth rate of the biostrome. Water depth thereby increased above the biostrome to more than approximately 20 m. This is calculated in Chap. 6 to be the maximal depth where hermatypic corals could survive in the sediments investigated. Corals in the biostrome therefore died. The effect of the eustatic rise in the *Planula* Chron that is shown in Fig. 4 in GYGI (1986) was that shallow-water sedimentation stepped back landward from Olten. Concomitant deepening of the water near Olten is documented in section RG 21 in the quarry on Mt. Born by the superposition of ammonite facies in the Kimmeridgian over the Oxfordian coral biostrome of the Olten Member (GYGI 1969a, Pl. 18). Production of calcareous ooid sand and growth of coral biostromes and of some small coral bioherms within the sand bank resumed later in the Kimmeridgian Age near Balsthal (Fig. 1.5). The vertical and lateral shift of the basinward boundary of shallow water sedimentation is outlined in Fig. 1.5 by a gray band.

4.10   Depositional Slopes with Truncation Surfaces, Debris Flows, and Turbidites

Depositional slopes in the sediments investigated evolved in the following way. Sedimentation at the turn between the Middle and the Late Jurassic proceeded at a rate that was on average very low. There was even omission (non-deposition) at this time between Auenstein, Villigen, and Oberehrendigen in Canton Aargau. Evidence was presented above that the sea floor between northwestern Switzerland and northeastern Canton Aargau was level at the time. Water depth at the beginning of the Late Jurassic in the proximal realm was about 80 m and approximately 100 m in the distal area in Canton Schaffhausen. The earliest Oxfordian sediment between Canton Jura and Canton Aargau was a thin iron oolite with a matrix of marl and with a macrofossil assemblage of mainly ammonites. Predominance of ammonites in a macrofossil assemblage indicates that water depth was in the pertinent habitat at least 35 m (Fig. 4.9). The rate of sedimentation in the northwest then increased and became normal, too much for iron ooid accretion. A sediment stack with the primary thickness of approximately 240 m thereby accumulated by aggradation and filled the basin up to sea level. Isostatic basement subsidence under the sediment load created the corresponding accommodation space of three times the initial depth of the basin of around 80 m (Fig. 4.15B). Endogenic basement subsidence during deposition of succession no. 1 was probably insignificant in the entire region investigated. The distal rim of proximal, thick succession no. 1 as well as the toe of the wedge, in which the aggradational part of succession no. 1 thins towards the adjacent starved basin, remained stationary during the entire time span of sedimentation of succession no. 1 (Fig. 1.5). The probable cause of this was approximate equilibrium between sediment supply and a net, essentially eustatic sea level rise. The primary inclination of the upper surface of the wedge-like part of proximal succession no. 1, in which the succession thins in the distal direction, steadily increased during sedimentation. This is visualized in Fig. 1.5.

The average sedimentation rate in the proximal and thick, aggradational part of succession no. 1 was greater than the average rate of relative, probably mainly eustatic sea level rise. Sedimentation in that area proceeded at a rate which was much greater than the rate of endogenic subsidence. The proximal, thick part of succession no. 1 is, therefore, shallowing-upward until sea level. Coeval sedimentation of the entire succession no. 1 on the level floor of the adjacent starved basin amounted to a measured thickness of on average about 5 m. Sedimentation of succession no. 1 began in the epicontinental basin of eastern Canton Solothurn and in Canton Aargau at a depth of probably around 80 m and ended at the top of the Birmenstorf Member, when water depth had increased to approximately 105 m mainly because of a net relative sea level rise (Fig. 4.15B). It was stated above that the primary inclination of the sediment surface increased during sedimentation of the wedge between the aggradational, thick part of succession no. 1 and the basin floor. This was counteracted by isostatic subsidence of the basement under the load of the thick, aggradational part of succession no. 1 (Fig. 4.15B). When the distal rim of aggradational succession no. 1 rose to close below sea level at the end of sedimentation of the succession, the declivity of the top of the Pichoux Formation was at its maximum of between less than 0.5 and about 4°. The growing inclination of the sediment surface depended on the varying width of the belt of the Pichoux Formation between the basinward rim of the carbonate platform of the St-Ursanne Formation and the Birmenstorf Member on the basin floor (see below). The outer rim of the St-Ursanne Formation rose to at least 100 m above the top of the Birmenstorf Member, when sedimentation of the St-Ursanne Formation was completed at the end of the *Transversarium* Chron (Fig. 4.15B).

Gravitational sliding of unconsolidated sediment could be documented to have occurred for the first time at an early stage of sedimentation of succession no. 2 above the depositional slope of the Pichoux Formation. The lowermost, level part of aggradational succession no. 2 in the proximal realm of the Vorbourg Member was at this time in the peritidal zone. Evidence of this is a stromatolite with dewatering cracks and shrinkage pores or bird's-eyes in the Vorbourg Member at Liesberg. A photograph of a polished cross-section of the stromatolite is Fig. 4.6. About coeval with this stromatolite from the upper intertidal zone is a turbidite in the lowermost Effingen Member which crops out on the territory of the township of Rüttenen above the toe of the Pichoux Formation. The turbidite is at the head of the ravine called Schofgraben on the northern slope of Mt. Weissenstein, north of the city of Solothurn. The turbiditic bed with much siliciclastic silt is in Schofgraben below the upper rim of a landslide, which is about 3.5 km west of the village of Günsberg. This township is mentioned in Fig. 1.5. The author of this study found the first two starfish weathered out from this turbidite in the drift in 1963 and gave them to the echinoderm specialist H. Hess for identification. Hess then located the thin turbiditic bed in the outcrop, from which the starfish had weathered out. He excavated many starfish from *in situ*, which after preparation proved to be in an excellent state of preservation. The specimens were assigned by Hess (1968) to a new species and even to a new genus in Asteroidea: *Pentasteria longispina* Hess, n. g., n. sp. Hess (1968, p. 612) concluded

Fig. 4.13 Subaqueous debris flow in polished cross-section, directly above the subaqueous truncation surface b–b that is shown in Fig. 4.12. The debris flow is bed no. 102 of section RG 37 east of Auenstein in Canton Aargau (GYGI 1969a, Pl. 17). Some of the large, cross-laminated and semi-consolidated clasts of pelsparite were plastically deformed within the moving flow and were burrowed after the flow came to rest. The light laminae include much silt-grade detrital quartz. Scale bar is 2 cm. Refigured from GYGI and PERSOZ (1986, Fig. 3)

Fig. 4.12 Synsedimentary fault (a) in the lower right of the figure. Movement along the fault was gradual until it died down. The shear plane is clear-cut, although it evolved in unlithified mud. Subaqueous truncation surface (b–b) above the fault. Temporary outcrop of the upper Effingen Member in section RG 37 in the quarry of Jakobsberg east of the village of Auenstein in Canton Aargau. Scale: see text in Chap. 10. Refigured from GYGI and PERSOZ (1986, Fig. 2)

from the lamination of the turbidite that one of the starfish figured by him was rapidly smothered. These starfish lived in a densely populated community at the surface of mixed calcareous and argillaceous mud. The silty turbidite above the community with a unique starfish assemblage was excavated in 1979 on a surface of 12 m². According to MEYER (1984), the excavation yielded 190 starfish specimens. GYGI (2003, p. 185) concluded that the starfish community must have been instantly smothered by the lowermost, silty part of a thick turbidite deeply enough that the animals could not free themselves and died. This turbidite in the lowermost part of the Effingen Member in Schofgraben came to rest at a water depth of approximately 100 m, to judge from Fig. 4.15B in this study.

Another depositional slope existed at a late stage of sedimentation of the Effingen Member in the upper part of succession no. 2 near Auenstein in Canton Aargau (Fig. 1.5). The Effingen Member was primarily a mixture mainly of calcareous and of argillaceous mud, which included some siliciclastic silt. The member is thick in eastern Canton Solothurn and in the greater part of northern Canton Aargau. It began to thin in the distal direction near Auenstein in *Hypselum* time (Fig. 1.5). A depositional slope at this locality was the precondition of failure of semi-consolidated, mud-grade sediment which then slid downslope in a small

debris flow (Fig. 4.13). A conspicuous truncation surface evolved at the base of the flow (Fig. 4.12). A thin turbidite with graded bedding in a probably approximately time-equivalent position in the upper Effingen Member was found in a core of the exploration well drilled in 1962 for a planned quarry in the small valley called Musital southwest of the village of Rekingen in northeastern Canton Aargau. The well log is the unpublished section RG 68, which is schematically represented in the upper synthetic section in GYGI (1969a, Pl. 19). A photograph of the turbidite is the lower part of Fig. 12 in Pl. 4 in GYGI (1969a). A small debris flow was recorded further up in the upper Effingen Member in a core from the same borehole RG 68. This debris flow, which was erroneously attributed by the author to the Geissberg Member, is represented in the block diagram of Fig. 6 in GYGI and PERSOZ (1987). The truncation surface left by a large debris flow in the upper Effingen Member is well exposed in the now disused quarry in Musital near Rekingen. It is probable that most or all of the debris flows in the upper part of succession no. 2 in eastern Canton Aargau graded in the distal direction into a turbidite.

A Recent counterpart of the debris flows mentioned above is in Lake Zürich in Switzerland the failure of the lakeshore that occurred near Horgen in 1875. According to HEIM (1932, p. 42), the lakeshore first subsided vertically and then moved downslope in a subaqueous debris flow towards the bottom of the lake. KELTS and HSÜ (1980) concluded that the debris flow of 1875 graded laterally into the turbidite that they found in a core at the bottom of the lake at the water depth of 136 m. Turbidites are consequently evidence of sedimentation in relatively deep water. It is known from other events in the Recent that turbidites can travel a long

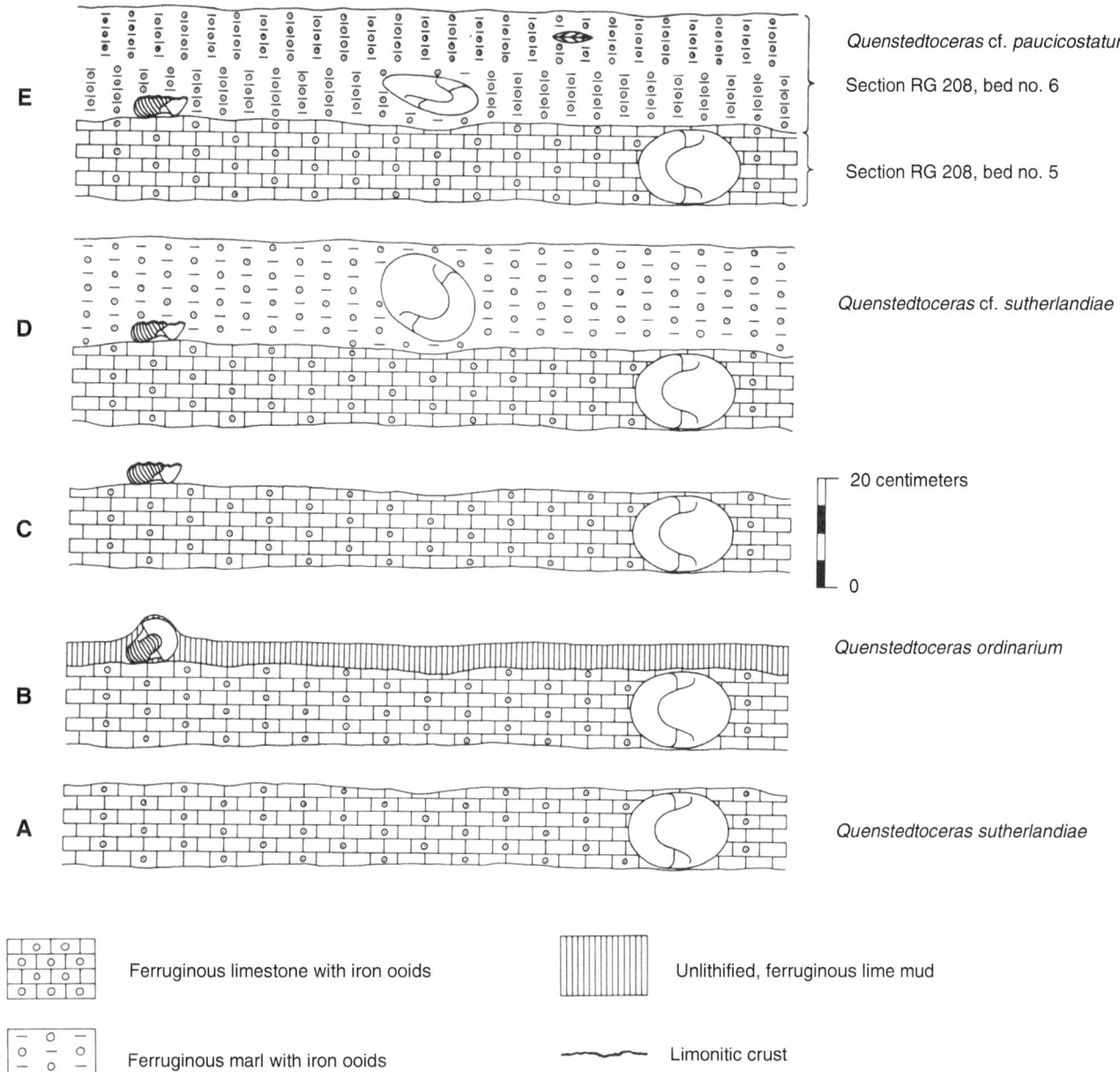

Fig. 4.14 Succession of depositional stages of ferriferous, iron oolitic limestone and marl in the upper Herznach Formation, directly below the boundary between the Middle and the Upper Jurassic Series in excavation RG 208 at locality Brunnrain southwest of the village of Ueken in Canton Aargau. (**A**) Rapid burial of the empty, near-globular ammonite shell of a *Quenstedtoceras sutherlandiae* (MURCHISON in J. de C. SOWERBY) in ferruginous lime mud with iron ooids during a storm, and prompt fossilization as a cast in the early-lithified bed. This part of the so-called *Lamberti* Bed must be a tempestite (storm layer) from a water depth of at least 35 m, according to the macrofossil assemblage with mostly ammonites. In fact, the depth of deposition of the *Lamberti* Bed was probably close to 80 m (see text). (**B**) A thin sheet of ferriferous lime mud smothered part of an empty ammonite shell. The part of this shell filled with sediment became early lithified as an incomplete cast. The mud around the cast remained to be unlithified. (**C**) Loose sediment which was laid down during stage B was churned up into suspension and was entirely removed by a unidirectional bottom current during another storm. (**D**) The lower part of marly bed no. 6 in excavation RG 208 is another tempestite which smothered an obliquely embedded ammonite shell of *Quenstedtoceras* cf. *sutherlandiae*. (**E**) The upper part of bed no. 6 with the ammonite cast of *Quenstedtoceras* cf. *paucicostatum* is the last sediment laid down in Middle Jurassic time. The entire bed no. 6 was subsequently thoroughly compacted under the weight of younger sediments with a total thickness of several hundred meters. Bed no. 6 is shown at the present, compacted state, as it was found in excavation RG 208. The successive, Upper Jurassic part of section RG 208 is represented in Fig. 8.2. Refigured from GYGI and MARCHAND (1982, Fig. 5) with permission of Geobios

way on very slight slopes or even over an almost level bottom. CREVELLO and SCHLAGER (1980, p. 1144) stated that a carbonate turbidity current traveled a distance of 100 km over the basin floor of Exuma Sound in the Bahamas, which has an average declivity of 0.5°. This is the order of inclination that probably existed at the toe of the advancing sigmoid clinoforms in the proximal Effingen Member, when the turbidite in Schofgraben mentioned above was sedimented at a water depth of about 100 m (Fig. 1.5).

Distinction of turbidites from tempestites (storm layers) can be difficult. Undubitable tempestites from *shallow* water are conspicuous and abundant in the uppermost Effingen Member in section RG 307 below the coral bioherms at the base of the Günsberg Formation in the quarry of La Charuque near Péry in Canton Bern (Gygi 2000a, Pl. 22, laminated beds with siliciclastic silt from no. 115 on upward to bed no. 157). Some tempestites issue laterally from the bioherms in this quarry, like the silty bed from which the rock sample of the polished slab photographed by Gygi (1986, Fig. 7) was taken from. There are several laminated tempestites with siliciclastic silt from shallow water in the uppermost, transgressive part of the Effingen Member above the Günsberg Formation in section RG 14 at the head of the landslide of Gschlief above the village of Günsberg (Fig. 1.5). These are bed nos. 171–183 of section RG 14 in Pl. 18 in Gygi (1969a).

The following two tempestites were formed in considerably deeper water, to judge from the macrofossil assemblages within them, in which ammonites prevail (Fig. 4.14). The earlier tempestite is bed no. 5 in Fig. 4.14A, the so-called *Lamberti* Bed in excavation RG 208 southwest of Ueken in Canton Aargau. This excavation was made a few hundred meters north of the Herznach iron mine. Bed no. 5 of excavation RG 208 includes some iron ooids. The cast of a well-preserved, undeformed cardioceratid ammonite of the taxon *Quenstedtoceras sutherlandiae* (Murchison in J. de C. Sowerby) was found in the excavation in this bed (Fig. 4.14A). Another well-preserved specimen of the taxon was found by the author in the *Lamberti* Bed in the Herznach iron mine. This ammonite was figured by Gygi and Marchand (1982, Pl. 1, Fig. 4). The thickness of the last whorl of the near-globular ammonite cast mentioned from bed no. 5 in excavation RG 208 is about equal to the entire thickness of the limestone bed in which the ammonite became fossilized. This is evidence that the *Lamberti* Bed at this locality was sedimented rapidly, most probably within hours during a storm. But it is important to note that the iron ooids within bed no. 5 in excavation RG 208 are not concentrated in laminae. They are evenly dispersed in the matrix of what was primarily calcareous mud. A depth of deposition of more than 35 m must be concluded from this bed, because ammonites prevail in the macrofossil assemblage (Fig. 4.9). Presence of iron ooids in the *Lamberti* Bed, not only in excavation RG 208 near Ueken, but for instance also in the *Lamberti* Bed no. 1 in section RG 32 near Küttigen (Gygi 1969a, Pl. 17), is evidence that the sedimentation rate was on average very low in the upper Herznach Formation in a vast area (see below). The calcareous *Lamberti* Bed in excavation RG 208 was not noticeably compacted because

of early lithification, which was the rule in calcareous beds or in ammonite casts in the ferriferous Herznach Formation.

The younger tempestite in excavation RG 208 is represented by the lower half of bed no. 6 (Fig. 4.14D). A large, near-globular ammonite cast of the taxon *Quenstedtoceras* cf. *sutherlandiae* was embedded obliquely into the iron-oolitic clayey marl of the lower half of bed no. 6 in the upper Herznach Formation, a short time before the close of the Middle Jurassic Epoch. Figure 4.14E shows the entire bed no. 6 after completion of sedimentation and at its present, compacted state, as it was found in excavation RG 208 with the deformed *Quenstedtoceras* cf. *sutherlandiae*. Compaction of bed no. 6 occurred under an overburden of younger sediments which once had a total thickness of probably several hundred meters. Sedimentation of bed no. 6 was completed at the time boundary between the Middle and the Late Jurassic Epoch. This is documented by the ammonite *Quenstedtoceras* cf. *paucicostatum* which was found in the upper half of bed no. 6 (Fig 4.14E). Very numerous ammonites were collected over the years in the neighboring iron mine west of the village of Herznach, when work in the mine was underway mainly during World War II. A choice of these ammonites was published by Jeannet (1951). The ammonites document in detail that the entire, iron-oolitic Herznach Formation was sedimented on average at a low rate in water which was more than 35 m deep (see above). A complete cast of an ammonite shell could only be formed, if the shell was filled with sediment soon after death of the animal within, and if sediment then smothered the filled shell rapidly and completely. The lower half of bed no. 6 in excavation RG 208 had a primary thickness of approximately 20 cm before compaction. This part of the bed was sedimented at a water depth of several tens of meters, probably within hours during a storm like the *Lamberti* Bed no. 5 below in the same excavation (Fig. 4.14). The iron ooids in the bed are evenly dispersed in the matrix. There is no lamination in the bed like in typical tempestites from shallow water. The contiguous, upper part of section RG 208 is shown in Fig. 8.2.

4.11 Lateral Shift of Belts with Maximal Sedimentation Rate

The Oxfordian part of the uppermost Herznach Formation in northwestern Switzerland is a thin iron oolite with a matrix which was primarily mostly mud of clayey marl. The bed includes a macrofossil assemblage with mainly ammonites. The lithology and the macrofossil assemblage in the bed are similar both below the thick clayey marl of the *Renggeri* Member in northwestern Switzerland (see, for instance, bed

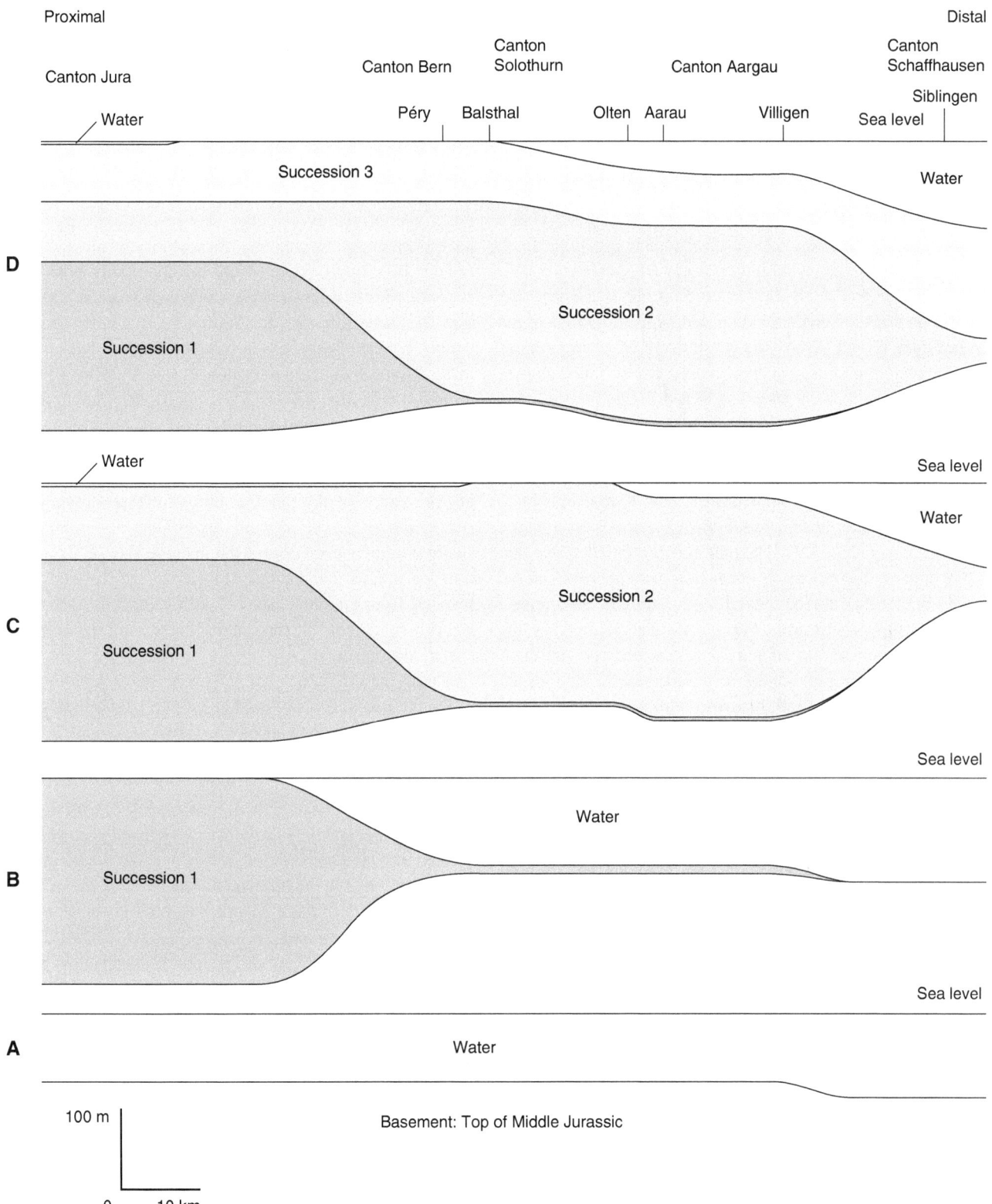

Fig. 4.15 Shifting of belts with maximal sedimentation during the Oxfordian Age. (**A**) Sea floor topography and water depth in the Rhodano-Swabian Basin of northern Switzerland at the beginning of the Late Jurassic. (**B–D**) Subsequent history of sedimentation of succession nos. 1–3 during a net relative sea level rise. Neighboring basement blocks subsided differentially, following lateral shift of belts with maximal sedimentation. Thicknesses are decompacted and partially recompacted using the nomogram by PERRIER and QUIBLIER (1974, Fig. 11) for argillaceous mud and according to figures in TERZAGHI (1940) for compaction of carbonate mud. Note the one hundredfold exaggeration of the vertical scale. Refigured from GYGI (2003, Fig. 187) with permission of Schweizerische Paläontologische Abhandlungen

no. 6 in section RG 280 near Liesberg in GYGI 1990a, Fig. 2) and beyond in the starved basin in eastern Canton Solothurn and northern Canton Aargau. The time equivalent of this bed in northwestern Switzerland is near Ueken in Canton Aargau, in excavation RG 208 directly north of the Herznach iron mine, the clayey marl with initially the double layer of calcareous concretions shown in Fig. 8.2A. The upper layer with iron oolitic nodules shown in the figure was subsequently incorporated in a multiphase process of stratigraphic condensation (see Sect. 8.2) into bed no. 8 in Fig. 8.2E. Each of the several stages in the process of condensation is visualized in Fig. 8.2B–E. Bed no. 8 in Fig. 8.2 is now the lower part of the Schellenbrücke Bed with ammonites of the *Cordatum* Chron which are positioned between nodules including ammonites of the much older *Scarburgense* Chron (Fig. 1.6). It is shown in Fig. 8.1 that stratigraphic condensation in the Schellenbrücke Bed could place ammonites with an age differing by more than one million years into direct contact.

A minimal depth of deposition of about 35 m is indicated, as soon as ammonites prevail in the macrofossil assemblage in a sediment (Fig. 4.9). Evidence was presented above that iron ooids in the Herznach Formation were *accreted* on the level floor of an epicontinental basin, and that the greatest water depth where iron ooids could be formed in the paleoenvironments investigated is approximately 100 m. A depth of sedimentation of 80 m was estimated above for the iron oolitic, earliest Oxfordian bed with mostly ammonites at the base of succession no. 1 in northwestern Switzerland (Figs. 1.5 and 4.15A). Uniform facies of both lithology and macrofossil assemblage in this bed in the upper Herznach Formation between northwestern Switzerland and northern Canton Aargau is evidence that water depth did not vary significantly during low-rate deposition of the thin unit in the entire area where it occurs (Fig. 1.5). The thick, marly Bärschwil Formation in the lower part of proximal succession no. 1 directly above the iron-oolitic bed at the base was, therefore, certainly not filled into a preexisting trough (Fig. 4.15A, B).

It is evident from Fig. 1.5 that the proximal, thick part of succession no. 1 was sedimented by aggradation. The Bärschwil Formation of mixed argillaceous and calcareous mud in the lower part of succession no. 1 was accumulated as a submarine bank of mud with a flat top. The formation is shallowing upward. Sedimentation began with a thin iron oolite with a macrofossil assemblage of mainly ammonites and ended with growth of biostromes of hermatypic corals in the Liesberg Member. It is calculated in Chap. 6, that probably all of the hermatypic corals in the area investigated are diagnostic of a maximal water depth of 20 m. The mud bank of the Bärschwil Formation was consequently a positive physiographic structure in a shallow epicontinental basin. The basinward rim of both this mud bank and the

carbonate platform of the St-Ursanne Formation directly above did not prograde. The basinward rims of the marly mud bank and the carbonate platform above are congruent (Fig. 1.5). It was stated above that this was probably the effect of equilibrium between sediment supply and predominantly gradual, net relative sea level rise which proceeded during sedimentation of the entire succession no. 1. The decompacted thickness of the proximal part of succession no. 1 was originally calculated by GYGI (1986, p. 470) to have been about 250 m. This thickness is recalculated here to have been approximately 240 m. The pertinent sediment stack filled the basin to close below sea level. A combination of massive exogenic, isostatic basement subsidence under the load of proximal succession no. 1 and of probably insignificant endogenic subsidence amounted to a total of around 140 m (Fig. 4.15B). The trough-like configuration of the base of the proximal, thick part of succession no. 1 evolved *after* sedimentation of the succession began. This is visualized in Fig. 4.15A, B. The "trough" was the *effect*, not the cause of the great primary thickness of proximal succession no. 1. The proximal part of succession no. 1 was shallowing upward, because the average rate of sedimentation was greater than the rate of relative sea level rise, which resulted from a combination of endogenic subsidence, isostatic basement subsidence under load, and a probably net eustatic sea level rise.

Hardly any marine accommodation space was left when sedimentation of the proximal, thick part of succession no. 1 was completed at the end of the *Transversarium* Chron. A massive supply of mixed calcareous and argillaceous, mainly mud-grade sediment with some siliciclastic silt of succession no. 2 began in the following *Schilli* Chron. It was essentially endogenic subsidence, which initiated and augmented the accommodation space for the lower part of succession no. 2 in the northwest. The peritidal environment into which the Vorbourg Member and the lower part of the Röschenz Member were sedimented persisted until the end of the *Stenocycloides* Chron. This condition changed when a small, gradual relative sea level rise began during the subsequent *Bifurcatus* Chron. This rise halted progradation of the carbonate platform of the Günsberg Formation. The timing of the change is documented by the *Perisphinctes (Dichotomoceras) bifurcatus* (QUENSTEDT) which R. ÉNAY found in 1964 in the type section RG 14 of the Günsberg Formation at the head of the landslide called Gschlief above the village of Günsberg in Canton Solothurn (GYGI 1969a, Pl. 18). This specimen was identified by R. ÉNAY, and it was figured by GYGI (1995, Fig. 17/2). The pertinent ammonite taxon is no. 19 in Figs. 1.5 and 1.6 in this study. A thin, but widespread deposit of calcareous oolite was sedimented until far into the interior of the shallow-water realm following the minor, but rapid relative sea level rise no. 6 in Fig. 30. This thin oolite separates the lower part of the Röschenz Member from the upper part. The oolitic intercalation is, for instance, bed no. 73 in section RG 359 near Bressaucourt (GYGI 2000a, Pl. 18), or bed no. 45 in section RG 402 near Röschenz (GYGI 2000a, Pl. 33). Much calcareous mud was precipitated probably in whitings (see below) in the same environment where calcareous ooids were accreted in the pertinent intercalation in the Röschenz Member. Currents in shallow water transported this mud across the sand bank of calcareous ooids which is now the Günsberg Formation. The mud was shed from there along with some calcareous mud which was produced on the Günsberg carbonate platform into the basin. Sedimentation of the Gerstenhübel Beds, a succession of micritic limestone from deeper water in Canton Aargau, was the result. The Gerstenhübel Beds with a thickness of more than 10 m are intercalated between marl of the Effingen Member. Therefore, the unit weathers as a prominent ridge out of marl where the pertinent strata are tectonically inclined. The Gerstenhübel Beds can be mapped in Canton Aargau at the scale of 1:25,000.

Running water transported from the early *Schilli* Chron on a great quantity of argillaceous mud and some siliciclastic silt from land into the vast peritidal realm of the lower Vellerat Formation. The terrigenous material was mixed in that environment with very much calcareous mud which was produced at the same time probably by whitings in the area of the Vellerat Formation (see below). Water currents driven by tides and by occasional storms carried most of this fine-grained sediment out of the area of the lower Vellerat Formation and across the carbonate platform of the Günsberg Formation. A great amount of mixed calcareous and argillaceous mud including some siliciclastic silt was thereby shed from the outer rim of the carbonate platform into deeper water of the adjacent Rhodano-Swabian Basin. This was because net sea level rise, endogenic subsidence, and resulting marine accommodation space were insignificant during sedimentation of the Vorbourg Member, the lower Röschenz Member, and the lower Günsberg Formation. The tidal delta in the uppermost Günsberg Formation at Péry shown in Fig. 5.2 is proof that water passing over the sand bank of calcareous ooids on the Günsberg carbonate platform sometimes transported a lot of fine-grained sediment. Water of the strong current which shed calcareous ooid sand into the lower part of this tidal delta was turbid with very much fines in suspension. This is documented by the thin section photograph shown in Fig. 5.3. Much silty mud settled as geopetal sediment in the interstices between ooids in the delta. The upper part of the interstices is filled with secondary cement of low-magnesian calcite spar. Time-equivalent with the tidal delta at Péry in Canton Bern is the sand wave near Liesberg in Canton Basel-Landschaft, which is represented in Fig. 5.5A. Neither silt nor mud was found in the interstices between pelletoids in the sand wave near

Liesberg. The thin section out of the sand wave which is shown in Fig. 5.5B is pure pelsparite.

Ample supply of mostly mud-grade sediment, insignificant endogenic subsidence, and a small net amount of eustatic sea level rise combined were the cause, that the belt with maximal sedimentation of succession no. 2 shifted by several tens of kilometers basinward in relation to the thick part of succession no. 1. This is evident from Fig. 1.5, and it is visualized in Fig. 4.15B, C. The initial accommodation space for proximal succession no. 1 as represented in Fig. 4.15A was less than the initial accommodation space for succession no. 2, for instance at Péry or near Balsthal (Fig. 4.15B). Proximal succession no. 2 near Balsthal in Canton Solothurn is, therefore, thicker than the proximal part of succession no. 1 in Canton Jura (Fig. 4.15C). Succession no. 2 is near Balsthal shallowing-upward like proximal succession no. 1, and it filled the initially starved basin entirely up to sea level (Fig. 4.15C).

4.12 Carbonate Platforms Versus Ramps

The top of carbonate *platforms* had some relief, which was normally of the order of a few meters. The situation could change after a *rapid* sea level rise. Rise no. 5 in Fig. 5.1 was rapid, and it created a lagoon above a sand bank of calcareous ooids in the internal part of the upper St-Ursanne Formation late in the *Antecedens* Chron. The greatest depth of the lagoon at an early stage of its evolution did not exceed 10 m. The Buix Member with an average thickness of approximately 30 m and with coral patch reefs was sedimented into the lagoon. The greatest water depth which could be documented to have evolved above any of the four carbonate platforms investigated of Late Jurassic age is 10 m. Relief of the sea floor in the lowermost part of the carbonate platform of the Günsberg Formation is imaged in the photograph of a coral bioherm in Fig. 19 in GYGI (1992). The domal top of this bioherm in the quarry of La Charuque at Péry shown in the photograph is 3 m above the surface of coeval sediment around the bioherm. The bioherm figured is one out of unit no. 160 of section RG 307 in Pl. 22 in GYGI (2000a).

Rapid sea level rise no. 6 during the *Bifurcatus* Chron (Fig. 5.1) created at Péry the shallow lagoon, into which the oncolite of the Grüne Mumienbank Bed with some hermatypic corals was sedimented. The oncolite is bed no. 163 with a thickness of 4.1 m in section RG 307 (GYGI 2000a, Pl. 22). The thickness of the bed is probably close to the initial depth of the lagoon, which evolved above a thin, supratidal bed with lenses of lignite coal and with gyrogonites of characean freshwater algae. A photograph of a polished slab of oncolite from bed no. 163 is Fig. 14 in GYGI (1992).

The time equivalent of this oncolite is in the interior of the shallow-water realm the intercalation of calcareous oolite mentioned above, which separates the lower half of the Röschenz Member from the upper half. The time equivalent of the oncolitic Grüne Mumienbank Bed is in the epicontinental basin the succession of micritic limestone of the Gerstenhübel Beds (Fig. 1.5). Time correlation of the pertinent units could be made with the mineral-stratigraphic correlations E and F by PERSOZ in GYGI and PERSOZ (1986, Pl. 1A).

The laterally shifting tidal delta of calcareous ooid sand with bed nos. 193–195 of section RG 307 in the uppermost Günsberg Formation at Péry, which is shown in Fig. 5.2, has a total thickness of 5.7 m. The delta was shed into water with a depth of about 6 m. This is the accommodation space that rapid sea level rise no. 7 in Fig. 5.1 provided for in the uppermost part of the carbonate platform of the Günsberg Formation in the *Hypselum* Chron. The transgressive surface at the base of bed no. 193 of section RG 307 is not represented in GYGI et al. (1998, Fig. 2). The environment of the tidal delta at Péry was probably similar to that of the Recent delta of calcareous ooid sand between two of the Joulters Cays north of Andros Island in the Bahamas, which is shown in Fig. 4.8.

The effect of rapid relative sea level rise no. 8 in Fig. 5.1, which occurred early in the *Berrense* Chron, is evidence of an extremely flat topography in the uppermost part of the carbonate platform of the Günsberg Formation and in the adjacent interior of the shallow-water realm, in the coeval part of the uppermost Röschenz Member. Rise no. 8 was rapid, but the amount of the rise was modest. This sea level rise caused a transgression over peritidal and even supratidal facies in proximal succession no. 2, which can be followed from Péry in Canton Bern over a palinspastic distance of more than 30 km perpendicularly to depositional strike to Bressaucourt near Porrentruy in Canton Jura. The sea level rise was rapid enough to augment water depth substantially even far in the interior of the shallow-water realm, where the mainly marly upper part of the Röschenz Member was sedimented. Enough fully marine accommodation space was created by the rise that water circulation became sufficient for hermatypic corals to advance or to retrograde, respectively, to the inner part of the carbonate platform of the upper Günsberg Formation. Growth of the coral biostrome in the upper Günsberg Formation in Court Gorge near Moutier was made possible by rapid sea level rise no. 8 early in *Berrense* time (GYGI 2000a, Pl. 28, bed nos. 39 and 41). Coral biostromes became established at this time even far beyond in the interior of the shallow-water realm of the uppermost Röschenz Member, like for instance north of Bressaucourt (3.5 km southwest of Porrentruy, as mentioned by SCHNEIDER 1960, p. 7), and elsewhere thanks to the rapid sea level rise. Hermatypic corals in the shallow-

water realm of the upper Röschenz Member did not survive for long. Sediment filled in the accommodation space soon after the rapid, but small-scale sea level rise no. 8. Some of the corals in the upper Röschenz Member are indicated in Fig. 1.5 by the single cross symbol for hermatypic corals under the word "Porrentruy".

A horizontal plane results when the relief at the top of a carbonate platform is averaged. The mineral-stratigraphic correlation C in Pl. 1A by PERSOZ in GYGI and PERSOZ (1986) is an indication that the top of the carbonate platform of the St-Ursanne Formation was indeed flat, when sedimentation of the formation was completed. Evidence of this is indirect. It is provided by stromatolites of the upper intertidal zone in the peritidal Vorbourg Member directly above the St-Ursanne Formation. A small bedding plane in the Vorbourg Member with mud-cracked prisms with a diameter of about 10 cm was visible for some time on the territory of Asuel in Canton Jura beside the road leading from St-Ursanne upward to Les Malettes. The outcrop is where the road intersects the Vorbourg Member at the foot of the slope called Côte du Frêne. The photograph of a polished cross-section through another such mud-cracked stromatolite in the middle Vorbourg Member on the crest of Chestel ridge south of the village of Liesberg in Canton Basel-Landschaft is represented in Fig. 4.6.

The upper part of the wide, middle Oxfordian carbonate platform of the St-Ursanne Formation was rimmed against the adjacent epicontinental basin by what was originally the sand bank of calcareous ooids of the Tiergarten Member. The bulk of the subsequent, narrow carbonate platform of the Günsberg Formation is calcareous oolite which delimits the mostly peritidal realm of the Vorbourg and the Röschenz Member from deeper water in the Rhodano-Swabian Basin. The Steinibach Member of the lowermost Balsthal Formation was initially a sand bank of calcareous ooids which rimmed the shallow lagoon with oncoids of the Hauptmumienbank Member against deeper water in the distal basin. The oolite of the Steinibach Member grades toward deeper water laterally into the coral biostrome of the Olten Member with a matrix of what was primarily calcareous mud. This is evidence that the Olten Member with mainly plate-like coral colonies was sedimented below the paleobathymetric range of between 0 and 6 m where calcareous ooids could be accreted. Water depth at the base of the biostrome of the Olten Member was east of Olten probably about 20 m like at the base of the boundary beds with corals between the uppermost Effingen Member and the lowermost Günsberg Formation near Péry. Coral bioherms with an elevation from base to top of as much as 20 m project out of the biostrome of the Olten Member, where this member crops out on the slope above the eastern end of the bridge connecting the city of Olten with Winznau in Canton Solothurn. The mainly micritic Courgenay Formation was sedimented into a very shallow lagoon, which was rimmed by the wide shoal of calcareous ooid sand of the coeval Balsthal Formation against deeper water in the adjacent Rhodano-Swabian Basin. The distal part of the Reuchenette Formation between Moutier and Egerkingen is a carbonate platform. There are coral biostromes and small coral bioherms in calcareous oolite in the Reuchenette Formation at Balsthal (Fig. 1.5). All of the carbonate platforms of Jurassic age investigated in this study were at least temporarily rimmed by banks of calcareous sand, either bioclastic or ooidal, not by coral reefs like in the Recent.

The *ramp* of the Pichoux Formation laterally replaced the sand bank of calcareous ooids with minor coral reefs, which is now the Tiergarten Member in the marginal, basinward part of the upper St-Ursanne Formation. The bedded, micritic limestone of the Pichoux Formation was sedimented upon the gently sloping upper surface of marly mud of the distal part of the Bärschwil Formation, which thins basinward in the wedge-like lower part of succession no. 1. The inclination of the depositional slope steadily increased in the course of sedimentation of the wedge-like Pichoux Formation (Fig. 1.5). The declivity of the upper surface of the Pichoux Formation reached a maximum at the end of sedimentation of the unit and amounted to approximately 0.5° in the transect between Mervelier and Günsberg. The palinspastic width of the ramp of the Pichoux Formation perpendicularly to depositional strike was about 20 km in that transect. The primary width of the facies belt of the Pichoux Formation diminishes very much to the north toward Basel.

The ramp of the Pichoux Formation is approximately 2 km wide between locality Titterten east of the village of Erschwil and locality Hohrüti south of the village of Meltingen. Both villages are townships in Canton Basel-Landschaft. The width of the ramp of the Pichoux Formation can be approximately quantified between the villages of Erschwil and Meltingen, because the depositional strike of the ramp southwest of Meltingen crosses the axis of a Jura fold at the angle of at least 40°. The primary inclination of the sediment surface at the top of the ramp could be calculated to have been about 4° in this case, when exogenic, load-induced subsidence of the basement and compaction were taken into account according to Fig. 4.15B. The difference between the horizontal top of the carbonate platform of the St-Ursanne Formation and the average declivity of the top of the adjacent ramp of the Pichoux Formation was therefore everywhere very slight. Nevertheless, the *platform* can be distinguished from the distally adjacent *ramp*. This subdued sea floor topography is the consequence that global sea level variations in the Late Jurassic were modest. It is shown in Fig. 5.1 that rapid relative sea level rises did not exceed 10 m in this time span. The seaward margin of Recent carbonate platforms with barrier reefs is fundamentally different. Sea level fluctuated during the later part of the Pleistocene with a

period of approximately 100,000 years by up to between 120 and 130 m. The probable consequence of this is that for instance the seaward margin of the Belize barrier reef drops off steeply and partially vertically to a depth of more than 100 m (James and Ginsburg 1979, Figs. 3–4 on p. 30).

The ramp of the Pichoux Formation was at the time of its sedimentation primarily inclined compared with the initially flat-lying, thin iron oolite of the earliest Oxfordian underneath (Fig. 1.5). This bed of iron oolite in the uppermost Herznach Formation is at the base of both the Bärschwil Formation and the Wildegg Formation. The Pichoux Formation was as well in a primarily inclined position in relation to the lagoonal Hauptmumienbank Member above. This mainly oncolitic unit was sedimented in a perfectly level position like the adjacent and coeval, equally flat-lying calcareous oolite of the Steinibach Member. The sand bank of calcareous ooids of the Steinibach Member rimmed the shallow lagoon of the Hauptmumienbank Member with its abundant and spherical, centimeter-size calcareous oncoids which originally floated in calcareous mud. Both the Hauptmumienbank Member and the Steinibach Member are sediments from very shallow water. The Pichoux Formation is in a diagonal position between the primarily horizontal Herznach Formation from deeper water below and the perfectly flat-lying Hauptmumienbank Member from shallowest water above. Because of this position, the Pichoux Formation separates the older Bärschwil Formation *laterally* from the younger Effingen Member (Fig. 1.5). This is why the argillaceous Bärschwil Formation and the marl of the Effingen Member were thought in the past to be coeval.

Figure 1.5 is the outcome of a detailed investigation of the sedimentology and the paleobathymetry as well as of the flat sea floor topography upon which the thin and widespread, earliest Oxfordian iron oolite of the uppermost Herznach Formation with a macrofossil assemblage of mostly ammonites was sedimented in comparatively deep water between northwestern Switzerland and Canton Aargau. This was supplemented by measuring and averaging the thicknesses of mainly the marly Bärschwil Formation and the Effingen Member in the Wildegg Formation. The resulting picture was checked by establishing, principally on the strength of ammonites, the time equivalence of the lagoonal Buix Member in the upper St-Ursanne Formation with the upper Pichoux Formation on the slope, and of the lower, distal Pichoux Formation with the Birmenstorf Member, which was sedimented upon the perfectly level floor of the adjacent epicontinental basin (see Sect. 7.4 on time correlation). It was stated above that the *primary* declivity of the Pichoux Formation was *secondarily* enhanced by isostatic basement subsidence under the load of the subsequent, thick Effingen Member in the proximal part of the Wildegg Formation (compare Fig. 4.15B with Fig. 4.15C). The apparent, strong declivity of the top of the Pichoux Formation in Fig. 1.5 is the consequence of the hundredfold exaggeration of the vertical scale as compared with the horizontal scale of the figure. Both scales are represented in the lower left part of Figs. 1.5 and 4.15.

LATHUILIÈRE et al. (2005) claimed, according to their Fig. 7, to have discovered in an outcrop of the time equivalent of the Swiss St-Ursanne Formation below the village of Bonnevaux-le-Prieuré in the French Jura an "Oxfordian reef tract" with a steep fore-reef slope extending from the intertidal or shallow subtidal zone down to a water paleodepth of around 40 m. The village is 16 km southeast of the city of Besançon. The water depth of 40 m at the base of the supposed "reef" was alleged on p. 553 of the paper by comparison with Recent coral communities with mainly plate-like colonies. This is to be compared with the result of the semiquantitative calculation in Chap. 6 of this study, that hermatypic corals lived in the Oxfordian Age in the environments investigated by the author in water probably nowhere deeper than approximately 20 m. This is repeated in main conclusion no. 10 of this study, and it was previously stated on p. 481 in GYGI (1986).

The outcrop studied by LATHUILIÈRE et al. (2005) is along the road leading from the small city of Ornans upward through a small valley to the village of Bonnevaux-le-Prieuré. The strata exposed are subhorizontal and begin with the unit called Calcaires argileux de Bonnevaux including some ammonites. Hermatypic, mainly plate-like corals first appear at the base of the Calcaires siliceux de Dole directly above. A *Perisphinctes (Perisphinctes) parandieri* DE LORIOL was found at the base of this unit. Corals occur to the top of the Oolithe coralliene de Pagnoz. According to Fig. 2 in the paper, the thickness from the base of the Calcaires siliceux de Dole to the top of the oolitic Pagnoz unit is 60 m. Above is the Calcaire de Clerval that the authors assigned to a peritidal environment. The level strata investigated are intersected by the steep flank of a small valley. The more or less horizontal stratification of Oxfordian units in the section below Bonnevaux-le-Prieuré is corroborated by level, coeval limestone units in good outcrops in the close neighborhood of the village. The shallowing upward trend of the succession was recognized by LATHUILIÈRE et al. (2005, p. 557), and they presumed the top of their bed no. 15 in Fig. 2 to be a sequence boundary. They related this bedding plane to sequence boundary O4 in GYGI et al. (1998, Fig. 2) at the base of the St-Ursanne Formation. This distinctive bedding plane in France is much more probably time equivalent with the transgressive surface within the Swiss St-Ursanne Formation which is indicated in the assembled section in the left part of Fig. 5 in GYGI et al. (1998). It corresponds to the "transgression argovienne" of ÉNAY (1966, p, 284) and is the effect of the rapid sea level rise late in the *Antecedens* Chron which is shown in Fig. 4 in GYGI (1986). The configuration of the supposed "reef tract" which was imagined by LATHUILIÈRE et al. (2005) is shown in their Fig. 7.

The succession below Bonnevaux-le-Prieuré is in the marginal part of what is in adjacent Switzerland the carbonate platform of the St-Ursanne Formation. The succession in France is, except for thicknesses, almost identical to the assembled section 1 km southwest of the village of Vellerat in Swiss Canton Jura. This detailed section is shown in Pl. 24 by GYGI (2000a). Clean outcrops in the Sornetan Member (previously Terrain à chailles) near Vellerat were excavated by the author in trenches. The pertinent sections RG 373 and 388 were shown by GYGI and MARCHAND (1993) in Fig. 2, and many cardioceratid ammonites as well as a *Perisphinctes (Otosphinctes) paturattensis* DE LORIOL complete with lappet were shown in Pls. 1–3 of the paper. The time equivalent of the upper Sornetan Member at Vellerat is the Calcaires argileux de Bonnevaux near Bonnevaux-le-Prieuré. The Liesberg Member with its partially chertified macrofossils at Vellerat is the time equivalent of the Calcaires siliceux de Dole in France. The entire St-Ursanne Formation and the peritidal Vorbourg Member directly above were measured at the cliff of Peute roche southwest of Vellerat on the rope in section RG 451. Large parts of the St-Ursanne Formation are easily accessible in two unpublished sections north of Vellerat, RG 296 along the road to Forêt de la Cendre and RG 367 along the road to the village of Courrendlin. Vellerat is indicated on the map of Fig. 1.4 and in the assembled section of Fig. 1.5 in this study. The basinward boundary of the carbonate platform of the St-Ursanne Formation against the Pichoux Formation south of Vellerat is shown in Fig. 1.5 with a bold black line and in the map of Fig. 1.4 by the dashed boundary line no. 1. This platform boundary continues from northwestern Switzerland almost due west to the region southeast of Besançon. When attention is paid to the vertical and lateral scales in the lower left of Pl. 1A in GYGI and PERSOZ (1986) and in Fig. 1.5 of this study, it becomes apparent that the declivity of the depositional slope at the top of the Pichoux Formation (abbreviated PIC in Fig. 1.5) was very slight.

It is important to note in this context that the paleobathymetric range of Late Jurassic hermatypic corals in the region of what is now the Jura Mts. was fundamentally different from that of Recent reef corals. The colony shape of Recent coral taxa was used by LATHUILIÈRE et al. (2005) without further consideration in order to interpret the environment of middle Oxfordian coral colonies with a similar shape. The water depth of 40 m in Fig. 7 of their paper was alleged on p. 553 "by analogy with Recent coral communities with nearly exclusive platelike paucispecific non-incrusting assemblages." The regional facies relationships of the outcrop below Bonnevaux-le-Prieuré with coeval subhorizontal strata in the neighbourhood were ignored by LATHUILIÈRE et al. (2005). The authors wrote on p. 546 that "The general organisation of sedimentary units is very close to the well-known frame proposed by GYGI and PERSOZ (1986) for Switzerland." The basinward margin of the carbonate platform equivalent to the St-Ursanne Formation is not at

Fig. 4.16 Tremalith of primary, low-magnesian calcite of a disintegrated coccosphere which was embedded into what was initially mainly aragonitic ooze. The calcareous mud recrystallized into micritic limestone of the Wangental Member. Upper Villigen Formation near Immendingen on the Danube River in southern Germany. Scale bar is 1 μm. Refigured from GYGI (1969a, Pl. 11, Fig. 41) with permission of Beiträge zur Geologischen Karte der Schweiz, Neue Folge

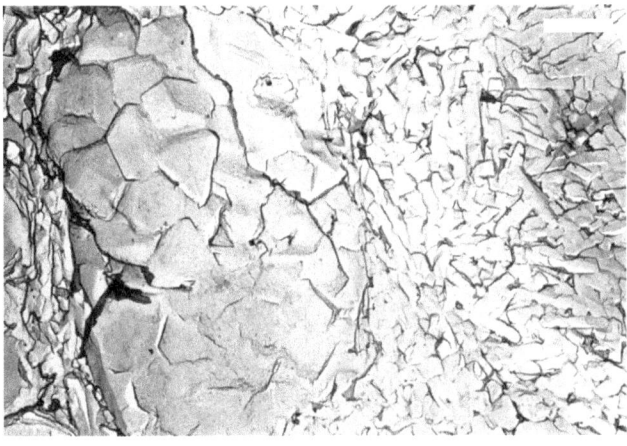

Fig. 4.17 Single-cell organism, unidentified. The cell began to divide and died in the process of mitosis. A framboidal aggregate of rhombic dodecahedral pyrite crystals replaced the decayed organic matter of the dividing cell. The aggregate of pyrite crystals is surrounded by probably disintegrated rods of primary, low-magnesian calcite. From sample of bed no. 102 in an intercalation of micritic limestone within the Effingen Member in section RG 14. The section is in the landslide of Gschlief above the village of Günsberg in Canton Solothurn (GYGI 1969a, Pl. 18). Scale bar is 1 μm. Refigured from GYGI (1969a, Pl. 5, Fig. 17) with permission of Beiträge zur Geologischen Karte der Schweiz, Neue Folge

Bonnevaux-le Prieuré. It is a short distance to the south where the depositional slope with a slight declivity toward the basin begins. The thickness of strata with corals below the peritidal Calcaire de Clerval is 60 m. Following the calculation by GYGI (1986), step no. 15 on p. 472, the basement subsides under the load of sediments isostatically by two thirds of the compacted sediment thickness. Sedimentation of the Calcaires siliceux de Dole consequently began at the water

depth of approximately 20 m, and this is the maximal depth at which corals could survive in France like in adjacent Switzerland at the base of the coeval Liesberg Member. LATHUILIÈRE et al. (2005) did not take into account, or more probably were unaware of isostatic basement subsidence under load which is concluded to be prompt even under subtle loads in Sect. 11.3 in this study. The authors obviously misinterpreted the steep slope incised by Quaternary erosion into a subhorizontal, shallowing upward succession of Jurassic sediments to be an exhumed fore-reef slope.

4.13 Lateral Shift of Carbonate Platform Margins

The paleogeographic position of the basinward margins of the carbonate platforms of the early Bathonian Spatkalk Member, of the early Callovian Dalle nacrée Member, of the middle Oxfordian St-Ursanne Formation, and of the late Oxfordian Balsthal Formation was represented in a map of northern Switzerland in GYGI (1986, Fig. 1). It is evident from the assembled cross-section of Fig. 1.5 in this study that the basinward shift of the rims of the carbonate platforms of the Günsberg Formation and of the subsequent Balsthal Formation proceeded over basinward advancing, sigmoid clinoforms of mixed calcareous and argillaceous mud of the Effingen Member below. The direction of progradation of these platform rims can be read from the map in Fig. 1.4. Progradation was controlled by the sea floor topography which is shown in cross-section in Fig. 4.15B in this study, and by the rate of supply of mud (either mainly argillaceous or predominantly calcareous, according to climate) in relation to sea level. Figure 4.15 is a slightly revised version of Fig. 3 in GYGI (1986).

The comparatively great thickness and the high rate of sedimentation of the partially argillaceous Effingen Member in eastern Canton Solothurn and in northern Canton Aargau are evidence that the coeval carbonate platform of the Günsberg Formation, and mainly the uppermost part it, evolved when a great quantity of terrigenous mud along with some siliciclastic silt was supplied from land in the northwest. The terrigenous weathering products bypassed the platform and its coral bioherms in suspension in water currents with a net basinward component, and which were strong enough to transport a mixture of calcareous and argillaceous mud as well as some siliciclastic silt. Currents were such that no significant tidal delta of calcareous ooids was shed beyond the outer rim of the Günsberg carbonate platform into deeper water where the Effingen Member was sedimented. It is probably the high concentration of clay minerals in suspension in the water passing over the platform, which reduced the rate of carbonate production on the carbonate platform of the Günsberg Formation, and that thereby limited the width of the platform. Nevertheless, the great amount of clay minerals that currents in very shallow water transported across the Günsberg carbonate platform could not inhibit carbonate production on the platform altogether. This platform is very narrow at its base. Then it grew wider, but it shrank again and became narrow in its uppermost part (Fig. 1.5), when the rate of supply of siliciclastic minerals bypassing the platform was at its acme. The carbonate platform of the Günsberg Formation is much narrower than the earlier carbonate platform of the St-Ursanne Formation and than the subsequent carbonate platform of the Balsthal Formation. The wide carbonate platforms of the St-Ursanne Formation and the Balsthal Formation evolved at times, when supply of terrigenous mud from land was minimal because of relatively dry climate.

4.14   Origin of Calcareous Mud

Calcareous mud was quantitatively the principal initial component of most of the rocks investigated. A great amount of calcareous mud was incorporated into the Bärschwil Formation, the St-Ursanne Formation, the Pichoux Formation, the Effingen Member, and into the Courgenay and the coeval Villigen Formation.

4.14.1   Different Sources of Calcareous Mud

One source of calcareous mud which was sedimented in deeper water of the epicontinental basin is nannoplankton like coccolithophorids (Fig. 4.16), or remains of unidentified organisms like the one which is represented in Fig. 4.17. Coccolithophorids contributed submicroscopic particles of stable, low-magnesian calcite to sediments in the Rhodano-Swabian Basin. GYGI (1969a) scanned many micritic limestone samples of the Upper Jurassic Series for nannoplankton and showed in his Fig. 1 which of the samples investigated by him included coccolithophorids and which did not. Only limestones were then scanned for nannoplankton. No entire coccospheres were found. Only isolated tremaliths, which are the individual elliptic shields protecting the algal cell within a coccosphere, occur in the micritic limestones investigated. One such tremalith was figured by GYGI (1969a, Pl. 11, Fig. 41), and it is reproduced here as Fig. 4.16. The elements of a tremalith are micellae. These are stubby, slightly conical rods which were probably primarily a single crystal of calcite. The micellae radiate obliquely from the center of the tremalith. The approximate length of a micella is one micron.

Spherical, individual cells of an unidentified organism in the limestones which were scanned by the author for

nanoplankton in 1963 had a diameter of 5 μm. When organic matter decayed after death of the cell, it was replaced by a spherical, framboidal aggregate of euhedral pyrite crystals. The pyrite could be identified in photographs made with a transmission electron microscope, because it crystallized in the form of rhombic dodecahedra (Fig. 4.17). The uniform diameter of the individual pyrite crystals is about 1 μm. The two spherical, framboidal aggregates of pyrite crystals which are shown in Fig. 4.17 are partially coherent. They are probably the replica of a cell that began to divide and died in the process of mitosis. This photograph was published for the first time by GYGI (1969a, Pl. 5, Fig. 17). It is reproduced here, because it apparently passed unnoticed by subsequent authors. A pure accumulation of randomly oriented rods of calcite surrounds the framboidal spheres. GYGI (1969a) surmised that the framboidal, spherical aggregates of pyrite crystals might be the replica of the organic matter of the algal cell in coccospheres, and that the rods around the aggregate of pyrite crystals were micellae of tremaliths, which fell apart after death of the algal cell. This is unlikely, because the diameter of the figured, spherical aggregates of pyrite crystals is substantially smaller than that of known, coeval coccospheres. The length of the rods of calcite around framboidal aggregates of pyrite crystals is about one micron like that of a micella of a coccolith, but the unidentified rods are not conical, and they are more slender than micellae. Isolated, pure accumulations with a diameter of 10–12 μm of the unidentified, randomly oriented rods of primary calcite without a framboidal nucleus of pyrite crystals were embedded into what was primarily probably almost pure aragonitic ooze. This is now recrystallized into a panxenotopic mosaic of micritic, low-magnesian crystals of calcite. Such an accumulation of rods was figured by GYGI (1969a, Pl. 4, Fig. 16) from an intercalation of micritic limestone in the Effingen Member.

All of the hard parts of nannoplankton which were figured by GYGI (1969a) from Oxfordian limestones were embedded into primarily aragonitic ooze which is now calcitic micrite (see below). The mode of formation of such micrite was investigated in great detail by GYGI (1969a, p. 39–42). The skeletal elements of nanoplankton were not recrystallized like the probably mostly aragonitic, mud-grade particles around them in the course of diagenesis, because they consist primarily of stable, low-magnesian calcite. This is the reason why it was easy to identify remains of coccoliths and other nannoplankton with the electron microscope where they occur in micritic limestone. Parts of coccoliths and unidentified nannoplankton were seldom found in the micritic limestones investigated. Unidentified organisms of nannoplankton like that represented in Fig. 4.17 were found only in the Effingen Member. The proportion of primarily calcareous mud which was produced by nannoplankton in micritic limestones from deeper water in the Rhodano-

Fig. 4.18 Two different kinds of whitings in shallow marine water over the northwestern Great Bahama Bank, as seen from the air. The two whitings in the foreground are stationary and are probably caused by schools of fish feeding on the bottom. The fish thereby stirred up carbonate mud into suspension. A tidal current carries the mud in suspension away in long tails. The photograph was made a short time after the direction of the tidal current changed. The much larger whiting in the background, behind a small atmospheric cumulus cloud, is of the type that was investigated by SHINN et al. (1989). The photograph was made by the author on May 31 in 1977 aboard a Douglas DC 3 plane cruising at an altitude of probably about 2,500 m

Swabian Basin was, therefore, probably negligible during the Oxfordian Age.

The Recent green alga *Penicillus capitatus* LAMARCK (figured in GYGI 1970, Fig. 16) produces calcareous mud in shallow water. When the tissue of this carbonate-secreting alga decays after death of the plant, an ooze of individual, clay-size needles of aragonite and of mammillate nanograins is left behind (MACINTYRE and REID 1992, Figs. 1F and 2). According to SHINN et al. (1989, p. 159), the standing crop of calcifying green algae of all types is not more than 1.5 plants per m^2 on the Great Bahama Bank northwest of Andros Island. The authors concluded on p. 156 that carbonate derived from algae may constitute only a small portion of either carbonate mud on the bottom or mud in suspension in whitings northwest of Andros Island.

Bioerosion of carbonate rock or of skeletal material mainly in shallow water can produce a considerable amount of fine-grained carbonate sediment. NEUMANN (1966) calculated the rate of erosion by boring sponges of the genus *Cliona* DE LAUBENFELS on a limestone coast in Bermuda, GYGI (1969b, 1975) and OGDEN (1977) that of parrotfish on coral reefs, and OGDEN (1977) quantified erosion of coral reefs by sea urchins. Much of the white, very pure calcareous sediment between the coral reefs in the Buix Member of the upper St-Ursanne Formation is possibly the product of bioerosion. Most of the fine-grained inter-reef sediment of the Buix Member is porous and resembles chalk, but no coccoliths were found in it to date. Chert nodules occur in

the uppermost Buix Member in the unpublished section RG 342 at Courchavon, and very large such nodules were found in that member in the unpublished sections RG 343 and 344 at Courtemaîche, all in Canton Jura. Empty rhombohedra of what were primarily euhedral dolomite crystals occur in the uppermost Buix Member dispersed in bed no. 8 of section RG 344. The source of silica and magnesium could not be identified.

Whitings are widespread patches in surficial sea water, which are turbid and colored white by clay-size carbonate particles in suspension. SHINN et al. (1989) concluded on p. 159 that whitings produce a large part of the lime mud which is sedimented on the Great Bahama Bank, and that part of this mud is exported to deeper water. MILLIMAN et al. (1993, p. 589) stated that inorganically precipitated aragonite needles are apparently produced everywhere on the Great Bahama Bank, and that much of this material is exported to deeper waters. A controversy arose about whether whitings are caused by inorganic precipitation of aragonite needles and possibly of some magnesian calcite from surficial water oversaturated with $Ca(CO)_3$, or whether schools of fish feeding on the shallow bottom of the Great Bahama Bank stir up carbonate mud from the sediment surface into suspension. SHINN et al. (1989) could not find schools of fish feeding on the bottom during their thorough research on whitings on the northwestern part of the Great Bahama Bank. They consequently denied that bottom-feeding fish stir up carbonate mud into suspension (p. 159). SHINN et al. (1989, p. 158) found whitings over sandy and even over rocky bottoms and concluded "if precipitation of aragonite and Mg calcite in whitings is a major source of sediment on the northwest portion of the Great Bahama Bank, this would then imply that the bottom sediment is derived from whitings, rather than the converse". MACINTYRE and REID (1992) showed that aragonite needles chemically precipitated in whitings (Fig. 1A) and in bottom sediment (Fig. 1B) west of Andros Island in the Bahamas have a different size and shape than aragonite needles secreted by common algae like *Halimeda incrassata* (ELLIS), *Udotea flabellum* (ELLIS and SOLANDER), *Rhipocephalus phoenix* (ELLIS and SOLANDER), and *Penicillus capitatus* LAMARCK (Fig. 1C–F).

The author of this study made the photograph shown in Fig. 4.18 during a flight aboard a Douglas DC 3 plane on May 31, 1977, which was a calm day with almost no atmospheric clouds over the northwestern part of the Great Bahama Bank. The aerial photograph was taken from an altitude of presumably about 2,500 m. This was the normal fairweather cruising altitude of a DC 3 over level ground or water. The flight was from Andros Town, Bahamas, to Miami, Florida, on return from a field trip to the Bahamas that was organized by the 3rd International Coral Reef Symposium at Miami, Florida. The two whitings in the foreground of Fig. 4.18, both with a drifting tail, can hardly be anything else than carbonate mud being stirred up from the bottom by two schools of fish. The whiting with a great areal extent in the upper part of Fig. 4.18 is one which was caused by inorganic-chemical precipitation of calcium carbonate from surficial water. This whiting covers a much greater area than the two whitings in the foreground that fish stirred up from the bottom. Whitings stirred up into suspension by bottom-feeding fish are stationary, and water current carries the mud away in a long tail. Whitings caused by fish are probably not as common as inorganically precipitated whitings, because the two small whitings with tails shown in Fig. 4.18 were the only ones of the kind which were seen by the author during the flight from Miami to Andros Town on May 28 and during the return flight on May 31. Several large whitings of inorganically precipitated calcium carbonate of the kind studied by SHINN et al. (1989) were seen during both of these flights. The temperature of sea water off Andros Island was measured on May 28, 1977, to be 24°C.

4.14.2 Export of Calcareous Mud from Carbonate Platforms

Only part of the St-Ursanne Formation is represented in Fig. 1.5. Much of the calcareous mud which was produced on the platform of the St-Ursanne Formation remained there and was incorporated into the Grellingen Member and into the Buix Member. The rest was exported to the adjacent depositional slope on which the Pichoux Formation was sedimented. When the volume of the calcareous mud that was laid down in the Pichoux Formation is compared with the partial volume of the carbonate platform of the St-Ursanne Formation which is represented in cross-section in Fig. 1.5, then the volume of the Pichoux Formation appears to be modest. The total width of the carbonate platform of the St-Ursanne Formation was possibly much greater than the part which is represented in Fig. 1.5. Most of the Pichoux Formation was initially calcareous mud. Spicules of siliceous sponges (GYGI 2000a, Fig. 11) and skeletal detritus of other macrofossils can be abundant in the proximal part of the formation, but the average content of such material is limited. The wedge-like configuration of the Pichoux Formation in cross-section perpendicularly to depositional strike is evidence that most of the lime mud in the formation was derived from the carbonate platform of the St-Ursanne Formation. The volume of the Pichoux Formation is consequently indicative of how much calcareous mud was exported from the St-Ursanne Formation.

The average width of the ramp of the Pichoux Formation was less than what can be concluded from the cross-section in Fig. 1.5. The belt on which the Pichoux Formation was sedimented is only about 2 km wide southeast and south of

Basel. It grew wider toward the southwest. The belt in the transect between Mervelier and Günsberg or in the transect between Sornetan and Péry has a palinspastic width of about 20 km (Fig. 1.5). The *average* width of the belt cannot be estimated, but it was certainly less than 20 km. The conclusion from this is that the volume of calcareous mud that was exported from the carbonate platform of the St-Ursanne Formation was probably a small fraction of the total volume of mud which was produced on that platform. It follows from the wedge-like geometry of the Pichoux Formation in cross-section perpendicularly to depositional strike and from the slight thickness of the coeval Birmenstorf Member on the level basin floor that production of calcareous mud *by whatever process* was insignificant in the deep part of the Rhodano-Swabian Basin during the *Antecedens* and the *Transversarium* Chron. Parts of coccoliths were found in the Birmenstorf Member (GYGI 1969a, Fig. 1), but they are quantitatively negligible as rock-forming constituents.

The Günsberg Formation as it is represented in Fig. 1.5 is a narrow carbonate platform which evolved in a belt running between the landward, shallow-water facies realm of the coeval lower Vellerat Formation and the facies realm of the coeval Effingen Member on the distal side, which was sedimented in deeper water of the Rhodano-Swabian Basin (GYGI 1990c, Fig. 6). The platform of the Günsberg Formation was much narrower than the preceding platform of the St-Ursanne Formation. The quantity of calcareous mud which was produced on the carbonate platform of the Günsberg Formation by carbonate-secreting algae, by bioerosion of coral reefs and biostromes, or by whitings must, therefore, have been much less than the quantity of such mud which was previously produced on the carbonate platform of the St-Ursanne Formation. Only a small part of the great quantity of calcareous mud which was sedimented in the Effingen Member can, therefore, have been produced on the carbonate platform of the Günsberg Formation.

The Effingen Member was sedimented mostly out of suspension of a varying mixture of calcareous mud and terrigenous, argillaceous mud including some silt-size detrital quartz and feldspar. The average carbonate content of the Effingen Member in Canton Aargau can be calculated to be about 60%, based on the many analyses that were made by cement companies of samples from the quarry east of Auenstein (section RG 37, GYGI 1969a, Pl. 17) and of samples taken from the quarry west of Villigen (unpublished technical report). PERSOZ in GYGI and PERSOZ (1986, p. 431) arrived at the same result based on his own analyses. The author of this study made additional, visual estimates of the carbonate content of samples from the Effingen Member by comparison with the numerous samples that he had given to Holderbank Cement Company to be chemically analyzed, when he was doing consulting for that company. The result of the visual estimates of samples from the Effingen Member in proximal section RG 14 near Günsberg (GYGI 1969a, Pl. 18) and in the distal section RG 87 in the landslide of Bleiche on the western slope of Mt. Eichberg northwest of Blumberg in southern Germany (GYGI 1969a, Pl. 16) was as follows. The average carbonate content of the entire Effingen Member is everywhere about the same as in Canton Aargau, where the mean thickness of the member is approximately 200 m. The fact that the total average carbonate content of the member does not vary substantially between the most proximal and the most distal localities investigated, irrespective of varying thickness of the entire member, is evidence that most of the calcareous mud in the Effingen Member was transported into the environment of final sedimentation from an area landward of the Günsberg carbonate platform like the terrigenous, argillaceous mud.

It is evident from Fig. 1.5 that the volume of the mostly marly Effingen Member with its content of primarily about 60% of calcareous mud is several times greater than that of the Pichoux Formation, which is all limestone. The volume of initially calcareous mud within the Effingen Member was, therefore, much greater than that in the limestone of the Pichoux Formation. When the volume of calcareous mud which was exported from the wide carbonate platform of the St-Ursanne Formation was modest, then the amount of calcareous mud which was derived from the narrow carbonate platform of the Günsberg Formation and that was sedimented in the Effingen Member must have been very small. The volume of calcareous mud which was sedimented in the mixed calcareous and argillaceous Effingen Member was much greater than the volume of limestone in the entire Günsberg Formation. This is *proof* that most of the calcareous mud in the Effingen Member was produced in the shallow-water realm landward of the carbonate platform of the Günsberg Formation, where the Vorbourg Member and the Röschenz Member were sedimented.

Nannoplankton like coccolithophorids (Fig. 4.16) and unidentified organisms (Fig. 4.17) occur according to Fig. 1 in GYGI (1969a) in intercalations of micritic limestone in the Effingen Member. The unidentified organisms were found exclusively in the Effingen Member (GYGI 1969a, Pl. 4, Fig. 16, and Pl. 5, Fig. 17). Green algae like Recent *Penicillus,* which secrete clay-size needles of aragonite, have no potential of becoming fossilized. It is consequently unknown whether similar algae existed in the Late Jurassic. To judge from SHINN et al. (1989, p. 159), such green algae could not have produced the great quantity of probably aragonitic ooze which was incorporated into the Effingen Member, even if they existed during the Late Jurassic. Bioerosion of coral reefs and biostromes in the Günsberg carbonate platform certainly produced some fine-grained calcium carbonate. A few small coral biostromes or low bioherms also existed in the uppermost, internal part of the Röschenz Member, for instance, near Bressaucourt. The

volume of coral communities in the Günsberg Formation and the Röschenz Member combined was so small, that bioerosion of these coral biostromes and bioherms was by far insufficient to supply the great amount of calcareous mud which was sedimented in the coeval Effingen Member.

All of the skeletal elements of calcareous nannoplankton which were found in Oxfordian limestones in the Jura Mountains are as a rule well preserved. Such elements were, therefore, easy to identify with the electron microscope. They are now embedded in a panxenotopic mosaic of micritic calcite grains, as it was imaged with a transmission electron microscope by GYGI (1969a, Pl. 4, Fig. 16). The micritic mosaic of the limestone's groundmass was probably formed, when ooze with mostly needles of aragonite was compacted and then recrystallized (BATHURST 1959, p. 366). Skeletal elements of nannoplankton are so rare in the limestone intercalations in the Effingen Member that they are negligible as rock constituents. It can be concluded from this that much or most of the carbonate mud in the Effingen Member was primarily aragonitic ooze. Only micritic limestones were scanned for nannoplankton during the investigations for the Ph. D. thesis of GYGI (1969a). Later it turned out that calcareous nannoplankton is more abundant, diverse, and better-preserved in sediments with a high content of primarily argillaceous mud (see GRÜN and ZWEILI 1980). The same is the case in palynomorphs which are rare for instance in the predominantly calcareous Birmenstorf Member (GHASEMI et al. 1999). Even if nannoplankton was much more abundant in the more argillaceous parts of the Effingen Member than in the pure, micritic limestone of the Wangental Member, as it was mentioned above, it was possibly quantitatively insignificant as a rock constituent in marl as well.

Since only a small quantity of calcareous mud can have been produced on the narrow carbonate platform of the Günsberg Formation and have been exported into the Effingen Member, and because it could be documented above that production of calcareous mud in the realm of deeper water of the epicontinental basin was slight, it is concluded below that whitings are the most probable source of the greater part of the calcareous mud which was sedimented in the Effingen Member. The pertinent whitings must consequently have occurred in the vast shallow-water realm landward of the carbonate platform of the Günsberg Formation, where the peritidal Vorbourg and the Röschenz Member were sedimented. From there, calcareous mud was transported by tidal or occasionally by storm-driven currents across the Günsberg carbonate platform and into the epicontinental basin.

An unsolved problem is why the thickly bedded and almost pure, micritic limestone of the peritidal Vorbourg Member in the lowermost Vellerat Formation of northwestern Switzerland possibly grades laterally into the coeval, but much more argillaceous, lowermost part of the Effingen Member in the adjacent epicontinental basin. Time correlation between the two sedimentary units could only be made according to the mineral-stratigraphic correlation C by Persoz in Pl. 1A by Gygi and Persoz (1986), because there are no ammonites in the Vorbourg Member. A conceivable solution of this correlation problem can be concluded from section RG 411 that is photographed in Fig. 11.1. The uppermost tongue of the St-Ursanne Formation in the eastern wall of Combe des Geais north of Grandval thins within the figured section over the distance of 100 m basinward to almost one-half of its thickness. The probable time equivalents of the Vorbourg Member in the lowermost Günsberg Formation directly above pinch out basinward in this formation and are not represented in the lowermost Effingen Member. To judge from Fig. 4.19A, B, the Vorbourg Member should be assigned to the St-Ursanne Formation rather than to the Vellerat Formation, as it is currently done.

4.14.3 Whitings, their Localization and Quantitative Relevance

The Courgenay Formation is a bedded, micritic limestone with few macrofossils, which was sedimented into a shallow lagoon. This lagoon was rimmed against the Rhodano-Swabian Basin by the wide sand bank of calcareous ooids in the carbonate platform of the Balsthal Formation. The volume of carbonate mud which was produced on the Balsthal carbonate platform and that was exported from there into the adjacent epicontinental basin was probably minimal (see below). The basinward time equivalent of the Balsthal Formation is the Villigen Formation, which was sedimented in deeper water. The Villigen Formation is a bedded, micritic limestone. It resembles in this the Courgenay Formation, but macrofossils are in the Villigen Formation more abundant. Macrofossils in the Villigen Formation of Canton Aargau are mainly bivalves and some brachiopods. Ammonites prevail in the macrofauna of this formation in Canton Schaffhausen where water was deeper (Fig. 1.5). The great volume of calcareous mud which was sedimented in the Villigen Formation and in its distal time equivalents in southern Germany as well as in the coeval lower Quinten Formation further south cannot possibly all have been produced on the Balsthal carbonate platform. Gygi (1969a) stated that well preserved remains of calcareous nannoplankton are generally rare in the basinal calcilutites that he had investigated. Coccolithophorids were found to be most abundant in sediments from deepest water, as for instance in bed no. 101 of the Wangental Member in the unpublished section RG 91 near Immendingen in southern Germany (Fig. 4.16), where

Gygi (1969a, p. 24) calculated the abundance of coccolithophorids to be 0.004% of the rock volume. The problem was revisited by Pittet and Strasser (1998). They concluded on p. 161 that nannofossils probably contributed only a small quantity to the carbonate mud which was sedimented in deeper water. Much of the calcareous mud in the Villigen Formation and in its distal time equivalents was probably derived from whitings which occurred in the lagoon in which the Courgenay Formation was sedimented. The facies realm of the Courgenay Formation may have been similar to Recent Florida Bay, in which whitings occur according to Shinn et al. (1989, p. 147).

Indirect evidence from the geometry and sedimentology of sedimentary bodies as well as from time correlation by means of ammonites and the clay mineral kaolinite leads to the *general conclusion* (see main conclusion no. 13), that most of the lime mud sedimented in the deeper part of the epicontinental basin was produced by inorganic chemical precipitation in whitings mainly in shallowest, near-shore water. The greatest amount of calcareous mud in the sediments represented in cross-section in Fig. 1.5 was incorporated into the Effingen Member where it is thickest in eastern Canton Solothurn and northern Canton Aargau. The bulk of this member is marl with an average carbonate content of approximately 60%. This is evident from section RG 37 at Auenstein in Gygi (1969a, Pl. 17). Comparison of the thick Effingen Member in the epicontinental basin with the coeval, peritidal and much thinner lower Vellerat Formation landward of the carbonate platform of the Günsberg Formation is instructive. An excellent section of the entire Vellerat Formation could be measured by the author near Röschenz, now in Canton Basel-Landschaft, on a day in August 1983. The section is shown in detail in Pl. 33 in Gygi (2000a). Much terrigenous material and nutrients were transported into deeper water in the epicontinental basin from the *Schilli* Chron on until *Hypselum* time. Only an insignificant portion of the calcareous mud which was laid down in the Effingen Member can have been produced on the narrow carbonate platform of the Günsberg Formation. Nannoplankton certainly contributed close to nothing to carbonates in the Effingen Member. Most of the calcareous mud in the Effingen Member was consequently produced in the vast marginal marine, proximal facies realm of the lower Vellerat Formation (Fig. 1.2). Large-scale production of calcareous mud in that environment was probably possible only by inorganic chemical precipitation of calcium carbonate. A perhaps even greater quantity of calcareous mud was produced in the shallow lagoon, into which the micritic limestone of the Courgenay Formation was sedimented. The surplus of mud that was produced, but did not remain in this lagoon, was exported.

Little or no calcareous mud at all was produced on the very widespread calcareous ooid sand bank of the Hauptrogenstein Formation. Proof of this is the large tidal delta of cleanly washed ooid sand at the basinward margin of the carbonate platform of this formation which is now very well exposed at Auenstein and Veltheim in Canton Aargau. The source in the uppermost part of the carbonate platform of the Günsberg Formation from which, according to Fig. 5.3, much lime mud spilled into the tidal delta near Péry which is shown in the photograph of Fig. 5.2 must have been local and exceptional. For instance the coeval, large sand wave near Liesberg represented in Fig. 5.5A is cleanly washed pelsparite which is shown in Fig. 5.5B. The situation on the widespread bank of calcareous ooid sand of the Balsthal Formation was different. Cleanly washed oosparite is uncommon in the Holzflue Member and in the Verena Member. This is evidence that water above the sand bank of the Balsthal Formation was possibly permanently a turbid suspension including clay-size calcareous particles in high concentration. Almost total absence of cross-bedding in the two members is an indication that tidal currents rolling ooids were rather slack. There are even intercalations of pure micrite, for instance in the Verena Member in beds no. 228b and 234 in section RG 307 at Péry (GYGI 2000a, Pl. 22) and in the Holzflue Member in bed no. 13 of section RG 440 at Mt. Rüttelhorn above Rumisberg and in bed no. 44 of section RG 438 in Steinibach Gorge north of Balsthal (GYGI 2000a, Pl. 43 and 44). When rapid sea level rise no. 10 in Fig. 5.1 interrupted for some time export of calcareous mud from the shallow lagoon of the Courgenay Formation across the oolitic sand bank of the Balsthal Formation to the Villigen Formation, omission and subsolution were the effect in deeper water in the Villigen Formation, for instance at Villigen. This is shown in the photograph of Fig. 12 in GYGI et al. (1998). This is additional evidence that most or all of the lime mud sedimented in the Villigen Formation was produced in the lagoon of the Courgenay Formation and was transported across the widespread bank of calcareous ooid sand of the Balsthal Formation in between.

Conversely, most of the wide sand bank of calcareous ooids of the Steinibach Member was initially cleanly washed ooid sand. Part of the Steinibach Member is cross-bedded, and large wave ripples were found in the oosparitic member in the unpublished section RG 448 at Egerkingen in Canton Solothurn. It can be read from Fig. 1.5 that the sand bank of the Steinibach Member rimmed the shallow, very widespread lagoon of the mainly oncolitic Hauptmumienbank Member against the deeper water realm where the coeval Geissberg Member was laid down. Almost pure calcareous mud was sedimented in the Geissberg Member in Canton Aargau. The bulk of the lime mud in the Geissberg Member was most probably produced by whitings in the shallow lagoon with abundant calcareous oncoids of the Hauptmumienbank Member. The spherical oncoids with a diameter of up to 4 cm in that member were initially floating in loose calcareous mud. This is evidence that the oncoids

were rolled on the lagoon floor from time to time by currents with the minimal velocity of 1.5 m/s, according to the diagram in HJULSTRÖM (1935, Fig. 18). Currents strong enough to roll calcareous oncoids with a diameter of several centimeters at the sediment surface must have been episodic and driven by occasional storms, not by tides at regular intervals, or else lime mud could not have settled out of whitings between oncoids and partially have become permanent sediment in the lagoon. *Occasional*, storm-driven currents churned calcareous mud up from the sediment surface into suspension in the lagoon of the Hauptmumienbank Member. These currents removed part of the mud in suspension in water of the lagoon and exported lime mud across the bank of calcareous ooid sand of the Steinibach Member into deeper water. *Regular* tidal currents in shallowest water above the sand bank of the Steinibach Member were strong enough to prevent lime mud to settle and remain on the shoal between the lagoon of the Hauptmumienbank Member and deeper water of the epicontinental basin, where the mud could settle out of suspension and was sedimented into the Geissberg Member.

4.15   Deepening-Upward and Shallowing-Upward Successions

4.15.1   Deepening-Upward Successions

When accommodation space in a given, shallow-marine environment is augmented by a *rapid* relative sea level rise, then both the source of terrigenous sediment and the source of calcareous mud, which is produced by whitings in very shallow-marine water, recede landward. The effect in a more distal environment in deeper water can be a particularly low sedimentation rate or even nondeposition (omission, see below). A hiatus evolves, when omission continues over an extended time span. The quantitatively most important example of such a situation is section RG 226 between Auenstein and Veltheim in Canton Aargau. The entire section was measured by the author in the deep road cut leading from the quarry of Unteregg down to the quarry of Jakobsberg east of Auenstein (GYGI 1973, Fig. 3). MANGOLD and GYGI (1997, Fig. 2) published an enlarged part of section RG 226 in the road cut which shows details of the boundary beds between the Middle and the Upper Jurassic Series. Unit no. 25 of section RG 226 is the marginal part of the thin carbonate platform of the early Bathonian *Spatkalk Member*. The member is on the eastern wall of the road cut a tidal delta of calcarenite with a total thickness of 3.2 m and with distinct foresets. The top of the delta is covered with a crust of goethite, which documents sea level or possibly emersion. Above the crust is a thin seam of marl, bed no. 26. The following bed nos. 27 and 28 above the marl are iron oolite with a matrix of what was primarily ferriferous calcilutite.

The aggregate thickness of the iron-oolitic bed nos. 27 and 28 is 25 cm. The photograph of a longitudinal, polished section through a drill core of these beds is shown in GYGI (1969a, Pl. 2, Fig. 4). The macrofossil assemblage in the iron oolite is mainly bivalves and rhynchonellid brachiopods with few ammonites of middle Bathonian age. Some of the ammonites from bed nos. 27 and 28 were figured by MANGOLD and GYGI (1997, Figs. 3/3, 5, 6, and 8). GYGI (1992, Fig. 28) figured abundant, large ostreids from bed no. 28. Brachiopods like *Rhynchonelloidella alemanica* (ROLLIER), which are the *Rhynchonella varians* of earlier authors, can locally be abundant in the iron oolite. The macrofossil assemblage in the Bathonian iron oolites near Auenstein and near Holderbank in Canton Aargau indicates a water depth of between 20 and 30 m, to judge from Fig. 4.9 in this study. Figure 4.9 shows the interrelation of the composition of fossil assemblages in sediments of *Late Jurassic* age with depth of deposition. The macrofossil assemblage in the iron oolite of Bathonian age with subordinate ammonites and with large ostreid bivalves, but without hermatypic corals, is similar to the macrofossil assemblage in the Geissberg Member of the late Oxfordian in Canton Aargau. This is an indication that the two units were sedimented in the same bathymetric range. The rate of sedimentation of the two units was certainly different. Thin and widespread iron oolite in the deep subtidal zone is a sediment diagnostic of a very low average sedimentation rate, whereas calcilutite of the Geissberg Member was sedimented at a normal rate. The similar macrofossil assemblage in the two sediments with a great difference in lithology, mode of formation, and age is proof that the composition of the macrofossil assemblage in a sediment depends to a large extent on the depth of deposition (see main conclusion no. 24). The further conclusion appears to be permissible that the environmental requirements of the investigated organisms fossilized, mainly of hermatypic corals and of ammonites, did not noticeably vary in the considerable time span between the middle Bathonian and the late Oxfordian.

The rich macrofossil assemblage of mostly benthic animals in the Bathonian iron oolites is evidence that the corresponding organisms lived in well-aerated water. Nevertheless, many of the ferriferous ooids in the uppermost part of the iron oolite of bed no. 28 in section RG 226 are dark green and probably consist of berthierine. The ooids are colored green by ferrous iron. The predominantly green iron ooids of presumably berthierine in the coeval iron oolite of bed no. 2 in section RG 51b near Oberehrendingen in eastern Canton Aargau were initially soft and were plastically deformed. It is likely that deformation of the soft green iron ooids was caused by animals burrowing in the embedding carbonate mud soon after the iron ooids were accreted.

The deformed iron ooids were photographed by GYGI (1969a, Pl. 3, Fig. 8). The iron-oolitic bed near Oberehrendingen was dated with a *Tulites (Tulites) cadus* BUCKMAN to be of middle Bathonian age by MANGOLD in MANGOLD and GYGI (1997, Fig. 5/4).

The upper bedding plane of the iron-oolitic bed no. 28 of middle Bathonian age in section RG 226 east of Auenstein is covered with a domal microbialite that can be as thick as 8 mm. This microbialite was figured by GYGI (1992, Fig. 29), and again by RAIS (2007, Chap. 2, Fig. 6B and appendix). The ferriferous stromatolite is partly green with berthierine and partly brown with goethite. The stromatolite was probably formed after a long time span of nondeposition at the end of the early Oxfordian *Cordatum* Chron. This is concluded for two reasons. First, the ammonite *Cardioceras (Cardioceras) persecans* (BUCKMAN) was found *below* the corresponding microbialite in bed no. 1 in section RG 39, which is at locality Berg 2 km south of the village of Scherz. This *Cardioceras* of *Cordatum* age was figured by GYGI and MARCHAND (1982, Pl. 9, Fig. 1a–b). Section RG 39 is 2.5 km east of section RG 226, and 400 m east of section RG 276 in the quarry of Chalch east of Holderbank. Second, the cast of the *Cardioceras* from near Scherz was formed in a small pocket of Oxfordian iron oolite within a bed of iron oolite of middle Bathonian age. The lithology of the early Oxfordian contents of the pocket could not be discerned from the lithology of the encasing Bathonian iron oolite. When the author measured section RG 39 in 1962, he did not recognize the pocket to be a much younger, Oxfordian inclusion in a bed of probably Bathonian iron oolite, because he did not find any Bathonian ammonites in bed no. 1 in the small outcrop of section RG 39. Nevertheless, the iron-oolitic bed with an inclusion of a *Cardioceras* of the Oxfordian *Cordatum* Chron near Scherz is probably coeval with bed no. 6 in section RG 276 in the quarry of Chalch near Holderbank 400 m to the west, in which numerous ammonites of Bathonian age were found. Some of the Bathonian ammonites from bed no. 6 in section RG 276 were figured by MANGOLD and GYGI (1997, Fig. 3/9 and Figs. 4/1, 2, and 4). Bed nos. 27 and 28 in section RG 226 east of Auenstein are consequently coeval with bed no. 6 in section RG 276 east of Holderbank.

The same early Oxfordian microbialite which is above the Bathonian iron oolites near Auenstein, Holderbank, Scherz, Villigen, and Oberehrendingen is continuous above the early Oxfordian iron oolite of the Schellenbrücke Bed in western and in northern Canton Aargau. This and the *Cardioceras* from Scherz is why GYGI (1969a, Pl. 2, Fig. 4) assigned bed no. 3 in his section RG 37a east of Auenstein to the Schellenbrücke Bed of the *Cordatum* Chron. Section RG 37a was measured in 1962 on the cores of an exploration well which was drilled in that year in order to plan the road cut, in which section RG 226 could be measured in 1972. Bed no. 3 of section RG 37a became in 1972 bed no. 28 of section RG 226, out of which the Bathonian ammonites as figured by Mangold and Gygi (1997) were collected.

The lowermost, stratigraphically condensed bed of the Birmenstorf Member of middle Oxfordian age is directly above the microbialite which covers bed no. 28 in section RG 226. Cauliflower pellets of glauconite and some brown, goethitic iron ooids occur at this locality together in the lowermost bed of the Birmenstorf Member, which is the time equivalent of bed no. 10 in section RG 208 near Ueken (see above). The occurrence of both iron ooids and glauconite pellets in the condensed, lowermost bed of the Birmenstorf Member is evidence of a depth of sedimentation of approximately 100 m. The base of this marker bed in section RG 226 is only 25 cm above the top of the carbonate platform of the early Bathonian Spatkalk Member which is covered with a crust of goethite indicating water depth of around zero. This documents that the relative sea level rise which occurred between the beginning of the middle Bathonian and the beginning of the middle Oxfordian was no less than approximately 100 m in the region between Auenstein and Holderbank in Canton Aargau. The rise occurred in the time span of about 7.5 million years. Only iron-oolitic sediment of mainly middle Bathonian age with a total thickness of 0.25 m was laid down in section RG 226 during this long period of time. There was consequently no sedimentation at all during most of the time of 7.5 million years. This was the effect of omission (or nondeposition, see Sect. 8.3), not of subaqueous or even of subaerial erosion.

The Bathonian *Calcaire roux sableux Member* in north-western Switzerland is carbonate sediment from very shallow water that drowned late in Bathonian time. The boundary beds between the Calcaire roux sableux Member and the lowermost sediments of Callovian age directly above were excavated by R. and S. Gygi in 1985 as section RG 427 in the clay pit of Hinter Chestel south of the village of Liesberg in Canton Basel-Landschaft. The section is represented as Fig. 1 in Dietl and Gygi (1998). The uppermost bed of the Calcaire roux sableux Member, bed no. 1 of section RG 427, is a bioturbated, bioarenitic wackestone with some detrital quartz as well as with large ostreid bivalves and small rhynchonellid brachiopods. Neither hermatypic corals nor ammonites were found in the bed. The upper surface of bed no. 1 is hummocky. It is the vast, inclined bedding plane with deep grooves up to several meters long and 0.4 m wide which bounds the former clay pit to the south.

The Callovian Age begins with the *Keppleri* Chron. The *Keppleri* Chron is named after the ammonite *Kepplerites keppleri* (Oppel). Two specimens of this taxon were collected from bed no. 2 in excavation RG 427 in the clay pit of Hinter Chestel. The bed is at the base of the inappropriately named *Macrocephalenkalke* (in Dietl and Gygi 1998, Fig. 1).

Figures of the two *Kepplerites* from bed no. 2 of excavation RG 427 were published in Pl. 1 by Dietl and Gygi (1998). A coeval ammonite cast from bed no. 2 of excavation RG 427 is a *Bullatimorphites (Kheraiceras)* transitional between the taxa *hannoveranus* (Roemer) and *bullatus* (D'Orbigny). This ammonite was figured on Pl. 4, Fig. c–d, in Dietl and Gygi (1998). According to the ammonites figured, the Callovian succession at Liesberg begins with bed no. 2 of section RG 427. Bed no. 2 is marl with carbonate concretions. The bed yielded 110 macrofossils, 65 of which are ammonite casts (59%), 43 are bivalves, mainly of the genera *Pholadomya* and *Pleuromya* (39%), and two are pleurotomariid gastropods (2%, see Dietl and Gygi 1998, p. 249). Under the assumption that ammonites of the earliest Callovian had the same ecological requirements as those living in the Late Jurassic, it is concluded according to Fig. 4.9 in this study that bed no. 2 of excavation RG 427 was sedimented at a water depth of slightly more than 35 m. This must have been substantially deeper than the depth of deposition of the Calcaire roux sableux Member of Bathonian age directly underneath, in which no ammonites were found. The major, rapid relative sea level rise no. 2 in Fig. 5.1 is, therefore, presumed to have caused non-deposition and to have created the substantial accommodation space of at least 35 m, before sedimentation resumed near Liesberg in earliest Callovian time. The rise had probably a eustatic component. This would explain, why a hiatus evolved both between the Bathonian Calcaire roux sableux Member from shallow water and the bed of earliest Callovian age directly above with mainly ammonites near Liesberg in northwestern Switzerland, and in the southeast between iron oolite of middle and early late Bathonian age from somewhat deeper water and the middle Oxfordian Birmenstorf Member between Auenstein and Oberehrendingen in Canton Aargau (Mangold and Gygi 1997, Fig. 2).

The *Herznach Formation* is a thin, but conspicuous deepening-upward succession of iron oolite including a macrofossil assemblage of mostly ammonites. The formation is defined above in the text of Sect. 4.8.2 and in Fig. 4.10. Sedimentation of the formation began in northwestern Switzerland before the end of the early Callovian after the carbonate platform of the Dalle nacrée Member had drowned. The best section of the Herznach Formation in western Switzerland was measured by the author in 1980 in the quarry of La Charuque near Péry in Canton Bern. The formation in that section is above a tidal delta with foresets at the basinward margin of the Dalle nacrée carbonate platform (Gygi 1990a, Fig. 4). The Herznach Formation in section RG 307 near Péry in Gygi (2000a, Pl. 22) is represented by bed nos. 11–20 with a total thickness of 1.5 m. The age of the formation ranges at that locality from the latest early Callovian to the earliest Oxfordian. The lowermost bed no. 11

of the formation in that section is 0.15 m thick and includes few iron ooids of goethite and berthierine.

The time equivalent of the carbonate platform of the Dalle nacrée Member is in section RG 280 in the clay pit of Andil southwest of Liesberg in Canton Basel-Landschaft (GYGI 2000a, Pl. 30) the bioclastic calcarenite of unit no. 2 with hummocky cross-stratification (term after HARMS et al. 1975). This calcarenite was sedimented in the shallow subtidal zone some distance off the basinward rim of the carbonate platform of the Dalle nacrée. The Herznach Formation above the calcarenite at Liesberg has a similar iron-oolitic lithology and a macrofossil assemblage of mostly ammonites like at Péry and a total thickness of 1.05 m. An enlarged part of section RG 280 with the calcarenite and the Herznach Formation is shown in Fig. 2 in GYGI (1990a). The Herznach Formation begins in that section with bed no. 3 with a thickness of 10 cm. The bed includes nodules of iron sulfide, and it is covered by a thick crust of goethite. No iron ooids were found in the bed near Liesberg, but there are abundant bivalves, gastropods, and brachiopods. B. HOSTETTLER found a few hermatypic corals in the upper-most part of the bed which indicate a maximal water depth of sedimentation of approximately 20 m (see Chap. 6). He also found in bed no. 3 the well-preserved and near-complete ammonite *Sigaloceras (Catasigaloceras) enodatum* (NIKITIN) of latest early Callovian age. Ammonites make their first appearance in bed no. 3, but they are few. STÄUBLE (1959, Fig. 2) called this bed at Liesberg Lumachellenbank. The coquina of BITTERLI (1979, Fig. 4) is a synonym. The time equivalent of this bed in section RG 307 at Péry is bed no. 11. These beds at the base of the Herznach Formation are part of a very widespread, fossiliferous marker bed. Sedimentation of the Herznach Formation began at Péry at the water depth of around zero. Depth of deposition at Liesberg was initially of the order of 5 m and increased to approximately 80 m at the end of sedimentation of the formation.

Sedimentation of the Lumachellenbank Bed in the central part of the Jura Mountains ceased at the close of the early Callovian at the water depth of 20 m, to judge from the ammonite that B. HOSTETTLER found in the bed near Liesberg. The time equivalent of this marker bed in the Central Jura Mountains is in Canton Aargau in the closed iron mine west of Herznach the lowermost bed with iron ooids above the Kornberg Sandstone Member. This is bed no. 12 with a thickness of 10 cm in Fig. 4.10 in this study, or bed A5 in Fig. 2 in JEANNET (1951). Three *Sigaloceras (Catasigaloceras) enodatum* (NIKITIN) were figured by JEANNET (1951, Pl. 25, Figs. 8–10) from his bed A5. The entire deepening-upward succession of the Herznach Forma-tion above this regional marker bed is iron oolite near Herznach, Liesberg, and near Péry. Ammonites prevail in the macrofossil assemblage in the iron oolites of the

Herznach Formation everywhere above the marker bed at the base. Figure 4.9 in this study indicates a paleodepth of sedimentation of more than 35 m of the Herznach Formation above the basal marker bed in the entire region between Péry and Herznach.

The greater part of the rapid relative sea level rise which created a water depth of at least 35 m near Péry early in middle Callovian time must therefore have occurred just after the close of the *Enodatum* Chron, when sedimentation of the bulk of the iron-oolitic Herznach Formation began. A substantial, net increase in water depth occurred during sedimentation of the entire, thin and partially condensed Herznach Formation. Water depth increased during deposi-tion of the Herznach Formation near Liesberg from a few meters to a maximum of about 80 m at the time, when deposition of the Herznach Formation was completed after sedimentation of bed no. 6 in section RG 280 (GYGI 1990a, Fig. 2). The time equivalent of this bed no. 6 near Liesberg is near Péry bed no. 20 in section RG 307 (GYGI 1990a, Fig. 4). Ammonites of the early Oxfordian *Scarburgense* Chron were found in the upper part of this bed near Péry (GYGI 1990a, Pl. 3, Fig. 11). A clear-cut boundary between rock colors exists, both in section RG 307 near Péry and in section RG 280 near Liesberg, between the top of the brownish-gray, clayey marl with iron ooids in the upper-most Herznach Formation below, and the blue-gray, homogenous clayey marl without iron ooids of the *Renggeri* Member above in the lowermost Bärschwil For-mation. When sedimentation of the early Callovian carbon-ate platform of the Dalle nacrée Member in northwestern Switzerland was completed at a water depth of close to zero, then the relative sea level rise began and continued during sedimentation of the entire Herznach Formation. The sea level rise between late in early Callovian time and the beginning of the Oxfordian increased water depth in all of northwestern Switzerland to about 80 m. Much of this net rise was probably eustatic.

An instructive deepening-upward succession is that in the stratigraphically condensed *Schellenbrücke Bed* and in the lowermost, condensed bed of the Birmenstorf Member directly above, which was documented in excavation RG 208 near Ueken in Canton Aargau. Deepening of the water and processes of condensation at Ueken are explained in the text of Sect. 8.4. The different stages of condensation reached in the course of deepening of the water from approx-imately 80 m at the beginning of the Oxfordian to 100 m in *Densiplicatum* time in the middle Oxfordian are visualized in Fig. 8.2.

A deepening-upward succession of middle Oxfordian age is documented in the unpublished section RG 336 of the *St-Ursanne Formation* near St-Ursanne in Canton Jura. The section was measured at the railway station of St-Ursanne

Fig. 4.19 (**A**) General view of the St-Ursanne Formation east of the small medieval city of St-Ursanne in Canton Jura as seen from the road to Montmelon, looking to the northwest. (**B**) Close-up view of the formation 300 m east of the city, looking north across Doubs River from the road to Épauvillers. CHE Chestel Member, in which mainly massive hermatypic coral colonies and oncoids are disseminated in the lower part of the member. The upper part is in this section pure calcareous oolite. BUI Buix Member, porous, friable and very pure calcareous mudstone with two well-cemented coral bioherms. The lagoon reefs projects out of the interreef sediment because of differential weathering. The top of the reefs is approximately 18 m above their base. VOR Vorbourg Member, thickly-bedded peritidal, calcareous mudstone in the lower Vellerat Formation. Photographs by C. Lanz

and in the quarry directly above. A general view of the St-Ursanne Formation at St-Ursanne is Fig. 4.19A in this study. The lower part of Chestel Member in the lower St-Ursanne Formation in section RG 336 is a bioclastic limestone with isolated, mostly plate-like coral colonies and with oncoids. The limestone grades upward into pure calcareous oolite

with grainstone texture and without corals. Massive and thickly branching coral colonies appear in the section 1.2 m below the top of Chestel Member. Corals are between 15 and 20% of the oolitic rock volume directly below the top of the unit. Some of the uppermost, massive coral colonies are bored from the top. This indicates prolonged nondeposition, which was probably caused by rapid deepening of the water. Accretion of calcareous ooids ceased at this location in the platform interior altogether when sea level suddenly began to rise at a rate that was greater than the maximal production rate of calcareous ooid sand which was possible in this paleogeographic situation (GYGI 1986, p. 467, 2003, p. 178). This was the effect of rapid relative sea level rise no. 5 in Fig. 5.1.

The sand bank on which calcareous ooids were accreted in Chestel Member in the platform interior prograded in the basinward direction, before rapid relative sea level rise no. 5 in Fig. 5.1 began. Oolitic sand thereby spread over and buried the coral biostrome growing in somewhat deeper water in the distal part of the Grellingen Member (Fig. 1.5). This progradation is documented by the oolitic bed nos. 6–10 in section RG 400 near Corban (GYGI 2000a, Pl. 35) and by bed no. 2 in section RG 451 near Vellerat (GYGI 2000a, Pl. 24). Both of these sections are in Canton Jura. It can be read from Figs. 1.4 and 1.5 that the particularly instructive section RG 400 at Corban was measured in a position which is somewhat closer to the basinward rim of the carbonate platform of the St-Ursanne Formation (boundary no. 1 in Fig. 1.4) than section RG 336 at St-Ursanne. Outcrops are good and continuous in the lower half of section RG 400 at Corban. The section is proof that basinward progradation of the sand bank of Chestel Member began before the onset of rapid rise no. 5. The earliest oolite in section RG 400 is in bed no. 6. Oolite of bed no. 6 crops out beside the footpath leading along the creek through a small gorge. The footpath connects farm La Providence below with farm Nieder Fringeli above the gorge. The beginning of sea level rise no. 5 is documented in section RG 400 by the transgressive surface at the top of calcareous oolite of bed no. 10. This transgressive surface is indicated in Pl. 35 in GYGI (2000a). Above the surface is a coral bioherm. The base of this bioherm is probably coeval with the base of the lagoon reefs in the Buix Member at St-Ursanne (Fig. 4.19B).

It is evident from Pl. 35 in GYGI (2000a) and mainly from Fig. 1.5 in this study that accretion of calcareous ooids began at Corban above the coral biostrome of the Grellingen Member earlier than at the top of the Grellingen Member at Grellingen in the more basinward part of the carbonate platform of the St-Ursanne Formation. The base of the coral bioherms in the uppermost part of the Pichoux Formation in Pichoux Gorge near Sornetan, not far beyond the basinward margin of the St-Ursanne Formation, is probably younger than that of the base of the reefs in the Buix Member at St-Ursanne. Rapid sea level rise no. 5 is assumed to be documented in Pichoux Gorge by the marly bed no. 22 which is almost exactly in the middle of the Pichoux Formation in section RG 315 (in Pl. 21 in GYGI 2000a). This bed crops out in the roof of the northern entrance to the road tunnel in Pichoux Gorge. The bed is presumed to be coeval with the transgressive surface at the base of the Buix

Member in the middle of the St-Ursanne Formation at St-Ursanne, as it is shown in Fig. 4.19B.

The rate of sea level rise no. 5 was such that accretion of calcareous ooids was cut short in the interior of the carbonate platform at the top of Chestel Member. Conversely, ooid accretion continued from the beginning of the rise in the lower Tiergarten Member near the basinward margin of the platform. Ooid accretion was consequently possible in the lower part of the sand bank of the Tiergarten Member, although relative rise no. 5 in Fig. 5.1 amounted to a total of approximately 10 m (see below), and in spite of the fact that the maximal water depth where calcareous ooids can be accreted in the Recent is 6 m. This is proof that the rate of carbonate precipitation in the form of ooids was greater near the basinward rim of the carbonate platform of the St-Ursanne Formation than in the platform interior. Production of ooid sand on the bank of the Tiergarten Member kept pace with the high rate of sea level rise no. 5. This oolitic sand bank with coral reef knobs thereby began to rim a lagoon in the platform interior, which was rapidly deepening concomitant with sea level rise no. 5. The sand bank of the Tiergarten Member rimmed the lagoon against the epicontinental, Rhodano-Swabian Basin.

The Buix Member with coral patch reefs was sedimented into this lagoon (Fig. 4.19B). Growth of the lagoon reefs in the Buix Member began directly above the transgressive surface at the top of Chestel Member. Water circulation and animal food supply to corals were sufficient for vigorous reef growth from then on through much of the time span before the lagoon was entirely filled with sediment. The abrupt ending of calcareous ooid production at the top of Chestel Member and the onset of vigorous reef growth directly above the transgressive surface is proof that sea level rise no. 5 was rapid. The reefs grew to a maximal elevation from base to top of 18 m until their demise because of too shallow water and thereby of insufficient water circulation and food supply to the coral polyps. Evidence is given in Sect. 5.3 that sea level slightly fell toward the end of the *Transversarium* Chron. As a consequence, sedimentation of calcareous ooid sand began to spread landward from the Tiergarten Member into the uppermost part of the lagoonal Buix Member at Liesberg. Oolite in the uppermost part of the Buix Member at Liesberg is shown in section RG 306, bed nos. 114–119 on Pl. 31 in GYGI (2000a).

The lagoon in the platform interior, into which the Buix Member was sedimented, evolved on top of a sand bank of calcareous ooids, which was directly below mean sea level before the onset of sea level rise no. 5 (compare with Fig. 4.7). The lagoon was rapidly deepening at an early stage of its evolution. Several large casts of perisphinctid ammonites were found in the inter-reef sediment of the Buix Member. The casts are evidence that water in the lagoon deepened to a considerable amount. Some of these ammonites, which possibly lived permanently in the lagoon, were figured by GYGI (1995). The rock volume in the core of the lowermost part of lagoon reefs in the Buix Member at St-Ursanne can be as much as 50% of massive and thickly branching corals in micritic matrix. This indicates that conditions for reef growth were favorable in the lagoon at an early stage of its evolution. The sediment between the coral reefs in the Buix Member at St-Ursanne is white, very pure, and fine-grained, chalk-like calcium carbonate without coccolithophorids or ooids. The corresponding sediment in a quarry east of the railway station of Courtemaîche, north of Porrentruy in Canton Jura, in bed no. 1 of the unpublished section RG 343, is so porous and friable that it can be scratched with fingernails. Water depth in the lagoon of the Buix Member gradually diminished concomitant with inter-reef sedimentation. Water currents supplying animal food to the corals on the reefs thereby became increasingly slack. Ongoing sedimentation eventually choked the corals. When the lagoon was entirely filled with fine-grained calcareous sediment, deposition of the peritidal Vorbourg Member (VOR in Fig. 4.19B of this study) began in the *Schilli* Chron.

It can be read from Fig. 4.19B that the vertical facies change at the top of Chestel Member near St-Ursanne is conspicuous. The same vertical change of lithology could be observed in the disused limestone quarry in Chestel ridge south of Liesberg (GYGI 2000a, Pl. 31, section RG 306, boundary between bed nos. 110 and 111) and in the small gorge northeast of the farm La Providence near Corban (GYGI 2000a, Pl. 35, section RG 400, boundary between bed nos. 10 and 11). The vertical facies change is at the transgressive surface which is indicated between sequence boundaries O4 and O5 in GYGI et al. (1998, Fig. 2). The equivalent of this transgressive surface at the cliff of Peute Roche in section RG 451, 1 km west-southwest of the village of Vellerat in Canton Jura (GYGI 2000a, Pl. 24), is the top of the oolitic bed no. 2. Ooid sand of this bed is the outermost tongue of prograding Chestel Member, which smothered the coral biostrome of Grellingen Member at Vellerat. A similar situation in the Recent is imaged in the aerial view of Fig. 4.8. It is probable that marly bed no. 22 in the Pichoux Formation in section RG 315 in Pichoux Gorge near Sornetan, which GYGI (2000a) recorded in Pl. 21, and marly bed nos. 41–43 in the Pichoux Formation in section RG 307 in the quarry of La Charuque near Péry in Pl. 22 of the same study, were sedimented in deeper water on the slope at the same time as when the transgressive surface between Chestel Member and Buix Member evolved during rapid sea level rise no. 5 (Fig. 5.1).

A minor, and at an early stage gradual rise of sea level occurred in the *Bifurcatus* Chron. The gradual part of the rise was sufficient to halt progradation of the *Günsberg Formation*. The early, gradual part of the rise was such that the distal boundary of the lower part of the carbonate platform of the

Günsberg Formation became more or less vertical (Fig. 1.5). The eastern boundary of the outcrop of the landslide of Gschlief above the village of Günsberg in Canton Solothurn is close to the basinward boundary of the carbonate platform of the Günsberg Formation. This is indicated by the rapidly growing thickness of marly intercalations in the formation in the eastern half of the outcrop (Fig. 3 on p. 79 in GYGI (1969a)). No tidal delta of ooid sand was shed from the brink of the Günsberg carbonate platform into deeper water at this time. The vertical, distal boundary of the lower Günsberg Formation is best documented east of the gorge called Chatzensteg south of Ramiswil, which is the western fraction of the township of Mümliswil in Canton Solothurn. A succession of oolitic limestone of the lower Günsberg Formation in this gorge is in tectonically vertical position in the northern limb of a west-east running Jura fold. This pure oolite of the lower part of the formation rises in a prominent crest culminating in Point 827 west of the small creek, which flows northward through the gorge. The same crest rises again east of the creek and then disappears without a transition approximately 100 m east of the creek, probably because of a lateral facies transition into marl of the Effingen Member. The lateral lithologic boundary supposed is covered by vegetation. Tectonic shear is unlikely to be the cause of the abrupt eastward disappearance of the conspicuous limestone crest of the lower Günsberg Formation, because the younger, oolitic Steinibach Member in equally vertical position to the north forms a prominent crest, which continues toward the east without being disturbed by tectonic shear. The clear-cut, eastern boundary of the oolitic limestone crest of the lower Günsberg Formation at this locality is, therefore, most probably an abrupt lateral facies boundary at the basinward rim of a carbonate platform.

The gradual sea level rise which ended progradation of the Günsberg Formation became rapid during its latest stage. This rapid part of the rise created an environment of fully marine, but very shallow water above the perfectly flat-lying, mainly peritidal sediments of the lower half of the Röschenz Member in the interior of the shallow-water realm landward of the carbonate platform of the Günsberg Formation. The sea level rise caused the coastline to recede landward, from which argillaceous mud and some silt-grade siliciclastic minerals were shed into shallow sea water. As a consequence a thin, but very widespread sheet of calcareous ooid sand was formed approximately in the middle of the Röschenz Member (Fig. 1.5). This oolite is above the transgressive surface between O5 and O6 in GYGI et al. (1998, Fig. 2). The pertinent calcareous oolite is shown in the following plates in GYGI (2000a): Pl. 16, section RG 454 near Bure: bed no. 38, Pl. 18, section RG 359 near Bressaucourt: bed no. 73, Pl. 23, section RG 366 near Delémont: bed no. 47, and in Pl. 33, section RG 402 near Röschenz: bed no. 45. Time equivalents of the calcareous oolite in the middle Röschenz Member are the coral bioherms in the most proximal part of the Günsberg

Formation (Fig. 1.5). One of these coral bioherms is unit no. 33 in section RG 406 near Vermes, which is represented in Pl. 37 in Gygi (2000a), and unit no. 19 in section RG 404 near Mervelier, that is shown in Pl. 36 in Gygi (2000a). The rapid, but moderate increase in water depth at this time is documented most distinctly by the thickness of the oncolite of the Grüne Mumienbank Bed with corals in the Günsberg Formation, which is unit no. 163 above a thin seam of supratidal sediment with lignite and characean freshwater algae in section RG 307 near Péry (Gygi 2000a, Pl. 22). Age and time correlation of the pertinent units are stated in Sect. 5.1.

Bivalves prevail in the macrofossil assemblage in the Geissberg Member of Canton Aargau. There are no hermatypic corals in this lowermost member of the Villigen Formation between Schönenwerd and Villigen. Schönenwerd is a village west of the city of Aarau (Fig. 1.5). The Geissberg Member in the area east of Schönenwerd was, therefore, sedimented at a water depth of more than 20 m. Glauconite in the *Crenularis* Member directly above indicates a reduced rate of sedimentation because of the rapid, but small-scale relative sea level rise no. 9 in Fig. 5.1, which occurred late in the *Berrense* Chron. The glauconitic *Crenularis* Member is above the transgressive surface at the top of O7 in Fig. 2 in Gygi et al. (1998). The mass occurrence of bivalves of the genus *Pholadomya* in the *Crenularis* Member, which was found in the unpublished section RG 63 near Villigen in Canton Aargau, is evidence of a water depth approaching 30 m. Section RG 63 was measured on the rope at the cliff at the head of the ravine, which is 270 m northwest of the ruin of Besserstein castle above the village of Villigen. Ammonites are the main element in the macrofossil assemblage in the glauconitic Baden Member above the Villigen Formation at Schönenwerd, Möriken, Villigen, and Mellikon. Water depth must have increased gradually and substantially during sedimentation of mainly the Letzi Member to finally approximately 40 m at least when sedimentation of the Baden Member began. The rate of gradual, relatively rising sea level increased in the *Planula* Chron so much that growth of the coral biostrome in the Olten Member could no longer keep up with the rise. Demise of the coral biostrome occurred when water depth above the apparently slowly accumulating biostrome increased to more than approximately 20 m (Fig. 1.5).

4.15.2 Shallowing-Upward Successions and Walther's "Law"

Shallowing-upward sedimentation in ammonite facies began in northwestern Switzerland in earliest Callovian time at the base of what Stäuble (1959) called "Macrocephalenkalke". It was stated above that water was at the beginning of the Callovian Age at Liesberg deeper than 35 m. The Bathonian/ Callovian boundary beds were excavated by R. and S. Gygi on September 5 in 1985 in the then disused clay pit of Hinter Chestel south of the village of Liesberg. The pertinent section RG 427 was published as Fig. 1 by Dietl and Gygi (1998) along with photographs of the essential ammonites which were found in the excavation. Section RG 427 provided for a clean outcrop of the limestone nodules and beds in marl of the "Macrocephalenkalke" as well as of the lower part of the Callovienton Member above (see Fischer 1965, Fig. 5). Above the clayey marl of the Callovienton Member is the carbonate platform of the Dalle nacrée Member. The marginal part of this carbonate platform near Péry is unit no. 10 of section RG 307 with a thickness of 10 m (Gygi 2000a, Pl. 22). This unit is a prograding, biocalcarenitic tidal delta with foresets. The time equivalent of the Dalle nacrée Member in the clay pit of Andil southwest of Liesberg (1 km northwest of the clay pit of Hinter Chestel, see map in Fischer 1965, Fig. 1), was stated above to be a calcarenite with hummocky cross-stratification from the shallow subtidal zone just off the basinward edge of the carbonate platform of the early Callovian Dalle nacrée Member. The calcarenite in the clay pit of Andil at Liesberg is bed no. 2 of section RG 280 with a thickness of 3.2 m (Gygi 2000a, Pl. 30, see also Gygi 1990a, Fig. 2). The constituent sand grains of the rock are mostly fragments of skeletal elements of echinoderms. No ammonites were found to date at Liesberg in this calcarenite.

The boundary beds between the Middle and the Upper Jurassic Series in the proximal region of northwestern Switzerland were studied mainly in section RG 307 in the quarry of La Charuque near Péry and in section RG 280 in the clay pit of Andil near Liesberg. The Oxfordian beds in section RG 280 near Liesberg begin with the brownish-gray, iron-oolitic marl-clay of bed no. 6 (Gygi 1990a, Fig. 2). Iron ooids are most abundant at the bottom of the bed, but even there they are only about 5% of the rock volume. The slight percentage gradually diminishes upward, and iron ooids fade away and disappear at the top. The top of the slightly iron-oolitic bed no. 6 in section RG 280 at Liesberg, like the upper surface of bed no. 20 in section RG 307 near Péry, is the top of the Herznach Formation. The gradual fading of iron ooid accretion toward the top of the Herznach Formation in northwestern Switzerland was probably the effect of a slight increase in the rate of sedimentation of predominantly argillaceous mud. This corroborates the above conclusion that accretion of iron ooids in the deep subtidal zone at a water depth of several tens of meters was a slow process. Ammonites prevail everywhere in the macrofossil assemblage in the thin, iron-oolitic sediments of earliest Oxfordian age which were laid down in northern Switzerland. Water depth in northwestern Switzerland increased to a maximum of about 80 m at the end of sedimentation of the Herznach Formation in the early *Scarburgense* Chron.

The compacted, tectonically undisturbed thickness of the *Renggeri* Member in northwestern Switzerland can be as much as 60 m. To judge from the thicknesses of ammonite zones within the *Renggeri* Member near Liesberg which are shown in Fig. 3 in GYGI (1990a), the rate of sedimentation increased during deposition of the lower part of the member. The *Scarburgense* Zone is approximately 3 m thick in section RG 280 southwest of Liesberg. Above is the *Praecordatum* Zone with a thickness in that section of the order of 30 m, according to Fig. 3 in GYGI (1990a). Ammonite casts prevail in the macrofossil assemblage in the whole of the *Renggeri* Member in section RG 280 in the clay pit of Andil near Liesberg (GYGI 2000a, Pl. 30) as well as in the lower half of the Sornetan Member which follows above. The entire Sornetan Member could be studied in the neighboring section RG 306 in the clay pit of Hinter Chestel south of the village of Liesberg (GYGI 2000a, Pl. 31) at the time, when the clay pit was worked. Mostly large bivalves of the genus *Pholadomya* were found in the upper half of the Sornetan Member. This is the "étage Pholadomyen" of ÉTALLON (1862). ÉTALLON intended this to be a stage with a temporal meaning. Instead, the term is environmental, and it now stands for a marine soft ground of mixed calcareous and argillaceous mud at a water depth of between 20 and 30 m.

Large specimens of the bivalve genus *Pholadomya* were most abundant where water depth approached 30 m. Bivalves were replaced without a transition by a mass occurrence of platy, hermatypic corals in the lower part of the biostrome of the Liesberg Member, as soon as water depth diminished to less than approximately 20 m because of ongoing, aggradational sedimentation of calcareous and argillaceous mud at a rate which was greater than the rate of basement subsidence and relative sea level rise. This abrupt vertical change in the composition of the macrofossil assemblage is conspicuous in section RG 306 in the clay pit of Hinter Chestel, which is the type section of the Liesberg Member. Aggradational sedimentation of the proximal, thick part of succession no. 1 ended close below sea level at the top of the calcareous oolite with small coral bioherms of the Tiergarten Member in the marginal, uppermost part of the pure carbonate St-Ursanne Formation (Figs. 1.5 and 4.15B). The macrofossil assemblage with mainly ammonites in the lowermost thin, iron-oolitic bed in the proximal, thick part of succession no. 1, and calcareous ooids and hermatypic corals in the uppermost, basinward part of thick succession no. 1 are evidence that the proximal part of succession no. 1 is shallowing upward by aggradation.

Sedimentation of succession no. 1 in the deeper part of the Rhodano-Swabian Basin of Canton Solothurn and Canton Aargau began on a level sea floor with thin, early Oxfordian iron oolites of the uppermost Herznach Formation. Both litho- and biofacies of these iron oolites in Canton Aargau are nearly the same as the facies of the thin iron oolites at the base of succession no. 1 in northwestern Switzerland. Part of the early Oxfordian iron oolites in Canton Aargau is stratigraphically condensed, for instance in section RG 210 in Fig. 8.3. Glauconite pellets occur in this section in the condensed, lowermost bed of the Birmenstorf Member directly above. The bulk of thin succession no. 1 in the starved part of the Rhodano-Swabian Basin is represented by biostromes of siliceous sponges with abundant ammonites in the uncondensed, normal facies of the Birmenstorf Member (BIR in Fig. 1.5). The compacted, measured thickness of the entire succession no. 1 in the starved basin between Günsberg in Canton Solothurn (section RG 14, GYGI 1969a, Pl. 18) and Gansingen in northeastern Canton Aargau (sections RG 210 and RG 230 in GYGI 2001, Fig. 1) is between 5 and 6 m. Apart from very abundant siliceous sponges, ammonites prevail in the macrofossil assemblage in the Birmenstorf Member. The small total amount of sedimentation in succession no. 1 in the basin was outweighed by a combination of insignificant endogenic basement subsidence with a much greater, net sea level rise, plus the concomitant isostatic basement subsidence under the additional load of sediment and water. As a consequence, water depth in Canton Aargau increased from about 80 m at the beginning of the Oxfordian (Fig. 4.15A) to approximately 105 m at the end of sedimentation of succession no. 1 (Fig. 4.15B), as it was calculated in Sect. 4.2. Contrary to the proximal, shallowing-upward and aggradational part of succession no. 1 with a great thickness, the distal part of the same succession no. 1 is thin and deepening upward in the starved basin (Fig. 4.15A, B). The deepening-upward tendency in basinal, thin succession no. 1 became extreme in the most distal part of the succession in Canton Schaffhausen (Fig. 1.5). The entire succession is on average 0.5 m thick, for instance, near Gächlingen in bed nos. 11–15 of section RG 81b, which is shown in Fig. 8.3. Water depth in that region increased during deposition of sediment with a total thickness of 0.5 m from approximately 100 m at the beginning of the Oxfordian to 125 m when sedimentation of the unnamed glauconitic marl above the Mumienkalk Bed in Fig. 4.9 was completed.

The *vertical* succession of macrofossil assemblages with decreasing water depth in the proximal, thick and aggradational part of succession no. 1 in northwestern Switzerland has a counterpart in the *lateral* succession of coeval macrofossil assemblages of late *Antecedens* and *Transversarium* age. This lateral succession is between the Birmenstorf Member on the bottom of the epicontinental basin and the basinward rim of the carbonate platform of the St-Ursanne Formation. The biostromes of siliceous sponges with abundant ammonites in the uncondensed part of the Birmenstorf Member of Canton Aargau and of Canton Solothurn grew at a water depth of somewhat more than 100 m on the level basin floor. About halfway up on the depositional slope of what was initially calcareous mud in the upper Pichoux Formation (PIC in Fig. 1.5) are sporadic, large bivalves of the genus *Pholadomya* and some very large perisphinctid ammonites. Two casts of the pertinent ammonites were figured by GYGI (1995, Figs. 8 and 9). These fossils were formed on the depositional slope close below the upper surface of the wedge, in which the proximal, thick succession no. 1 thins toward the bottom of the epicontinental basin. The conspicuous coral bioherms in Pichoux Gorge near Sornetan in Canton Bern, in the proximal part of the uppermost Pichoux Formation, were photographed by BOLLIGER and BURRI (1970, Pl. 1, Fig. 1). One of these bioherms is schematically shown in Fig. 1.5 of this study. The bioherms in the proximal Pichoux Formation are about coeval with the coral bioherms within calcareous oolite in the adjacent Tiergarten Member.

The *vertical* succession of macrofossil assemblages in proximal and thick, aggradational succession no. 1 in northwestern Switzerland begins with mainly ammonite casts in iron oolite, which was sedimented at a water paleodepth of approximately 80 m. Almost exactly halfway up in the proximal, aggradational and shallowing-upward succession is the upper Sornetan Member with a macrofossil assemblage of mainly large bivalves of the genus *Pholadomya*. Hermatypic corals are the most abundant macrofossils in the bioherms in the Tiergarten Member near the basinward rim of the upper St-Ursanne Formation. The *vertical* succession of macrofossil assemblages in proximal succession no. 1 from relatively deep water with mainly ammonites at the base to assemblages from shallowest water with mostly hermatypic corals in calcareous oolite of the Tiergarten Member in northwestern Switzerland on the one hand, and the *lateral* succession of coeval macrofossil assemblages between the Birmenstorf Member with siliceous sponges and ammonites on the level basin floor, upward over the depositional slope of the Pichoux Formation to the coral bioherms of the Tiergarten Member in the upper St-Ursanne Formation on the other hand, are an instructive example of what Anglo-Saxon authors call WALTHER's law.

5.1 Rapid Relative Sea Level Rises: Timing and Quantification

Rise no. 1. The first rapid relative sea level rise represented in Fig. 5.1 could be documented in section RG 226 east of Auenstein in Canton Aargau (Mangold and Gygi 1997, Fig. 2). The rise drowned the thin, early Bathonian carbonate platform of the Spatkalk Member. Mainly the ammonite *Bullatimorphites (Bullatimorphites) polypleurus* (Buckman) of the *Subcontractus* Zone, which was figured by Mangold and Gygi (1997, Fig. 3/8a, b), is evidence that the rise occurred at the latest early in the middle Bathonian Age. The macrofossil assemblage of mainly bivalves with large ostreids and brachiopods, which is associated with this ammonite, is according to Fig. 4.9 indicative of a water depth of between 20 and 30 m, because there are no hermatypic corals and because ammonites are uncommon. Accordingly, the rise is estimated to have been about 25 m. This major rise and the following rise no. 2 cannot be correlated with the curve of short-term eustatic sea level fluctuations in Hardenbol et al. (1998, chart 1). Reference to that chart is, therefore, omitted in the text about the following sea level rises.

Rise no. 2. The second rapid relative sea level rise could be documented in excavation RG 427, which was made on September 5, 1985, in the disused clay pit of Hinter Chestel south of the village of Liesberg in Canton Basel-Landschaft. The section of the excavation is represented in Dietl and Gygi (1998, Fig. 1). The rise cut short carbonate sedimentation of the Bathonian Calcaire roux sableux Member with mainly brachiopods and with large ostreid bivalves at the top of bed no. 1 in section RG 427. The rise is dated indirectly by ammonites of the taxon *Kepplerites keppleri* (Oppel) of earliest Callovian age which were found in the following bed no. 2 in the excavation. According to these ammonites, rise no. 2 occurred directly before or at the beginning of the Callovian Age. Two specimens of *Kepplerites keppleri* from bed no. 2 of excavation RG 427 were figured by Dietl and

Gygi (1998, Pl. 2a, b). As stated above in Sect. 4.15.1, ammonites prevail in the macrofossil assemblage in bed no. 2 of excavation RG 427. To conclude from Fig. 4.9, bed no. 2 was sedimented at a water paleodepth of at least 35 m. The depth of deposition of the Calcaire roux sableux Member underneath may have been near Liesberg as much as 20 m, because there are no hermatypic corals at this locality in the member. A minimum of 15 m can, therefore, be assumed for rapid rise no. 2 (Fig. 5.1).

Rise no. 3. Rapid relative sea level rise no. 3 drowned the carbonate platform of the Dalle nacrée Member. The marginal part of this platform is unit no. 10 in section RG 307 near Péry in Canton Bern (Gygi 2000a, Pl. 22). Above the platform is the iron oolitic Herznach Formation. The lowermost bed of this formation is in northwestern Switzerland the Lumachellenbank Bed (Sect. 4.15.1 above). This is a regional marker bed which can be followed from Péry (bed no. 11 of section RG 307 in Gygi 2000a, Pl. 22) to Liesberg in Canton Basel-Landschaft (bed no. 3 of section RG 280, Gygi 1990a, Fig. 2). The time equivalent of the Lumachellenbank Bed at Herznach in Canton Aargau is bed no. 12 in section RG 260 in the iron mine (Fig. 4.10 in this study). Jeannet (1951, Fig. 2) labeled this marker bed A5 in his section of the Herznach iron mine and figured *Sigaloceras (Catasigaloceras) enodatum* (Nikitin) from the bed in his Pl. 25, Figs. 8–10. B. Hostettler recently found a well-preserved ammonite of this taxon at the top of bed no. 3 of section RG 280 in the clay pit of Andil southwest of Liesberg. Bed no. 3 in this section is the Lumachellenbank Bed above a calcarenite with hummocky cross-stratification. The calcarenite is unit no. 2 in Gygi (1990a, Fig. 2), which was sedimented off the basinward rim of the carbonate platform of the Dalle nacrée Member in the shallow subtidal zone (see above). The few hermatypic corals which B. Hostettler found at the top of bed no. 3 in section RG 280 indicate that water depth had increased during sedimentation of the bed at Liesberg to 20 m at most. Rapid relative sea level rise no. 3, which proceeded during the *Enodatum*

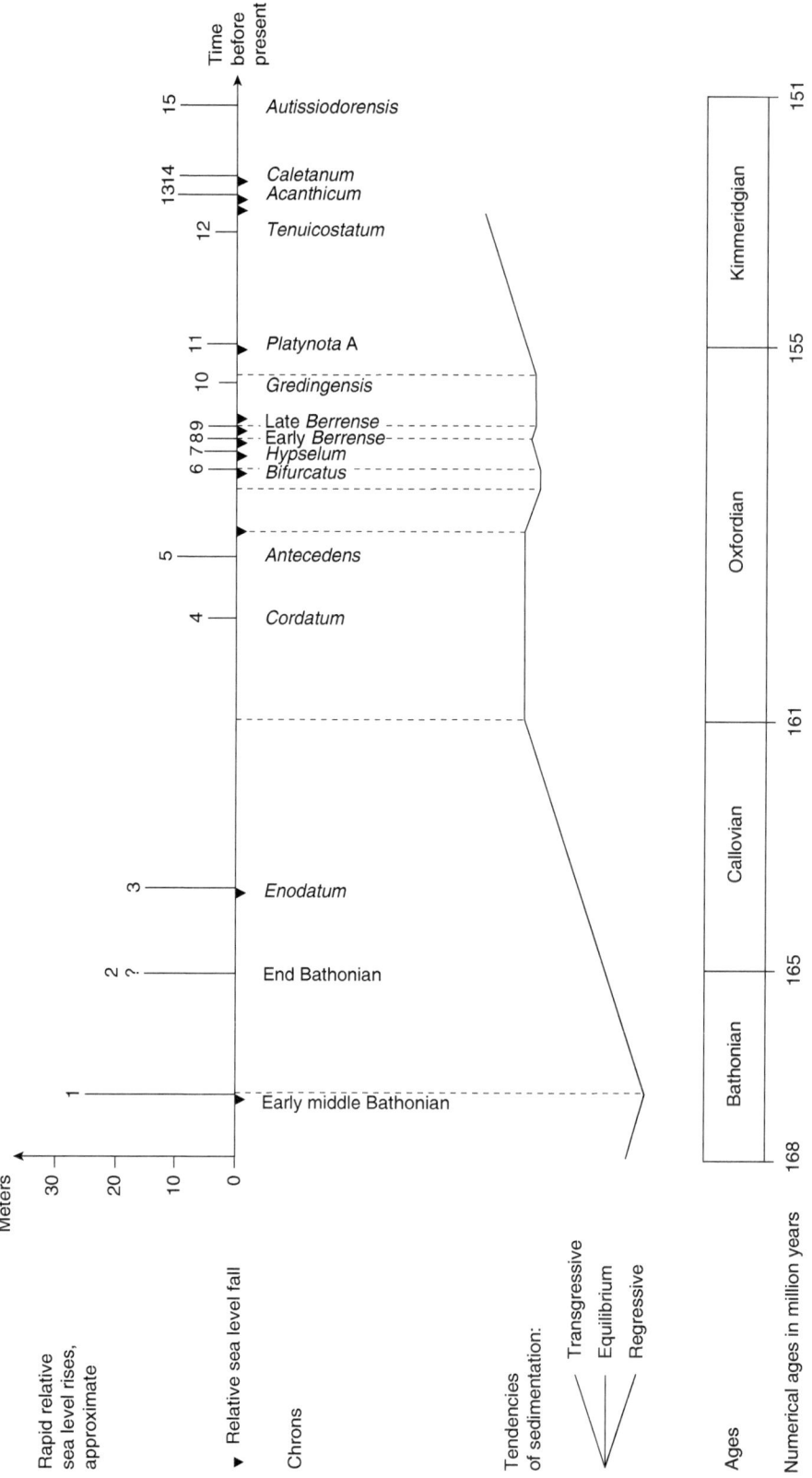

Fig. 5.1 Rapid relative sea level rises which occurred between early in the middle Bathonian and the latest part of the Kimmeridgian Age as well as transgressive and regressive tendencies of sedimentation. Rapid relative sea level rises in the Bathonian: (1) beginning of middle Bathonian, (2) latest part of Bathonian. Callovian: (3) *Enodatum* Chron. Oxfordian: (4) *Cordatum* Chron. (5) late in the *Antecedens* Chron. (6) *Bifurcatus* Chron. (7) *Hypselum* Chron. (8) early *Berrense* Chron. (9) late *Berrense* Chron. (10) late *Gredingensis* Chron. Kimmeridgian: (11) *Platynota* A Chron. (12) *Tenuicostatum* Chron. (13) *Acanthicum* Chron. (14) *Caletanum* Chron. (15) *Autissiodorensis* Chron

Chron until the end of sedimentation of bed no. 3 in section RG 280, can therefore be rated to have been not more than about 15 m. Ammonites prevail in the macrofossil assemblage in the entire Herznach Formation near Péry above the basal Lumachellenbank Bed no. 11 in section RG 307. Sea level must, therefore, have continued to rise in the area investigated from the beginning of the middle Callovian on. According to the calculation made above, a total relative sea level rise of approximately 80 m occurred in northwestern Switzerland during sedimentation of the thin, deepening-upward Herznach Formation, until deposition of the shallowing-upward Bärschwil Formation began in earliest Oxfordian time. No single, rapid relative sea level rises could be discerned in the long time span between the end of the early Callovian and the latest part of the early Oxfordian (Fig. 5.1).

Rise no. 4. The first rapid relative sea level rise which could be recognized to have occurred in the Oxfordian Age happened late in the *Cordatum* Chron (Fig. 5.1). Rise no. 4 is well documented in the middle of the Sornetan Member by a regional marker bed, which is either limestone or a layer of marl with calcareous concretions. The thickness of the marker bed is between 10 and 20 cm. The best sections in which the bed was recorded are compiled in Fig. 2 in GYGI and MARCHAND (1993). The distal part of the bed includes a rich macrofossil assemblage of mainly ammonites. GYGI and PERSOZ (1986) called this marker bed "Fossil Bed" in their Table 2 on p. 396–397. The marker bed in the distal part of the Sornetan Member crops out in several landslides between Fournet-Blancheroche in France (a village 7 km north of La Chaux-de-Fonds in western Switzerland) and Bärschwil in Canton Solothurn. The horizontal, non-palinspastic distance between Fournet-Blancheroche and Bärschwil is 53 km. The regional stratigraphic context of the fossil bed is represented in the following plates in GYGI (2000a): Pl. 21 of sections RG 314 and 315 near Sornetan in Canton Bern; Pl. 24 of sections RG 373, 389, and 451 near Vellerat in Canton Jura; and Pl. 34 of section RG 399 near Bärschwil in Canton Solothurn. Depth of sedimentation of the fossil bed was calculated semiquantitatively in Sect. 4.2 and is discussed in Sect. 11.4.3.

Ammonites are more than 80% of the macrofossils in the distal part of the bed (Fig. 4.9). There is an error about this in GYGI and MARCHAND (1993, p. 1003 and Fig. 4). Many ammonites from the bed were figured in Pl. 1–3 of that paper. The age of all of the ammonites figured is the *Cordatum* Chron. This is documented mainly by the *Cardioceras* (*Cardioceras*) *persecans* (BUCKMAN) from bed no. 46 in section RG 399 near Bärschwil, which was identified by D. MARCHAND and figured by GYGI and MARCHAND (1993, Pl. 2, Fig. 2). The high concentration of fossils in the marker bed is evidence of a particularly low rate of sedimentation, but not of stratigraphic condensation in the strict sense (see below).

According to GYGI and MARCHAND (1993, p. 1006), the bed was sedimented above a transgressive surface following rapid sea level rise no. 4 late in *Cordatum* time (Fig. 5.1). There are large bivalves of the genus *Pholadomya* in the upper Sornetan Member above the marker bed. Rise no. 4 was therefore probably of a small amount. The eustatic component of the rise was less than that, and rise no. 4 was certainly less than the eustatic rise that TALBOT (1973, p. 313) concluded from coeval sediments in southern England. Nevertheless, the minor, rapid relative rise no. 4 recorded in northern Switzerland must have been caused by a eustatic rise, because the effect of the rise was observed in the Jura Mountains in several sections more than 50 km apart, and because a coeval sea level rise was documented to have occurred in southern England.

Rise no. 5. This rapid relative sea level rise drowned the sand bank of calcareous ooids at the top of what is now the Chestel Member in the interior of the lower part of the carbonate platform of the St-Ursanne Formation. Only a rapid rise could cut short the process of ooid accretion on the internal part of the sand bank of Chestel Member and create the lagoon above the pertinent transgression surface, into which the Buix Member with coral reefs was sedimented (Fig. 4.19B). The rapid rise amounted to more than 6 m, or else it would not have drowned the oolitic sand bank in the internal part of Chestel Member, which probably resembled the Recent sand bank that is shown in Fig. 4.7. Rise no. 5 was eustatically driven, because it drowned the oolitic sand bank of Chestel Member independently of endogenic, synsedimentary tectonics, which occurred at distant localities. It is stated in Sect. 9.1 below, that a vertical displacement of as much as 60 m occurred during sedimentation of the St-Ursanne Formation between two sections which are palinspastically 5 km apart. The transgressive surface caused by rise no. 5 is in the quarry above the railway station of St-Ursanne shown in Fig. 4.19A at the top of bed no. 17 in the unpublished section RG 336. The same surface is clear-cut in the outcrop B approximately 550 m west of the quarry, above the cliff of Chestel Member shown in Fig. 4.19B. GYGI (2000a) observed the same transgressive surface at Delémont to be at the top of bed no. 9 of section RG 366 in Pl. 23, and at Liesberg at the top of bed no. 110 of section RG 306 in Pl. 31. The present, non-palinspastic distance between St-Ursanne in the west and Liesberg in the east is 20 km.

The amount of rise no. 5 can be calculated at St-Ursanne based on Buix Member in the following way. Figure 4.19A, B is a general view of the St-Ursanne Formation at St-Ursanne. The Buix Member with two coral reefs that grew in a lagoon is shown in Fig. 4.19B to be between the top of the calcareous oolite of Chestel Member below and the base of the peritidal Vorbourg Member above. Sedimentation of the Buix Member consequently began and ended at approximately sea level. The thickness of Buix Member was measured by the author to be 29.5 m in the unpublished section RG 336 in the quarry above the railway station of St-Ursanne. GYGI (1986, p. 471) calculated that the initial water depth of a basin can accommodate compacted sediment with a thickness that is equivalent to three times the original depth of the basin, until the basin is completely filled to sea level. Provided that the property of the pure carbonate, fine-grained lagoonal sediment between the early cemented, non-compactable coral reefs of the Buix Member was similar to the property of the equally pure carbonate, probably mostly biodetrital and non-compactable sediment between patch reefs in the lagoon of the Bermuda Atoll, from which the author of this study has taken several large samples, then the entire Buix Member was not compacted to a significant degree.

It can be concluded from the fact that sea level rise no. 5 was rapid that maximal depth of the lagoon of the Buix Member evolved at an early stage. The lagoon accommodated at St-Ursanne sediment with the thickness of 29.5 m. It is shown in Sect. 11.3 that basement reacts with isostatic equilibration by subsidence promptly and to even subtle loads. Endogenic basement subsidence was presumed in Sect. 4.10 to have been insignificant during the entire time of deposition of succession no. 1. Endogenic subsidence is therefore disregarded in the following calculation, because the Buix Member was sedimented during little more than one ammonite chron (mainly in the *Transversarium* Chron), or in a fraction of the 8 chrons in which succession no. 1 was laid down (Fig. 1.6). It can, therefore, be assumed that the initial, maximal depth of the lagoon was at St-Ursanne one-third of the thickness of the Buix Member of close to 30 m, or approximately 10 m. The maximal depth of the lagoon, which was created by sea level rise no. 5, is concluded to have been almost equivalent to the total amount of relative sea level rise no. 5, because the rise was rapid. The rise is, therefore, rated in Fig. 5.1 to have been 10 m. Rise no. 5 can be dated to have occurred late in the *Antecedens* Chron based on the ammonites *Perisphinctes (Dichotomosphinctes) dobrogensis* SIMIONESCU and *Perisphinctes (Perisphinctes) alatus* ÉNAY, which V. PÜMPIN found in the inter-reef sediment of the Buix Member at St-Ursanne. These ammonites and additional specimens from the lagoonal sediment were figured by GYGI (1995).

Rise no. 6. The first part of this sea level rise in *Bifurcatus* time was gradual. It was stated earlier in Sect. 4.15.1, that the rise thereby halted progradation of the carbonate platform of the Günsberg Formation (Fig. 1.5). The successive, minor sea level fall is documented, for instance, in section RG 307 at Péry by a wavy erosion surface at the top of the cross-bedded calcareous oolite no. 162. The depressions in the eroded surface are up to 10 cm deep. They are filled with marl including lenses of lignite coal, branches of trees, and gyrogonites of characean freshwater algae. Rise no. 6 then became rapid and thereby displaced the coastline of a very shallow sea landward.

Supply of terrigenous sediment to the peritidal environment of the Röschenz Member was thereby interrupted for a short time. A thin sheet of calcareous oolite was laid down following the rise over peritidal sediments in the middle Röschenz Member. The lagoonal oncolite of the Grüne Mumienbank Bed was sedimented at the same time in the Günsberg Formation over an intertidal stromatolite in Pichoux Gorge near Sornetan (GYGI 2000a, Pl. 21, unit no. 56 in section RG 315), and over the supratidal intercalation mentioned above at Péry (GYGI 2000a, Pl. 22, below base of bed no. 163 in section RG 307). The photograph of a polished slab cut across the oncolite of the Grüne Mumienbank Bed at Péry is Fig. 14 in GYGI (1992). Production of lime mud did not diminish in the shallow water realm of the Röschenz Member at the time, when almost no terrigenous sediment reached the area immediately after the rapid part of sea level rise no. 6. Much or most of this calcareous mud was transported across the carbonate platform of the Günsberg Formation and was deposited in deeper water of the epicontinental basin as the succession of micritic limestone of the Gerstenhübel Beds. Time correlation of all of the limestone units mentioned was made in GYGI and PERSOZ (1986) with the mineral stratigraphic correlations E and F by PERSOZ in Fig. 10 and in Pl. 1A. These correlations could be calibrated in the basinal Effingen Member with the ammonite *Perisphinctes (Dichotomoceras) bifurcatus* (QUENSTEDT) shown in Fig. 7.2C (see below). Correlations with the shallow water realm could be checked with specimens of the pertinent taxon no. 19 in Fig. 1.5 that were found in the Günsberg Formation. *Perisphinctes (Dichotomoceras) bifurcatus* (QUENSTEDT) from the Günsberg Formation were figured by GYGI (1995, Fig. 17/2) and HOSTETTLER (2006, Fig. 4).

The time equivalent of the supratidal erosion surface below the Grüne Mumienbank Bed at Péry is the equally wavy erosion surface at Moutier in section RG 381 in Court Gorge at the top of bed no. 14 which is a pelletoidal packstone. The probable time equivalents of the Grüne Mumienbank Bed with some hermatypic corals at Péry are in section RG 381 in Court Gorge bed nos. 18 and 19. Bed no. 18 in section RG 381 includes druses with a diameter of up to 10 cm lined with calcite crystals. The druses are probably the remains of massive coral colonies. The corals appeared in the environment after rise no. 6 had become rapid. The rise created the water depth and adequate water circulation required by hermatypic corals. Bed no. 19 above in Court Gorge is recorded in the author's field book no. 10, p. 11, to be a cross-bedded, calcareous-oolitic grainstone with overcrowded ooids at the base. The cross-bedding of the oolite was not indicated in Pl. 28 in GYGI (2000a) because of inadvertence when reading the field book.

Rise no. 6 was eustatically driven, because coeval sediments from very shallow water, which resulted from the rise, can be followed over a considerable distance in the Röschenz Member and the Günsberg Formation.

The minimal amount of the rapid, later part of rise no. 6 can be concluded from the marine accommodation space created by the rise above supratidal sediments. The thickness of the lagoonal oncolite of the Grüne Mumienbank Bed at Péry is 4.1 m. The rapid part of rise no. 6 is, therefore, rated to be 5 m in Fig. 5.1. This was sufficient to allow for hermatypic corals to reappear in the Grüne Mumienbank Bed at Péry, in bed no.18 of section RG 381 at Moutier in Gorges de Court, and in Pichoux Gorge near Sornetan in the upper part of bed no. 56 of section RG 315 (GYGI 2000a, Pl. 21). It can be read from Fig. 1.5 that the *entire* rise no. 6 occurred during the *Bifurcatus* Chron. This is documented by the specimen of *Perisphinctes (Dichotomoceras) bifurcatus* (QUENSTEDT), which B. HOSTETTLER found 2 m above the base of a coral bioherm in unit no. 160 in section RG 307 at Péry. The ammonite is figured in HOSTETTLER (2006, Fig. 4). The specimen of the conspecific taxon which R. ÉNAY found probably in the lowermost part of the Günsberg Formation near Günsberg in Canton Solothurn lived directly before progradation of the Günsberg Formation was halted at the beginning of the gradual part of rise no. 6. A specimen of *P. (D.) bifurcatus* kept in the ROLLIER collection in ETH Zürich is refigured in Fig. 7.2 C of this study. The specimen was found near Bözen (no. 19 in Fig. 1.5), most probably in the Gerstenhübel Beds. The pertinent ammonite lived *after* the rapid, second part of rise no. 6 occurred, when the Gerstenhübel Beds, the Grüne Mumienbank, and the thin intercalation of calcareous oolite in the middle of the Röschenz Member were sedimented.

Rise no. 7. Relative sea level rise no. 7 began gradually after a fall in *Hypselum* time. The sea level fall before the rise is documented by bed no. 177 in section RG 307 near Péry (GYGI 2000a, Pl. 22). Marly limestone bed no. 177 with a thickness of 30 cm is an intertidal stromatolite with parallel lamination. The marly limestone of bed no. 178 above is 1.3 m thick and is flaser bedded. The bed with flasers includes lens-like mud pebbles which are as much as 5 cm wide, as well as coaly twigs with a length of up to 10 cm and a diameter of 5 mm. The gradual sea level rise above is documented by a succession of limestone beds alternating with marl. Foraminifers of the genus *Alveosepta* found in four thin sections of the limestones are common (Pl. 22 in GYGI 2000a). The probable time equivalent of the intertidal stromatolite of bed no. 177 in section RG 307 at Péry is in section RG 315 in Pichoux Gorge near Sornetan the intertidal, crinkled stromatolite of bed no. 65 (GYGI 2000a, Pl. 21). A very large sample of this stromatolite about half a meter across is kept in the Basel Museum of Natural History. The photograph of a thin section through the stromatolite at Sornetan and the photograph of the crinkled surface of it are Figs. 11–12 in GYGI (1992). The upper surface of a coeval stromatolite, bed no. 42 in the unpublished section RG 417 north of Crémines in Canton Bern, is prism-cracked by

dewatering (Fig. 4.4). The photograph of an entire prism with margins upwarped by dewatering from a coeval stromatolite is Fig. 4.5. The pertinent stromatolite is bed no. 21 in the unpublished section RG 414 on the south slope of Mt. Raimeux north of Grandval in Canton Bern. A coeval, intertidal stromatolite near Vermes in Canton Jura was broken up in the process of dewatering into angular fragments. The fragments were then blackened to a varying degree. The eroded upper surface of this bed is planed, bored, and encrusted with ostreids. The photograph of a polished cross-section of the bed with lithoclasts is Fig. 5.4 in this study. The bed is no. 49 of section RG 406 in Pl. 37 in GYGI (2000a). The probable time equivalent of this bed near Vermes is in Court Gorge near Moutier bed no. 28 of section RG 381 in Pl. 28 in GYGI (2000a). Part of the eroded, planed, and bored upper surface of the bed with a crust of goethite and adhering ostreids is easily accessible on the south side of the road leading through Gorges de Court. This erosion surface is indicated to be a transgressive surface above sequence boundary O6 in Fig. 7 in GYGI et al. (1998). ZIEGLER (1956, p. 92, bed no. 10 of his section no. 49) drew attention to this surface for the first time.

Rise no. 7 turned to be rapid when sedimentation of bed no. 192 in section RG 307 at Péry began above a transgressive surface. Bed no. 192 includes an abundant and diverse macrofossil assemblage of rhynchonellid and terebratulid brachiopods, bivalves, parts of echinids, and fragments of coral colonies. No ammonite was found in the bed. Above is the tidal delta of bed nos. 193–195 with foresets of calcareous oolite and with a total thickness of 5.7 m (Fig. 5.2). The foresets of what was primarily calcareous ooid sand have a depositional inclination of up to 20°. Notwithstanding the strong water currents shedding ooid sand into the delta, the interstices between the tightly packed ooids were partially filled with calcareous silt and mud. The fines in the interstices settled geopetally. This is documented in the thin section photograph shown in Fig. 5.3. The thickness of the tidal delta at Péry is evidence that the second, rapid part of rise no. 7 created an accommodation space of several meters. This part of the rise must indeed have been rapid and eustatically driven, or else it could not have provided for the accommodation space in a vast area, into which steeply dipping foresets of calcareous sand were shed at many localities upon flat-lying sediments from very shallow water and partly even from a peritidal environment. In case that all of the rise no. 7 was slow and gradual, the accommodation space created by the rise would have been filled entirely with flat-lying sediments.

The top of uppermost bed no. 195 of the tidal delta at Péry is a hummocky erosion surface. The surface is covered with a crust of goethite. GYGI (2000a, Pl. 22) erroneously assigned the tidal delta of bed nos. 193–195 in section RG 307 at Péry to the Steinibach Member. Calcarenites with inclined

Fig. 5.2 Tidal delta of calcareous ooid sand in beds no. 193–195 with the aggregate thickness of 5.7 m in section RG 307 in the uppermost Günsberg Formation in the quarry of La Charuque south of Péry in Canton Bern. The delta is seen in cross-section at the northern wall of the quarry. The delta shifted laterally with time. The accommodation space for the delta was provided by rapid relative sea level rise no. 7 in the *Hypselum* Chron (in Fig. 5.1). The massive beds no. 198 and 199 above the delta are together 4.9 m thick and belong to the lagoonal

oncolite of the Hauptmumienbank Member. The lagoon, into which this member was sedimented, was created by rapid relative sea level rise no. 8 early in the *Berrense* Chron. The entire section RG 307 is shown in Pl. 22 in GYGI (2000a). Photograph by S. Dahint on the quarry floor, looking north. Refigured from GYGI (2003, Fig. 181 on p. 166), corrected, with permission of Schweizerische Paläontologische Abhandlungen

Fig. 5.3 Calcareous oolite with geopetal infillings of calcareous silty mud in interstices between ooids. Photograph of thin section from bed no. 193 in the tidal delta shown in Fig. 5.2, in section RG 307 at Péry in Canton Bern. Refigured from GYGI (2000a, Fig. 13 on p. 23)

bedding in a position probably coeval with the tidal delta near Péry occur in other sections in the shallow-water realm. Examples are the conspicuous bed no. 38 in section RG 381 in Court Gorge near Moutier (GYGI 2000a, Pl. 28), bed no. 84 in section RG 400 near Corban (GYGI 2000a, Pl. 35), bed no. 31b in section RG 404 near Mervelier (GYGI 2000a, Pl. 36), and the lowermost unit with a large, pelsparitic sand wave in section RG 441 near Liesberg, which is shown in this study in Fig. 5.5A, B. Section RG 441 is the most instructive of those which can be seen by the public. It is 1 km east of the village of Liesberg beside the cantonal highway, 60 m west-southwest of Point 377, where a small road branches off to the farm Greifel. The outcrop is along a parking strip and is, therefore, easily accessible.

The perisphinctid ammonite figured in GYGI (1995, Fig. 18) was found by A. and H. ZBINDEN in the quarry of La Charuque near Péry in the talus below bed no. 23 of section RG 435. This bed is the equivalent of bed no. 187

in section RG 307 below the tidal delta (GYGI 2000a, Pl. 22). The ammonite could only be identified approximately, but its age is most probably the *Hypselum* Chron. An adult and near-complete *Euaspidoceras hypselum* (OPPEL), which is the index taxon of the *Hypselum* Chron, was picked by D. KRÜGER from talus below the succession of limestone bed nos. 55–80 in the upper Effingen Member of section RG 37, at the foot of the western wall of the quarry of Jakobsberg east of Auenstein in Canton Aargau. This specimen was figured natural size by GYGI (2000a, Pl. 10, Fig. 1). It is refigured at the scale ×0.5 in Fig. 7.2D in this study. The position from which the ammonite cast fell is indicated in Fig. 1.5 with the encircled no. 23. This corresponds to no. 23 of the list in Fig. 1.6 of this study. The succession of micritic limestone of bed nos. 55–80 in section RG 37 is stratigraphically 60 m above the top of the Gerstenhübel Beds and about 70 m below the base of the Geissberg Member which follows above the top of the Effingen Member (this study, Fig. 1.5). The Geissberg Member is the time equivalent from deeper water of the Hauptmumienbank Member in the shallow-water realm, according to the mineral stratigraphic correlations H and I by PERSOZ in Pl. 1A in GYGI and PERSOZ (1986). The succession of bed nos. 55–80 in section RG 37 in the epicontinental basin with ammonite no. 23 in Fig. 1.5 of this study is, therefore, probably coeval with the calcarenites with inclined bedding below the Hauptmumienbank Member in the shallow-water realm. The succession of bed nos. 55–80 in section RG 37 was consequently sedimented after a small transgression like the Gerstenhübel Beds (see above). The rapid part of relative sea level rise no. 7 occurred in the *Hypselum* Chron. It amounted to an estimated 5 m, to judge mainly from the thickness of the tidal delta near Péry (Fig. 5.2).

Rise no. 8. Bed no. 196 of section RG 307 near Péry is marl with a thickness of 0.9 m (GYGI 2000a, Pl. 22). The bed was sedimented above the hummocky erosion surface with a crust of goethite at the top of bed no. 195, which is the uppermost bed of the tidal delta below. The base of bed no. 196 is a transgressive surface, because macrofossils are concentrated above the surface in the lower part of the bed. Most abundant in bed no. 196 is the small, smooth brachiopod *Zeillerina astartina* (ROLLIER). This is the *Zeilleria humeralis* of earlier authors. The rich macrofossil assemblage in bed no.196 also includes parts of echinids, belemnites, and an ill-preserved ammonite which could not be identified. The following succession of bed nos. 198–211 in section RG 307 is the Hauptmumienbank Member, which was not recognized as such by GYGI (2000a, Pl. 22). The Hauptmumienbank Member in section RG 307 is shown in Fig. 5.2 in this study. There are oncoids in a micritic matrix in bed nos. 199–202. *Cladocoropsis mirabilis* FELIX and pseudomorphs of calcite after sulfate minerals were found in micrite in a thin section of bed no. 203.

A time equivalent of bed no. 196 with abundant macrofossils in section RG 307 near Péry is bed no. 59 in the unpublished section RG 312 at the western end of the hill called Montaigu, which is 1 km southeast of the village of Souboz in Canton Bern. The marl of this bed no. 59 is 3.5 m thick and includes a rich assemblage of hermatypic corals. Associated with the corals are abundant *Zeillerina astartina* (ROLLIER), rhynchonellid brachiopods, bivalves, and echinids. A small coral bioherm in the bed is evidence that rise no. 8 was rapid and of the order of several meters. Time equivalent with the coral bioherm in bed no. 59 in section RG 312 near Souboz is the densely populated coral biostrome of bed nos. 39–40 in section RG 381 in the northern part of Court Gorge in the territory of Moutier (GYGI 2000a, Pl. 28). The coral biostrome in Court Gorge includes ostreid bivalves and abundant bivalves of the genus *Trichites*, gastropods, rhynchonellid brachiopods, serpulids, and parts of echinids. The coral biostrome is above an uneven erosion surface at the top of conspicuous bed no. 38 with foresets of calcareous oolite. The eroded top of bed no. 38 was covered with a crust of goethite, which was then settled by ostreid bivalves. Above the marl with plate-like and finely branching coral colonies of bed no. 41 in section RG 381 in Court Gorge is the Hauptmumienbank Member.

Rise no. 8 created sufficient accommodation space that hermatypic corals could advance far into the shallow-water realm landward of the carbonate platform of the Günsberg Formation. Hermatypic corals occur below the Hauptmumienbank Member north of Bressaucourt near Porrentruy (see Fig. 1.5, single cross below the Hauptmumienbank Member to the right of the vertical bar with the word Séquanien). The corals near Bressaucourt were mentioned for the first time by SCHNEIDER (1960, p. 7). The time equivalent of the patches with coral biostromes or with some small bioherms below the Hauptmumienbank Member is in the distal direction a bed, or several beds with glauconite below the Steinibach Member. The best outcrop of two successive glauconitic marker beds below the Steinibach Member is in Steinibach Gorge north of Balsthal (GYGI 2000a, Pl. 44, bed nos. 1 and 7 of section RG 438). The Steinibach Member west of Wangen near Olten was laterally replaced in water approximately 6 m deep by the lowermost part of the coral biostrome of the Olten Member (this study, Fig. 1.5). The lowermost part of the Olten Member graded laterally into the Geissberg Member, where water depth became greater than approximately 20 m. This time equivalence was not recognized by GYGI (1969a, Pl. 19, upper synthetic section). Appropriate time correlation became possible only when mineral stratigraphic correlation H by PERSOZ in GYGI and PERSOZ (1986, Pl. 1A) could be used, because there are no ammonites in the Steinibach Member.

The time equivalent of the glauconitic marker beds below the Steinibach Member is below the proximal part of the Geissberg Member a marly limestone with glauconite and pyrite. This is in section RG 21 in the quarry on Mt. Born, 2 km southwest of the city of Olten, the marly limestone of bed no. 11 with much glauconite, pyrite, and many bivalves (GYGI 1969a, Pl. 18). Glauconite and large specimens of the bivalve *Gryphaea* are abundant in the coeval bed nos. 3–7 in the unpublished section RG 25, which was measured in the old quarry at locality Heidenloch north of the village of Oftringen. The fossiliferous marker bed below the Geissberg Member could be followed from Mt. Born near Olten through the city of Aarau (well of section RG 30) as far as to the former quarry southeast of the village of Küttigen in Canton Aargau, in which the unpublished section RG 31 was measured in 1963. Bed no. 20 in section RG 31 is the marker bed. The quarry was the easternmost locality where the glauconitic marker bed below the Geissberg Member was found (see upper synthetic section on Pl. 19 in GYGI 1969a). The quarry near Küttigen is now entirely filled up with landfill.

The Geissberg Member grades northeast of section RG 69 at Baldingen in northeastern Canton Aargau into the lower Hornbuck Member. Bed nos. 10–25 in section RG 70 in the large quarry at Mellikon are part of a large sponge bioherm. They were assigned by GYGI (1969a, Pl. 17) to the *Crenularis* Member because of the slight content of glauconite in part of the bioherm. The bioherm at Mellikon is better assigned to the Hornbuck Member, which can be followed through Klettgau valley to the Randen Mountains in Canton Schaffhausen. Sponge bioherms are abundant in the Hornbuck Member. A transgressive surface time equivalent with rapid rise no. 8 could not be discerned at the base of the Hornbuck Member. A small specimen of the ammonite *Epipeltoceras berrense* (FAVRE) was found in the talus below the lowermost part of the Hornbuck Member at the foot of section RG 279 near Siblingen in Canton Schaffhausen. The section is shown in Fig. 3 in GYGI (1991b). It is because of this ammonite that the assumption can be made that rapid relative sea level rise no. 8 occurred early in the *Berrense* Chron.

Rise no. 9. This sea level rise flooded a paleosol above the Steinibach Member, bed no. 40B in the unpublished section RG 4 near Waldenburg in Canton Basel-Landschaft (Fig. 4.3), and another paleosol above the Steinibach Member, bed no. 3 in the unpublished section RG 15 at the head of Horngraben Gorge south of Aedermannsdorf in Canton Solothurn. The rise must have been of the order of at least several meters, because hermatypic corals were found in bed no. 189 directly above the calcareous oolite of the Steinibach Member with cross-bedding in section RG 14 above the landslide of Gschlief north of Günsberg in Canton Solothurn (GYGI 1969a, Pl. 18). The rise cannot have been more than a

few meters, because calcareous oolite follows in section RG 14 directly above bed no. 189 with corals, and above the paleosol in section RG 15, which is easily accessible beside a path at the head of Horngraben Gorge. Sea level rise no. 9 flooded a planed erosion surface on the upper bedding plane of bed no. 50 in the unpublished section RG 372 above the village of Roches, which is a township north of Moutier in Canton Bern. Unit no. 50 in section RG 372 is the uppermost bed of the Hauptmumienbank Member. Figure 5.6 is the photograph of a polished cross-section through the uppermost part of bed no. 50 with the planed top. The Hauptmumienbank Member in section RG 372 at Roches is intersected by the road leading from the fraction of Hautes Roches upward to the farm Le Trondai.

Rapid relative sea level rise no. 9 temporarily reduced the rate of sedimentation in deeper water in the epicontinental basin. This is documented by relatively abundant, but fine-grained glauconite in the *Crenularis* Member directly above the Geissberg Member in the Villigen Formation in Canton Aargau (GYGI 1969a, Pl. 17, section RG 62, bed nos. 32–37). The entire Geissberg Member with mainly bivalves, but without hermatypic corals, was sedimented at a water paleodepth of more than 20 m, to conclude from Fig. 4.9 in this study. There is a mass occurrence of bivalves of the genus *Pholadomya* in the *Crenularis* Member in section RG 62 and mainly in RG 63 near Villigen in Canton Aargau. Plentiful occurrence of large *Pholadomya* was concluded above to be indicative of a water paleodepth approaching 30 m. A depth of close to 30 m is not much more than the probable depth of sedimentation of more than 20 m of the uppermost part of the Geissberg Member in Canton Aargau. The small amount of rise no. 9 which was concluded from coeval sediments of the Balsthal Formation in the shallow-water realm is thereby corroborated in an environment of somewhat deeper water in Canton Aargau. Rise no. 9 occurred late in the *Berrense* Chron. This is documented by the *Epipeltoceras* transitional between the taxa *berrense* FAVRE and *bimammatum* QUENSTEDT (see below), which C. MOESCH found in the glauconitic *Crenularis* Member in what is now the unpublished section RG 36 west of locality Fahr east of Auenstein in Canton Aargau. The ammonite found by MOESCH was figured by GYGI (2000a, Pl. 10, Fig. 5). The only *Epipeltoceras berrense* (FAVRE) that the author could find in northern Switzerland is the small specimen mentioned above, which was picked from talus below the lower Hornbuck Member in section RG 279 near Siblingen in Canton Schaffhausen. The lower Hornbuck Member is the time equivalent of the Geissberg Member in Canton Aargau.

Rise no. 10. A rapid, eustatically driven relative sea level rise of very small scale is indicated to have occurred by the Knollen Bed. This is a thin and widespread marker bed of micritic limestone with some fine-grained glauconite from deeper water in the Villigen Formation. The Knollen Bed could be followed by the author from the unpublished section RG 27 above the cemetery at Schönenwerd (west of the city of Aarau) over a distance of 110 km to the northeast, to a quarry on Ortenberg hill near the village of Deilingen. The village is 12 km south-southwest of Balingen in southern Germany. The marker bed could very probably be tracked from Deilingen further to the east, but this was not done to date by German geologists. The Knollen Bed delimits in Canton Aargau the Wangen Member below from the Letzi Member above, and in Canton Schaffhausen the Küssaburg Member below from the Wangental Member above. The bed was named by MOESCH (1863). The position of the bed in the Villigen Formation was indicated by MOESCH (1867) on p. 169 in the text and in a section, which he figured on p. 176. The best outcrop of the Knollen Bed is in section RG 62 west of Villigen in Canton Aargau, which is shown in GYGI (1969a, Pl. 17). The outcrop is at the altitude of 525 m beside the road leading from the village of Villigen upward to the table mountain of Mt. Geissberg. A photograph of the outcrop is Fig. 12 in GYGI et al. (1998).

The top of the early lithified micritic limestone of the Wangen Member in section RG 62 was deeply corroded and partly dismantled by subsolution into nodules the size of several centimeters. This is evidence of a long period of omission. Knollen is the German word for nodules. The Knollen Bed no. 59 in section RG 62 is 0.3 m thick and is directly above the corroded surface mentioned earlier. P. R. VAIL and A. L. COE interpreted the corroded bedding plane in section RG 62 at the top of Wangen Member to be a transgressive surface above their sequence boundary O8 in GYGI et al. (1998, Fig. 2). Sea level rise no. 10 occurred late in the *Gredingensis* Chron. This can be concluded from a nearly complete adult specimen of the ammonite taxon *Wegelea gredingensis* (WEGELE), which is shown in Fig. 7.3 C. The specimen was found in bed no. 55 of the unpublished section RG 85 near Beggingen in Canton Schaffhausen, in the upper Küssaburg Member about 7 m below the Knollen Bed. Section RG 85 is schematically represented in the upper synthetic section of Pl. 19 in GYGI (1969a). *Subnebrodites planula* (QUENSTEDT) first appeared in Canton Schaffhausen directly above the Knollen Bed.

A surface coeval with the boundary between the well-bedded, micritic Wangen and the Letzi Member from deeper water in Canton Aargau cannot be discerned in the mostly massive calcareous oolite of the Holzflue Member in the distal part of the Balsthal Formation (Fig. 1.5). This is evidence that the amount of relative sea level rise no. 10 was less than the 6 m, which is the maximal water depth at which calcareous ooids can be accreted. The increment in water depth caused by the rapid rise above the shoal of ooid sand of the Holzflue Member was so small that it did not interrupt ooid accretion. This is why no trace of the rise could be found in the member. The rise was eustatically

driven, because it influenced the sedimentation budget in the very shallow lagoon, into which lime mud of the La May Member in the Courgenay Formation was laid down landward of the wide oolitic sand bank of the Balsthal Formation. It was shown in Sect. 4.14.3 that most of the calcareous mud accumulating in the micritic Villigen Formation was exported from the lagoon of the Courgenay Formation, where most of it was probably precipitated in whitings. No trace of sea level rise no. 10 was observed in the La May Member. The effect of the minor rise on the lagoon can only be concluded indirectly from evidence in the Villigen Formation from deeper water.

Sedimentation of the Knollen Bed near Villigen began after a considerable time span of omission. Nondeposition lasted long enough to allow for early lithification and subsequent deep subsolution of limestone at the top of micritic Wangen Member. Omission in parts of the Villigen Formation was the effect of some deepening of the very shallow lagoon of the La May Member because of sea level rise no. 10. Deepening of the lagoon at a time when a great quantity of calcareous mud was chemically precipitated in it was only possible, because the rise was rapid. The increase in water depth by a few meters in the lagoon must have been sufficient to substantially reduce, or possibly to interrupt for some time export of lime mud out of the lagoon, and to cause omission in deeper water in some parts of the Villigen Formation. This is evidence that little or no lime mud at all was produced on the oolitic sand bank of the Balsthal Formation in between. As soon as sedimentation of lime mud caused by ongoing precipitation in whitings had raised the lagoon floor of the La May Member to the level as it was before rise no. 10, export of calcareous mud out of the lagoon and transportation across the wide sand bank of calcareous ooids of the Balsthal Formation resumed. Sedimentation of the Knollen Bed began at this time in the Villigen Formation at a low rate. Some fine-grained glauconite in the Knollen Bed is evidence of the low sedimentation rate. It was concluded above that relative sea level rise no. 10 amounted to less than 6 m. To judge from ZIEGLER (1982, p. 106), the eustatic component was 60% of the rise. The amount of rise no. 10 was consequently minimal. Nevertheless, the rise had a noticeable effect in a vast area both in shallow and in deeper water of the epicontinental basin, because the rise was rapid and because sea floor topography in the shallow-water realm was very flat.

Rise no. 11. Rapid relative sea level rise no. 11 flooded the uneven, terrestrial erosion surface with a relief of up to 4 m, which is the top of the Balsthal Formation in the region of Balsthal in Canton Solothurn. The relief of the surface is the result of subaerial erosion of lithified pelletoidal packstone from shallow-marine water in the uppermost Balsthal Formation. A paleosol is above the erosion surface. A calcrete nodule from this paleosol near Balsthal is shown in Fig. 5.9. The thickness of the paleosol is between 0.1 and 0.6 m. The greatest thickness was measured by the author in 1989 on the rope in bed no. 7 of the unpublished section RG 450 at the cliff below Point 702 on Chluser Roggen south of Balsthal. The paleosol is local and occurs only around the village of Balsthal. It weathers in the cliff of Chluser Roggen as a deep, conspicuous notch. The relief of the erosion surface below the paleosol can be clearly discerned in the photograph of Fig. 37 on p. 64 in GYGI (2000a). The photograph was made from the brink of the cliff of Wannenflue on the opposite side of the cluse between Balsthal and Oensingen, looking northeast over a distance of 1.75 km. Sedimentation of shallow-marine pelletoidal packstone resumed above a well-defined boundary at the top of the paleosol. Absence of a transitional layer is evidence that sea level rise no. 11 was rapid. It can be concluded from the relief of the erosion surface of as much as 4 m at the top of the Balsthal Formation around Balsthal, from the thicknesses of the paleosol, and from the facies of the marine calcarenites below and above the paleosol that sea level rise no. 11 was of the order of at least 5 m. The rise could be dated to have occurred not earlier than the *Platynota* A Chron by the ammonite *Lithacosphinctes evolutus* (QUENSTEDT), which is photographed natural size in GYGI (1995, Fig. 19). B. MARTIN found the specimen in the equivalent of what subsequently became bed no. 9 of the unpublished section RG 439, 2.6 m below the base of the paleosol at Innere Klus, which is a fraction of the township of Balsthal (see above). Section RG 439 is located 800 m west of section RG 450 on Chluser Roggen.

Rise no. 12. A minor, rapid relative sea level rise probably occurred during the *Tenuicostatum* Chron. This is concluded from sediments in the shallow-water realm in northwestern Switzerland. Some fine-grained glauconite occurs, for instance, in bed no. 23 of section RG 340 in the quarry east of Banné hill at locality La Rasse south of Porrentruy (GYGI 2000a, Pl. 17), or in a limestone bed in the upper part of the small quarry of section RG 349 beside the cemetery of Fontenais, which is 2 km south of Porrentruy. The time equivalent of these glauconitic beds in the shallow-water realm near Porrentruy is in deeper water of the epicontinental basin the uppermost part of the glauconitic lower Baden Member in section RG 70 near Mellikon (GYGI 1969a, Pl. 17). Two *Crussoliceras tenuicostatum* (GEYER) were found by R. and S. GYGI in this part of the Baden Member near Mellikon. The lager specimen of these is shown in Fig. 7.4E. The four marly intercalations above glauconitic bed no. 23 in section RG 340 of the quarry at La Rasse near Porrentruy are probably coeval with the marl of bed no. 125 in section RG 70. There is no indication that rise no. 12 amounted to more than a few meters.

Rise no. 13. A major, rapid relative sea level rise which occurred during the *Acanthicum* Chron is documented in the unpublished section RG 341 in the quarry of L'Alombre aux Vaches near Courgenay in Canton Jura (see also JANK et al. 2006, Fig. 7/1). A small tidal channel was found within bed no. 32 of section RG 341 at the top of Paulin Member, which is the lowermost limestone member with a thickness of 45 m in the Reuchenette Formation (see Sect. 1.2 above and GYGI 2000a, Pl. 19). A stromatolite with crenulated lamination is above bed no. 32 in section RG 341. The stromatolite is covered with a crust of goethite. Ostreid bivalves settled adhering to the brown crust. The crenulated stromatolite with a thickness of 5–8 cm is from the upper intertidal zone. Directly above is the Banné Member which is a succession of marl and limestone. The total thickness of the member in this section is 7.65 m. The macrofossil assemblage in the Banné Member is characterized by a mass occurrence of specimens with a relatively low diversity. It can be read from Fig. 4.9 that bivalves and brachiopods prevail in the fossil assemblage in the member. Associated in section RG 341 are gastropods, some massive hermatypic coral colonies, and the ammonite *Aspidoceras* cf. *acanthicum*. This ammonite was found by A. and H. ZBINDEN and is the only one recorded from the Banné Member in section RG 341. The specimen is figured in GYGI (1995, Fig. 17/4). Foraminifers of the genus *Pseudocyclammina* and nerineid gastropods from a marginal marine environment occur in limestone bed no. 38 of section RG 341 directly above the Banné Member. This is evidence that most of the accommodation space created by rise no. 13 near Courgenay was filled in with sediments when bed no. 38 was laid down.

Rise no. 13 created marine accommodation space above intertidal sediments in many sections between Sornetan in the south and Ajoie region in the north. It can be estimated from the average, decompacted thickness of the Banné Member in the area of Ajoie and from presence of an ammonite in the member in section RG 341 near Courgenay that relative sea level rise no. 13 amounted to approximately 10 m. The rise must have been substantial, because it caused in deeper water of the epicontinental basin further east a vertical change from limestone sedimentation to deposition of blue-gray marl. The base of this marl is in the assembled section of the three quarries west of the village of Dielsdorf in Canton Zürich, which was published by NOTZ (1924), 63 m above the top of the Villigen Formation. The blue-gray marl at Dielsdorf is 4 m thick and includes numerous *Aspidoceras acanthicum* (OPPEL). A well-preserved, adult specimen of this taxon was found in 1958 by the author of this study in rubble of this marl immediately after a blasting in the middle quarry west of Dielsdorf. He gave the specimen to the Paleontological Institute of the University of Zürich. Another, equally well-preserved and complete adult of *Aspidoceras acanthicum* (OPPEL) from Dielsdorf is shown in this study in Fig. 7.5B.

Rise no. 14. The limestone succession of the Courtedoux Member between the marly Banné Member below and the marly Virgula Bed above in the Reuchenette Formation in the shallow-water realm of Ajoie region in Canton Jura, is approximately 30 m thick, according to sections that were measured by JANK (2004, Chap. 1, Fig. 4). The upper part of this succession was sedimented in a peritidal environment and includes west of the village of Courtedoux a bedding plane with very distinct footprints of dinosaurs. A stromatolite with prism cracks is 3 m below the Virgula Marl in the quarry in Combe de Varu, 1 km east of the village of Chevenez (JANK 2004, Chap. 1, Fig. 5). Rise no. 14 flooded this stromatolite, and some ammonites appeared in sediments above. JANK (2004, Fig. 11d) figured an *Orthaspidoceras schilleri* (OPPEL) from above the stromatolite and below the Virgula Marl, and he stated that *Aspidoceras caletanum* (OPPEL) occurs directly above the Virgula Marl. Two meters above the stromatolite is a limestone unit with a thickness of at least 1 m and with rounded black pebbles at the base. This unit in the western part of the quarry of Combe de Varu is a tidal delta with the same thickness and with steeply inclined foresets. The Virgula Marl directly above is in Combe de Varu about 1 m thick. It includes some glauconite and locally a mass occurrence of small ostreids (JANK 2004, Chap. 1, Pl. 1). There is some glauconite in a thin seam of marl 0.8 m above the Virgula Marl. The ammonite *Aspidoceras caletanum* (OPPEL) was found by A. and H. ZBINDEN stratigraphically a few meters above the Virgula Marl in the village of Alle, 10 km east-northeast of Chevenez. A photograph of this well-preserved specimen is shown in Fig. 26 in GYGI (1995). The presence of several ammonites and of hermatypic corals, which were mentioned by JANK (2004, Chap. 1, Fig. 5) to occur above the Virgula Marl, is an indication that sea level rise no. 14 was substantial, but that it probably did not amount to more than 10 m.

A bed of glauconitic marl occurs in the facies from deeper water of the epicontinental basin in the middle quarry west of the village of Dielsdorf in western Canton Zürich. This quarry in Kimmeridgian limestone at Dielsdorf is geographically (non-palinspastically) 110 km east of Chevenez in Canton Jura. The glauconitic bed near Dielsdorf is the Echinidenschicht of NOTZ (1924, p. 29). B. ZIEGLER (oral communication) found a specimen of the ammonite *Aulacostephanus eudoxus* (D'ORBIGNY) in the limestone bed directly below the Echinidenschicht at Dielsdorf. About 3 m above the Echinidenschicht in this quarry is another, thinner marly bed with glauconite, according to the drawing on p. 96 in the author's field book no. 2 that is kept in the Museum of Natural History at Basel. The same two glauconitic beds of marl as near Dielsdorf were drilled through in 1983 in the NAGRA exploration well near Weiach in northwestern Canton Zürich: the lower marl with a thickness of 1.35 m between

the depth of 274.80 and 276.15 m, and the upper marl with a thickness of 0.6 m between the depth of 271.70 and 272.30 m. A limestone succession 2.5 m thick is between the two glauconitic marls in this well (NAGRA 1988, Beilage 6.2a). According to the ammonites mentioned above, the Virgula Marl in Canton Jura, the Echinidenschicht of NOTZ (1924) in Canton Zürich, and the widespread Glaukonitbank of ALDINGER (1945, p. 112) in adjacent southern Germany must be coeval. A minor rapid relative sea level rise amounting to a total of probably 10 m at most consequently occurred in two phases during the *Caletanum* (or *Eudoxus*) Chron. The two-phase rise was sufficient to cause formation of a double glauconitic marker bed, which can be traced over a lateral distance of much more than 100 km, between sediments from very shallow water and deposits from deeper part of the epicontinental, Rhodano-Swabian Basin.

Rise no. 15. A substantial, rapid relative sea level rise occurred presumably at the beginning of the *Autissiodorensis* Chron in the region, where the city of Solothurn is now. This is concluded from the well-preserved ammonite *Aulacostephanus autissiodorensis* (COTTEAU) no. 10,842 in the Naturmuseum Solothurn, which is a cast of pure micrite. The ammonite was identified and figured by GYGI (1995, Fig. 24) and is shown again in Fig. 7.6 in this study. The specimen was found in the quarry Bargetzi in the northern outskirts of the city of Solothurn, which are called Kreuzen or Steingrueben. The Solothurn Turtle Limestone Member was quarried at that locality as a building stone. The turtle member is a thickly bedded pelmicrite, which can locally be packed with large nerineid gastropods. Nerineids are diagnostic of particularly shallow-marine water. The thickness of the Solothurn Turtle Limestone Member at Solothurn is approximately 7 m. The name of the member is derived from the numerous, well-preserved fossil turtles which were found over the years of quarrying at several localities in and around the city of Solothurn. The ultimate monograph of these turtles is that by BRÄM (1965). According to this author, *Platychelys oberndorferi* WAGNER probably lived in freshwater, and *Plesiochelys* and *Thalassemys* are thought to be marine genera. Above the turtle member in Bargetzi quarry at Solothurn is an erosion relict of a thinly bedded, pure micritic limestone unit with a preserved thickness at that locality of up to 2 m.

Both the Solothurn Turtle Limestone Member and the thinly bedded, unnamed micritic limestone unit directly above crop out in complete thickness in two quarries in the territory of the township of Oberdorf, which is a village 5 km northwest of the city of Solothurn. These sections were measured by the author in the summer of 1986. One of the quarries is north of locality Wäberhüsli, north of the railway station of Oberdorf (section RG 433, GYGI 2000a, Pl. 41). The other quarry is in the eastern part of the locality called Steingrueben, 2 km west-northwest of the village of

Oberdorf (section RG 434, GYGI 2000a, Pl. 42). It was mentioned above that the author found very well-preserved, circular footprints of sauropod dinosaurs with a diameter of 0.5 and 0.7 m, respectively, in the Solothurn Turtle Limestone Member at the upper surface of bed no. 17 of section RG 434 on July 24, 1986. These footprints were made on a tidal flat with a thin cover of soft calcareous mud above indurated substrate. The hard substrate is now bed no. 17 of section RG 434, and this bed includes a multitude of nerineid gastropods. The mass occurrence of nerineids in bed no. 17 is evidence of a marginal marine, very shallow subtidal environment. The Solothurn Turtle Limestone Member is consequently a peritidal sediment. More dinosaur footprints were uncovered on a bedding plane below, when bed no. 17 was broken up and removed by a large-scale blasting later in the summer of 1986.

The pure micritic matrix of the ammonite cast of *Aulacostephanus autissiodorensis* (COTTEAU) from Bargetzi quarry at Solothurn indicates that the specimen was found in the thinly bedded, micritic limestone unit directly above the Solothurn Turtle Limestone Member. The ammonite is from the lowermost part of the micritic succession, because the erosion relic of the unit preserved in Bargetzi quarry is at most 2 m thick. The corresponding unit of thinly bedded micrite is 6 m thick where it is complete in section RG 433 north of Wäberhüsli, and the entire succession is 9 m thick in section RG 434 in the eastern quarry of Steingrueben west of Oberdorf. The presence of an ammonite in the micritic limestone unit at Solothurn indicates the event of a rapid relative sea level rise of possibly as much as 10 m. The rise terminated peritidal sedimentation of the Solothurn Turtle Limestone Member. *Aulacostephanus autissiodorensis* is the index taxon of the *Autissiodorensis* Chron, which is the youngest chron of the Kimmeridgian Age *sensu gallico* that is used in northwestern Europe. The micritic, thinly bedded limestone unit directly above the Solothurn Turtle Limestone Member between the city of Solothurn and Oberdorf SO is the youngest sediment of the Kimmeridgian *sensu gallico* in the region, provided that the entire micritic unit represents the *Autissiodorensis* Zone. If this be the case, the thickness of the entire Kimmeridgian Stage near the city of Solothurn would be 52 m, according to Pl. 41 in GYGI (2000a).

5.2 Evidence that *Rapid* Relative Sea Level Rises Were Eustatically Driven

Rapid relative sea level rise no. 5 in Fig. 5.1 was dated with ammonites from the Buix Member near St-Ursanne to have occurred late in the *Antecedens* Chron. The amount of the rise could be calculated at St-Ursanne to have been 10 m.

When rise no. 5 is compared with coeval events in France and England, it becomes evident that the rise was not relative and local, but that it was a global, eustatic event. TALBOT (1973, p. 307) interpreted the "discordance antéargovienne" of ÉNAY (1966, Fig. 76) in southeastern France to have been caused by a eustatic sea level rise during the *Antecedens* Chron. In addition, he thought that the same eustatic sea level rise caused the unconformity, which ARKELL (1947, p. 99) recognized at the base of the Coral Rag, of the Wheatley Limestone, and of the Oakley Beds in southern England. TALBOT (1973, p. 310) estimated the eustatic component of this rise to have been about 13 m. Coincidence of discrete, major relative rises of sea level of the same order of magnitude in regions so far apart must have had a global cause. GYGI (1986, p. 474) estimated the relative sea level rise late in *Antecedens* time at approximately 10 m. This was confirmed in Sect. 5.1 with independent evidence. Provided that rise no. 5 was eustatically driven, then the eustatic component of it can be calculated according to ZIEGLER (1982, p. 106) to have been 60% of the total sea level rise, because isostatic basement subsidence under additional load is prompt (see Sect. 11.3). The eustatic sea level rise during the *Antecedens* Chron hence amounted to approximately 6 m, and the basement subsided under the additional load of water isostatically by around 4 m.

The thicknesses of coeval limestone members measured in the shallow-water realm vary from section to section. This becomes evident when the separate foldout plates in GYGI (2000a) are laid out side by side, and if the thicknesses of coeval carbonate members are compared. The substrate of the sediments was consequently unstable. This was because of endogenically driven, synsedimentary tectonics, because of regionally unequal compaction of sediments underneath, or possibly because of lateral, plastic flow of rock salt under unequal load within sediment of Middle Triassic age in the region investigated. Some sediments from very shallow water, which were laid down into the minor accommodation space created by a small-scale, rapid relative sea level rise, can be followed over a considerable distance in spite of the unstable basement. An example of this is the thin calcareous oolite of *Bifurcatus* age in the middle of the Röschenz Member, which was sedimented after the rapid part of rise no. 6 upon an almost exactly flat surface of peritidal deposits (Fig. 1.5). In subsequent *Hypselum* time, the tidal delta of calcareous ooid sand near Péry (Fig. 5.2) and its time equivalents of calcarenite with inclined bedding in other sections were shed after the rapid part of rise no. 7 upon the carbonate platform of the upper Günsberg Formation. The transgressive sediment which was laid down after rapid sea level rise no. 8 probably early in *Berrense* time in the uppermost part of the Günsberg Formation directly below the Hauptmumienbank Member includes at some localities abundant hermatypic corals. An example at Moutier is the coral biostrome of bed nos. 39–41 in section RG 381 in Court Gorge (GYGI 2000a, Pl. 28), or the small coral bioherm in bed no. 59 in the unpublished section RG 312, which is located 1 km southeast of the village of Souboz (see above). It was stated above that following the transgression caused by rapid sea level rise no. 8, corals could advance landward from the uppermost Günsberg Formation into the shallow-water realm of the uppermost Röschenz Member as far as Bressaucourt southwest of Porrentruy in Ajoie region of Canton Jura (see SCHNEIDER 1960, p. 7). The corals near Bressaucourt grew far landward from the basinward rim of the Günsbserg carbonate platform, above peritidal and even supratidal facies of the upper Röschenz Member. Sea level rise no. 8, which created the pertinent accommodation space, must consequently have been driven eustatically. The rise was rapid and of the order of at least several meters like the previous, rapid part of rise no. 7.

5.3   Falls of Sea Level in the Late Jurassic

Eustatic *falls* of sea level moved in the same direction as basement subsidence. Traces of sea level falls are, therefore, more difficult to recognize in a sedimentary succession than those of rises (HALLAM 2001, p. 24). A sea level fall could leave an unequivocal trace in the sediments investigated only on a carbonate platform. No undubitable traces of subaerial erosion were found at the vertical boundary between the Chestel Member below and the Buix Member above, or between the distally adjacent Grellingen Member and the Tiergarten Member within the St-Ursanne Formation, respectively. The age of these boundaries is the late *Antecedens* Chron. HALLAM (1999, p. 345) found evidence of subaerial erosion of consolidated rock of more than 1 m at the top of the Bencliff Grit on the Dorset coast in southern England, which occurred at about the same time.

Fall no. 1. The earliest, probably eustatic sea level fall which could be recorded with certainty in Oxfordian sediments is indicated in Fig. 5.1. The fall caused some subaerial erosion at the top of the St-Ursanne Formation at the upper surface of bed no. 3 in section RG 402 near Röschenz in Canton Solothurn (GYGI 2000a, Pl. 33). Some of the ammonite casts that were found over the years in the Buix Member near St-Ursanne were figured by GYGI (1995). They are numbered 9 and 10 in Pl. 1A by GYGI and PERSOZ (1986), and no. 10 in Fig. 1.5 of this study. These ammonites and the mineral stratigraphic correlation C by PERSOZ in Plate 1A by GYGI and PERSOZ (1986) are evidence that the sea level fall recorded near Röschenz occurred at the end of the *Transversarium* Chron. The fall must have been of minor order, because distinct traces of erosion at the top of the carbonate platform of the St-Ursanne Formation were found only in a few sections.

Fall no. 2. The second, probably eustatic sea level fall which could be recognized to have occurred in the Oxfordian Age left distinct traces in the Günsberg Formation. The fall occurred in the *Bifurcatus* Chron, before rise no. 6 turned to be rapid (Fig. 5.1). Distinct traces of the fall were found in:

1. Section RG 315 in Pichoux Gorge near Sornetan in Canton Bern (GYGI 2000a, Pl. 21), in bed no. 54. This is an intertidal stromatolite with parallel lamination below bed no. 56, which is the time equivalent of the Grüne Mumienbank Bed.

2. Section RG 430 in the quarry at Gänsbrunnen in Canton Solothurn (GYGI 2000a, Pl. 40). The uppermost part of bed no. 16 includes in a matrix of calcareous oolite gray to black, angular lithoclasts with a diameter of up to 8 cm. The edges of the lithoclasts are slightly abraded. The top of bed no. 16 is uneven and is covered with a crust of goethite.

3. Section RG 381 in the northern part of Court Gorge southwest of Moutier (GYGI 2000a, Pl. 28). The top of bed no. 14 is an uneven erosion surface with a relief of up to 15 cm. A crinkled, intertidal stromatolite filled in the depressions of the eroded surface. A small and ill-preserved cast of a perisphinctid ammonite was found by the author in 1958 in the upper part of bed no. 13 below. More ammonite casts and fragments were subsequently collected from the same bed at this locality by the author together with P. ALLENBACH. One of these ammonites is a *Perisphinctes* (*Dichotomoceras*) sp. of probably the late *Bifurcatus* Chron.

4. Section RG 307 in the quarry of La Charuque near Péry in Canton Bern (GYGI 2000a, Pl. 22). The uneven top of bed no. 162 with a relief of as much as 10 cm is an erosion surface covered with a crust of goethite. Above the crust are lenses of lignite coal and carbonized tree branches with a diameter of up to 4 cm, as well as gyrogonites (oogonia) of characean fresh-water algae. Above is the oncolite of the Grüne Mumienbank Bed.

Fall no. 3. The third and probably major sea level fall occurred before rise no. 7 became rapid in the *Hypselum* Chron. The fall left distinct traces in the following sections:

1. Section RG 406 in the ravine southeast of farm La Kohlberg, 2 km southwest of the village of Vermes in Canton Jura (GYGI 2000a, Pl. 37). Bed no. 48 of the section is a wavy-laminated, intertidal stromatolite with a thickness of 0.5 m. The top of the bed is an uneven erosion surface. Bed no. 49 directly above with a thickness of 0.15 m includes gray to black lithoclasts of the stromatolite underneath in a matrix of what was initially lime mud with thin dewatering cracks. Lamination is preserved in some of the larger, angular lithoclasts of the stromatolite with a diameter of up to 5 cm. The top of bed no. 49 is a planed and bored erosion surface

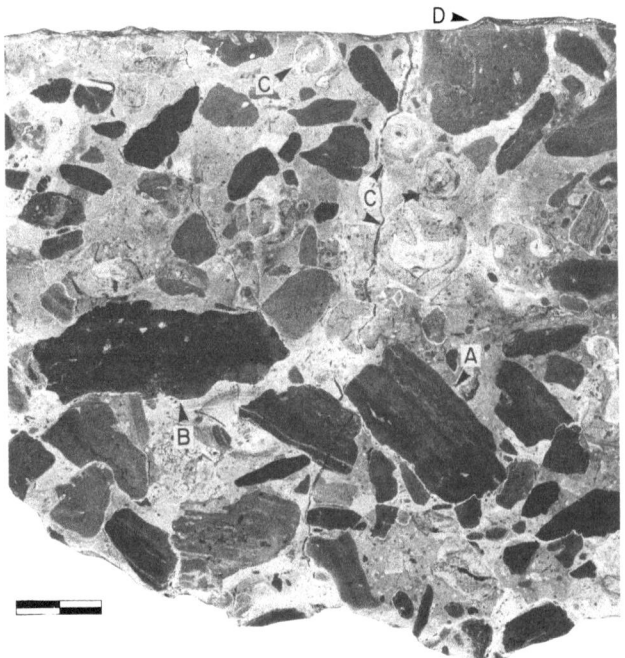

Fig. 5.4 Blackened lithoclasts in matrix of what was initially calcareous mud. Some of the clasts (A) were eroded from a laminated, intertidal stromatolite directly underneath. Clast B is conglomeratic. Nerineid gastropods are labeled C. The top of the bed is a planed and bored erosion surface encrusted with thick-shelled ostreid bivalves (D). Erosion of the surface was probably caused by sea level fall no. 3 in *Hypselum* time. This fall is presumed to be time-equivalent with the planed and bored top with adhering ostreids of bed no. 28 in section RG 381 in Court Gorge near Moutier (GYGI 2000a, Pl. 28). Polished cross-section of bed no. 49 in section RG 406 in ravine 2 km southwest of the village of Vermes in Canton Jura, shown in Pl. 37 in GYGI (2000a). Refigured from GYGI and PERSOZ (1986, Fig. 5)

encrusted with thick-shelled ostreid bivalves. A polished cross-section of bed no. 49 is shown in Fig. 5.4.

2. Section RG 419 (unpublished) measured along the forest road above the gorge south of Bächlen at Seehof in Canton Bern. Formation of lithoclasts by dewatering of an intertidal, laminated stromatolite in the upper Günsberg Formation in bed no. 74b of the section is documented in Fig. 9 in GYGI (1992). This partially brecciated stromatolite is probably coeval with bed no. 48 in section RG 406 near Vermes.

3. Section RG 315 in Pichoux Gorge near Sornetan in Canton Bern (GYGI 2000a, Pl. 21). Bed no. 61 is a wavy-laminated, intertidal stromatolite with prism cracks. The top of the bed is an uneven erosion surface intersecting laminae. Photographs of the crinkled stromatolite with characean gyrogonites of bed no. 65 above are shown in GYGI (1992, Figs. 11 and 12).

4. Section RG 414 (unpublished) north of Grandval in Canton Bern. Bed no. 21 is a prism-cracked stromatolite with characean gyrogonites within the cracks. A single prism with margins upwarped by dewatering is shown in Fig. 4.5.

5. Section RG 417 (unpublished) north of Crémines in Canton Bern. The prism-cracked upper surface of the laminated stromatolite of bed no. 42 is shown in Fig. 4.4.

6. Section RG 390 in Moutier Gorge at Moutier in Canton Bern (GYGI 2000a, Pl. 27). The top of bed no. 80 is a seam of lignite coal with a thickness of up to 3 cm.

7. Section RG 430 in the quarry at Gänsbrunnen in Canton Solothurn (GYGI 2000a, Pl. 40). The top of bed no. 25 is locally a thin seam of lignite coal.

8. Section RG 381 in Court Gorge at Moutier in Canton Bern (GYGI 2000a, Pl. 28). Bed no. 28 with a thickness of 0.9 m includes druses with a diameter of up to 10 cm lined with coarse calcite crystals, which are the remains of massive coral colonies. The top of the bed is a planed and bored erosion surface encrusted with goethite. The closely spaced borings descending from the planed surface have a diameter of up to 1 cm. A small part of the surface encrusted with ostreid bivalves is uncovered on the south side of the road.

9. Section RG 307 in the quarry of La Charuque at Péry (GYGI 2000a, Pl. 22). Bed no. 177 of the section is a stromatolite with flat laminae and a thickness of 30 cm. Carbonized twigs with a length of up to 10 cm and a diameter of 5 mm are common in bed no. 178 above.

Fall no. 4. The fourth sea level fall occurred after rapid sea level rise no. 7, probably late in the *Hypselum* Chron. H. ZBINDEN found the ammonite *Orthosphinctes?* aff. *rhodanicus* (DUMORTIER) in a fallen block which probably tumbled, to judge from its lithology, from a bed 5 m above bed no. 178 in section RG 307 at Péry with carbonized wood mentioned above. The ammonite was identified and figured by GYGI (1995, Fig. 18). The effect of fall no. 4 was, depending on the outcrop, an uneven or a planed erosion surface, respectively. This surface is on top of the widespread calcarenites with inclined bedding in the uppermost Günsberg Formation. These calcarenites are time equivalents of limestone succession no. 55–80 in the upper Effingen Member of section RG 37 east of Auenstein in Canton Aargau (GYGI 1969a, Pl. 17). The ammonite *Euaspidoceras hypselum* (OPPEL) shown in Fig. 7.2D was picked by D. KRÜGER from talus below succession no. 55–80 in section RG 37. Erosion caused by fall no. 4 is conspicuous in the following sections.

1. Section RG 441 (unpublished) along the cantonal road 1.7 km east of the village of Liesberg in Canton Basel-Landschaft. A planed erosion surface truncates a large sand wave (Fig. 5.5A) of pelsparite (Fig. 5.5B). Directly above is the Hauptmumienbank Member.

2. Section RG 414 (unpublished) north of Grandval. The lower and the upper part of unit no. 35 with a thickness of 3.1 m is cross-bedded, oolitic grainstone with some massive, hermatypic corals. The top of the unit is a hummocky erosion surface encrusted with goethite.

Fig. 5.5 (A) Base (*gray*): Pavement of cantonal road 1.7 km east of the village of Liesberg in Canton Basel-Landschaft. Above the road is a sand wave (B) in the uppermost Günsberg Formation. The accommodation space for the sand wave was created by rapid relative sea level rise no. 7 in the *Hypselum* Chron (in Fig. 5.1). The sand wave is truncated by a planed erosion surface which was caused by sea level fall no. 4 late in *Hypselum* time. The sand wave is the lowest unit of section RG 441. *Above*: oncolite of the Hauptmumienbank Member which is massive (C) in the upper part of the section. Scale bar is 1 m. Outcrop west of Point 377, at the turnoff of a small road to the farm Greifel. (B) Thin section photograph of pelsparite, sample taken from circle B in Fig. 5.5A. Scale bar is 200 µm. (C) Oncomicrite in the massive part of the Hauptmumienbank Member. Polished slab of sample taken from circle C in Fig. 5.5A. Dark rods within the largest oncoid are calcite pseudomorphs after calcium sulfate minerals. Scale bar is 1 cm. Figure. 5.5A, B, C refigured from GYGI and PERSOZ (1987, Fig. 2), with permission of Neues Jahrbuch für Geologie und Paläontologie, Abhandlungen

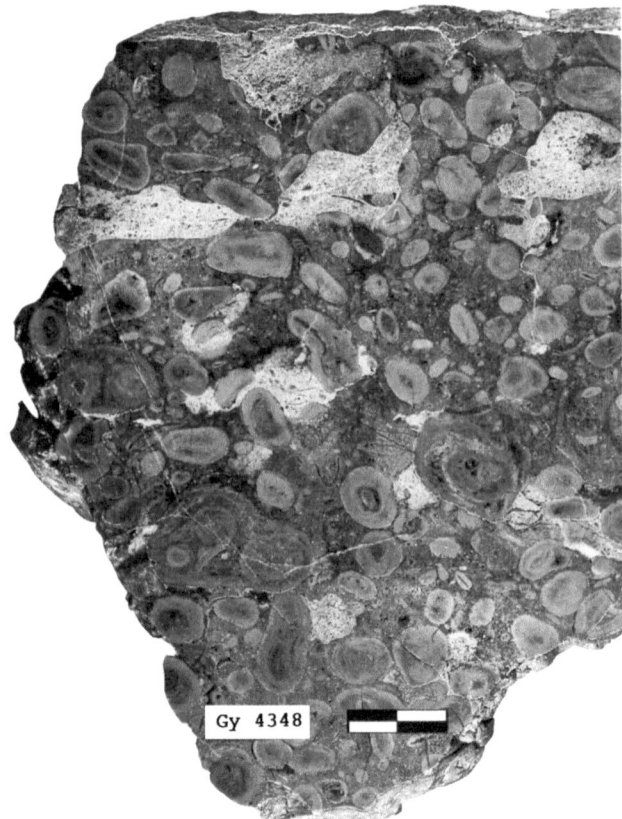

Gy 4348

Fig. 5.6 Oncolite of the Hauptmumienbank Member which was sedimented into a vast lagoon that was created by rapid relative sea level rise no. 8 early in the *Berrense* Chron. Polished slab from the uppermost part of the member in the township of Roches north of Moutier in Canton Bern. The slab vertically intersects the planed erosion surface at the top of the member. Both the erosion surface and the curved borings descending from it intersected oncoids in the lithified rock. The planed surface is the top of bed no. 50 in the unpublished section RG 372 which was measured above the village of Roches along the road leading from the fraction of Hautes Roches to the farm Le Trondai. Scale bar is 2 cm. Refigured from GYGI (2003, Fig. 183 on p. 168) with permission of Schweizerische Paläontologische Abhandlungen

3. Section RG 381 in Court Gorge in the territory of Moutier (GYGI 2000a, Pl. 28). Unit no. 38 in the section with a thickness of 1.8 m is a tidal delta of oolitic grainstone with conspicuous foresets dipping 15–20° to the north. The relief of the eroded surface at the top of the unit is 10 cm. Erosion of the surface occurred in the supratidal zone, because ooids directly below the hummocky bedding plane are thoroughly overcrowded. Rainwater percolating through unlithified sand of calcareous ooids caused overcrowding of aragonitic ooids by intergranular solution of the grains during exposure. Few large ostreid bivalves settled on the crust of goethite at the top of bed no. 38.

4. Section RG 307 in the quarry of La Charuque at Péry (GYGI 2000a, Pl. 22). The top of the uppermost unit no.

195 of the oolitic, laterally shifting tidal delta shown in Fig. 5.2 is a hummocky erosion surface encrusted with goethite. The oncolitic Hauptmumienbank Member above begins with bed no. 198.

Fall no. 5: This sea level fall occurred before rapid sea level rise no. 9 late in *Berrense* time. Erosion caused by the fall was not ubiquitous. It truncated the upper surface of the oncolitic Hauptmumienbank Member and of the distal time equivalent, the oolitic Steinibach Member (Fig. 1.5), mainly at the following localities.

1. Section RG 372 (unpublished) above the village of Roches north of Moutier, along the road leading from the fraction of Hautes Roches to the farm Le Trondai. The upper surface of the uppermost, oncolitic bed no. 50 of the Hauptmumienbank Member is planned by erosion and bored (Fig. 5.6). Erosion truncated oncoids and, therefore, occurred after lithification of the oncolite was completed. The borings descending from the erosion surface into the oncolite are bent and hence resemble burrows. Nevertheless, they are borings, because they intersect both primarily hard oncoids and micritic matrix.

2. Section RG 4 (unpublished) in the cliff of Brocheni Flue, 1.5 km west-southwest of Waldenburg in Canton Basel-Landschaft. Aggregates of radiating, acicular calcite crystals grew in marl of a paleosol directly above the top of Steinibach Member (Fig. 4.3).

3. Section RG 15 (unpublished) in the uppermost part of Horngraben Gorge south of Aedermannsdorf in Canton Solothurn. Bed no. 3 directly above the top of Steinibach Member is marl of a paleosol with a thickness of 0.3 m with calcrete nodules. The nodules are similar to, but smaller than those in the paleosol at the base of the Reuchenette Formation at Balsthal (see Fig. 5.9).

The sea level fall which is documented by the erosion surfaces at the top of the Hauptmumienbank Member and at the top of the coeval Steinibach Member must have been eustatic, because it caused erosion at localities that are far apart. The fall was probably of minor scale, because erosion of the top of the two members was not ubiquitous. The fall occurred some time before the close of the *Berrense* Chron. This is concluded from the mineral stratigraphic correlation I by PERSOZ in Plate 1A in GYGI and PERSOZ (1986). Correlation I could be traced from the top of the Hauptmumienbank Member in the lower Balsthal Formation to the top of the Geissberg Member of the lower Villigen Formation in Canton Aargau. Sedimentation of the Geissberg Member ended probably a short time before the close of the *Berrense* Chron, to judge from the ammonite of the genus *Epipeltoceras* that C. MOESCH found in the glauconitic *Crenularis* Member directly above the Geissberg Member in the quarry west of locality Fahr, in which the unpublished section RG 36 was measured east of Auenstein in Canton Aargau (see above).

Fig. 5.7 Blackened lithoclasts above a local, hummocky subaerial erosion surface within the Verena Member. Polished slab cut across bed no. 233 in section RG 307 near Péry, Canton Bern (GYGI 2000a, Pl. 22). Scale bar is 2 cm. Refigured from GYGI (2003, Fig. 186 on p. 194) with permission of Schweizerische Paläontologische Abhandlungen

A *relative* sea level fall is documented in section RG 307 near Péry by the uneven, subaerially eroded surface at the top of the shallow-marine limestone bed no. 232 in the upper Verena Member (GYGI 2000a, Pl. 22). The bed was lithified before erosion occurred. The gray and black, angular lithoclasts shown in Fig. 5.7 occur in a *local* lens directly above the erosion surface. The diameter of the largest clasts is as much as 12 cm. The lens with lithoclasts is intersected by the eastern wall of the quarry, but *at the present time* the lens does not extend down to the quarry floor. The lens was temporarily accessible in 1980 over a cone of blasted limestone rubble which then reached from the quarry floor up to above the lens. The wavy-laminated stromatolite that B. MARTIN found in the upper Holzflue Member on Mt. Chamben north of Günsberg in Canton Solothurn, at coordinates ca. 609,950/235,850, is in about the same stratigraphic position below the base of the Reuchenette Formation like the lens with black lithoclasts near Péry. The stromatolite on Mt. Chamben is 25 km east-northeast of the possibly coeval erosional surface with black lithoclasts above it in section RG 307 near Péry. It is doubtful whether the subaerially eroded surface below the blackened lithoclasts near Péry and the intertidal stromatolite on Mt. Chamben north of Günsberg are the effect of a eustatic sea level fall, because the two relative falls could be recorded nowhere else. Nevertheless, it should be noted that the vertical boundary between the La May Member and the Porrentruy Member in the Courgenay Formation of Ajoie region in Canton Jura is approximately in the same position as the erosion surface within the Verena Member in the Balsthal Formation near Péry and the stromatolite on Mt. Chamben north of Günsberg. This can be read from Fig. 1.5 in this study.

Fall no. 6. Sea level rise no. 11 was preceded by a sea level fall which caused formation of an intertidal stromatolite with laminae and bird's-eye pores 3.5 m above the base

of the type section of the Reuchenette Formation at Péry in Canton Bern. This stromatolite is coeval with an intertidal, laminated stromatolite 3.5 m above the base of the Reuchenette Formation on Mt. Rüttelhorn at Rumisberg, and a paleosol at the base of the Reuchenette Formation at Balsthal. Fall no. 6 occurred during the *Platynota* A Chron. The fall is documented mainly in the following sections.

1. Section RG 307 at Péry (GYGI 2000a, Pl. 22). The stromatolite is bed no. 236. The photograph of a polished slab cut vertically through this stromatolite is Fig. 5.8.
2. Section RG 440 is 100 m east of the summit of Mt. Rüttelhorn on the southern slope, northwest of the village of Rumisberg in Canton Bern (GYGI 2000a, Pl. 43). The section is located 8 km west-southwest of Balsthal in Canton Solothurn. The base of bed no. 20 in section RG 440 is 3.5 m above the base of the Reuchenette Formation. Bed no. 20 with a thickness of 0.6 m is a wavy stromatolite with bird's-eyes. The top of this intertidal stromatolite is a hummocky erosion surface with a relief of 15 cm which truncates the laminae in the uppermost part of the stromatolite.
3. Fall no. 6 is documented by a paleosol in the following three sections around Balsthal: RG 438 in Steinibach Gorge (Pl. 44, bed no. 52 in GYGI 2000a), RG 439 at Innere Klus (Fig. 5.9 in this study), and RG 450 at the cliff below Point 702 on Chluser Roggen (Fig. 37 on p. 64 in GYGI 2000a). Sea level fall no. 6 was going on probably concomitant with *local* tectonic uplift in the region near Balsthal. The two processes combined caused emersion of marine calcarenite and prompt lithification when meteoric water percolated. Subsequent subaerial erosion of rock produced a relief of up to 4 m. A paleosol was formed above the erosion surface. A broken calcrete nodule from the paleosol is shown in Fig. 5.9.

Fig. 5.8 Intertidal stromatolite in polished cross-section, 3.5 m above the base of the Reuchenette Formation in the formation's type section east of fraction of La Reuchenette in the township of Péry, Canton Bern. Polished slab from bed no. 236 in section RG 307, which is shown in Pl. 22 in GYGI (2000a). Scale bar is 2 cm. Refigured from GYGI (2003, Fig. 184 on p. 169) with permission of Schweizerische Paläontologische Abhandlungen

5.4 Why Stillstand, Progradation, and Backstepping of Basinward Rims of the Shallow-Water Sedimentary Units Investigated Are Diagnostic of Eustasy

Fig. 5.9 Calcrete nodule with acicular calcite crystals radially arranged below the broken surface. Nodule out of marly matrix of the paleosol with a thickness of 45 cm at the base of the Reuchenette Formation at Innere Klus near Balsthal in Canton Solothurn, bed no. 15 in the unpublished section RG 439. Scale bar is 1 cm. Refigured from GYGI (2003, Fig. 175 on p. 162) with permission of Schweizerische Paläontologische Abhandlungen

There are no means to quantify eustatic sea level falls that occurred during the Late Jurassic in the region investigated. Nevertheless, it can be said that the falls must have been of the order of not more than a few meters. They certainly did not amount to tens of meters as it was concluded by SAHAGIAN et al. (1996, Fig. 11 on p. 1453) from falls which occurred during the same time span on the Russian platform and in Siberia.

A plan view of the long-term pattern of seaward advance and landward retreat of the distal rims of shallow-water sedimentary units between the early Bathonian and the late Oxfordian is the map of Fig. 1 in GYGI (1986). More detail in the pattern of progradation of the basinward rims of Oxfordian carbonate platforms is shown in plan view in Fig. 1.4 of this study and in cross-section in Fig. 1.5. Maximal *regression* was reached late in the early Bathonian, when the basinward rim of the carbonate platform of the Spatkalk Member with a tidal delta in front had prograded to Auenstein in Canton Aargau. From that time on, shallow-water sedimentation stepped back or receded landward. Maximal *transgression* occurred in the region investigated in the Oxfordian Age at the end of the *Transversarium* Chron, when the greatest water depth of approximately 125 m was reached in Canton Schaffhausen, or about 105 m at the top of the Birmenstorf Member at Auenstein in Canton Aargau (Fig. 4.9). The gray band in Fig. 1.5 highlights *long-term* vertical and lateral shifts of the basinward margin of shallow-water sedimentation during the earlier part of the Late Jurassic. Tendencies of transgression, stillstand, or regression are distinctly documented in the Oxfordian and early Kimmeridgian sediments investigated in spite of synsedimentary play of normal faults. Such faults displaced deposits vertically by as much as 60 m (see Chap. 10).

The pattern of lateral shifting of shallow-water sedimentation in the earlier part of the Late Jurassic is diagnostic of eustatic variation in sea level, because sediments of the Oxfordian and part of the Kimmeridgian were laid down in the region investigated upon a comparatively stable shelf. Stillstand of the basinward rim of the thick, aggradational sediment stack through the entire time of deposition of proximal succession no. 1 was, therefore, the effect of an almost exact equilibrium between an essentially gradual relative sea level rise and supply of fine-grained siliciclastic and calcareous sediment. No traces of substantial sea level fluctuations were found in succession no. 2. The only exception is the small-scale sea level rise no. 6 in the *Bifurcatus* Chron that halted progradation of the Günsberg carbonate platform. The rapid part of the probably slightly greater sea level rise no. 7 in *Hypselum* time (Fig. 5.1) was the cause of minor backstepping of the basinward rim of the Günsberg carbonate platform, which is shown in Fig. 1.5. Substantial eustatic sea level rise resumed and proceeded at an accelerating rate during sedimentation of succession no. 3. Relative stability of the shelf ended in the earlier part of the Kimmeridgian Age when the rate of *endogenic* subsidence became much greater in Ajoie region of Canton Jura than it was in the region around the city of Solothurn. According to the sections which were measured by JANK (2004, Chap. 1, Fig. 4), the Kimmeridgian Reuchenette Formation is approximately 140 m thick in Ajoie region. The uppermost part of the Kimmeridgian Stage is not preserved in that area because of pre-Tertiary erosion. It was stated above that the thickness of the entire Kimmeridgian Stage is slightly more than 50 m around the city of Solothurn. Endogenic basement subsidence in Ajoie region during the Kimmeridgian was consequently at least 90 m greater than in the area of the city of Solothurn.

5.5   The Pattern of Global Sea Level Variation in the Late Jurassic

5.5.1   Long-Term Variations

The main trends of global sea level variation in the earlier part of the Late Jurassic can be read from data in the preceding section. GYGI (1986, Fig. 4) made an attempt to devise a curve of eustatic sea level variation in the Oxfordian Age. The long-term interaction between supply of fine-grained siliciclastic and calcareous sediment to the area investigated and variation in sea level is documented Fig. 1.5. Some of the relative sea level rises during the Oxfordian Age were gradual and some of them were rapid (GYGI 1986, Fig. 4, and Fig. 5.1 in this study). As stated above, the distal rim of the aggradational sediment stack of proximal, thick succession no. 1 did not shift noticeably in the lateral direction during

sedimentation of the entire succession. This is an indication that sea level rise was gradual most of the time from the beginning of the Oxfordian Age to the end of the *Transversarium* Chron. The short-term, rapid relative rises no. 4 and 5 in Fig. 5.1 only embossed the probably otherwise smoothly rising curve of long-term, gradual sea level rise during sedimentation of succession no. 1. Pronounced progradation during sedimentation of the older part of the Günsberg Formation and mainly of the Steinibach Member indicates approximate stillstand of sea level. This was interrupted by the minor rapid rises in *Bifurcatus* time (no. 6 in Fig. 5.1) and in *Hypselum* time (no. 7 in Fig. 5.1, see also GYGI 1986, Fig. 4). Gradual sea level rise resumed during sedimentation of succession no. 3. The rise continued to be gradual at a progressively increasing rate in the *Planula* Chron and thereby caused the demise of the apparently slowly growing coral biostrome of the Olten Member. This is visualized in Fig. 1.5.

5.5.2   Short-Term Variations

A detailed and probably complete record of small-scale, short-term sea level variation is provided by sediments on carbonate platforms. This is the case with rapid sea level rises no. 5–9 which are represented in Fig. 5.1. Rise nos. 6–9 are documented in sediments of the Günsberg Formation and the lower Balsthal Formation directly above. They occurred at more or less regular time intervals of approximately 400,000 years. Each of these rises amounted to 5 m at most. Neither can the falls of sea level be quantified, which are indicated in Fig. 5.1 to have occurred mainly before the rapid rises, nor can the rate of falling be estimated. Sea level falls moved in the same direction as endogenic basement subsidence. The rate of sea level falls could have been slow, and it may have been similar to the rate of endogenic subsidence (see below). This is a possible cause why some of the sea level falls in Oxfordian time left indistinct or no traces at all in the sedimentary succession investigated. It can be read from Fig. 5.1 that rapid sea level rise no. 5 was about twice as much as rise nos. 6–9. Rise no. 5 was with 10 m the quantitatively most important rapid rise that occurred in Oxfordian time. Proof was presented above that the rate of sea level rise no. 5 late in *Antecedens* time in the middle Oxfordian was rapid and that the rise was eustatically driven. The eustatic component of the rise was 6 m. Rise no. 5 is separated from rise nos. 6–9 by a much longer time span than the almost regular period of approximately 400,000 years which elapsed between each of the rise nos. 6–9. Rapid sea level rise nos. 5–15 were of a varying amount, and they occurred at widely different time intervals.

Assemblages of hermatypic corals occur in the sediments investigated either in the form of widespread biostromes or as individual, minor bioherms. The most widespread coral biostromes are those in the Liesberg Member with a matrix which was primarily a mixture of calcareous and argillaceous mud. Coral biostromes in a matrix of primarily pure calcareous mud occur at the basinward margin of the St-Ursanne Formation and in the Olten Member at the basinward rim of the Balsthal Formation. Coral colonies in the lowermost part of biostromes are almost exclusively plate-like microsolenids or *Isastraea* with probably a low diversity. The initially substantial content of clay minerals in the muddy matrix at the base of the Liesberg Member diminished in the course of sedimentation upward in the member. The upper boundary of the member to the limestone of the St-Ursanne Formation is, therefore, transitional. Growth of coral assemblages at the basinward margin of the carbonate platform of the St-Ursanne Formation was uninterrupted from base to top. The growth form of these coral assemblages was probably a biostrome at all stages of growth. The best place to see this is section RG 404 in the gorge of the Scheulte River 1 km southeast of the village of Mervelier in Canton Jura, which is shown on Pl. 36 in GYGI (2000a). These biostromes probably grew at a water depth of more than 6 m below the bathymetric range in which calcareous ooids were accreted in the coeval Tiergarten Member. Most of the Olten Member at the basinward margin of the carbonate platform of the Balsthal Formation was a coral biostrome growing in deeper water off the wide oolitic sand bank of the Steinibach Member. The Olten Member is intersected in a natural outcrop in the small gorge north of the village of Wangen west of Olten and above the eastern end of the bridge across the Aare River between Olten and Winznau.

Individual coral bioherms grew in the Buix Member and the Tiergarten Member of the upper St-Ursanne Formation, in the proximal, uppermost part of the Pichoux Formation, and in the boundary beds between the uppermost Effingen Member and the lowermost Günsberg Formation above (Fig. 1.5). The largest individual coral bioherms growing in the sediments investigated were those in the lagoon of the Buix Member. This was an environment of very pure limestone sedimentation. The lagoon reefs in the Buix Member are between 20 and 60 m wide and as much as 18 m from base to top (Fig. 4.19B). The elevation of the domal top of the reefs above the surface of the surrounding sediment was of the order of a few meters. This can be read from Fig. 64 in PÜMPIN (1965), which shows a reef in the Buix Member in the quarry above the railway station at St-Ursanne. Coral reefs in this quarry are close together, but the average interspace between reefs at St-Ursanne is much greater. The distance between the westernmost bioherm in the quarry and the next bioherm in the Buix Member further west above the outskirts of St-Ursanne is approximately 600 m. This interspace is shown in Fig. 4.19A. The coeval bioherms in the proximal part of the uppermost Pichoux Formation were located just off the basinward margin of the carbonate platform of the St-Ursanne Formation in somewhat deeper water (Fig. 1.5).

The knob-like coral bioherms in the boundary beds between the uppermost Effingen Member and the lowermost Günsberg Formation grew very close together in front of the basinward margin of the prograding sand bank of calcareous ooids of the carbonate platform of the Günsberg Formation (Fig. 1.5 in this study and GYGI 2000a, Pl. 22, unit no. 160 of section RG 307 at Péry). The lateral interspace between the bioherms at Péry could not be measured. It was estimated looking from the floor of the northern part of the quarry of La Charuque, using as scale the elevation from base to top of the two bioherms at Péry which could be measured. Judging from the figures measured in the section, the lateral interspace between the bioherms at Péry is 10–30 m. One of the bioherms and part of the adjacent sediment were photographed in the quarry of La Charuque by GYGI (1992, Fig. 19). The top of the bioherm photographed is 7 m above its base. When after nucleation the bioherm grew higher and wider, the structure subsided under its increasing weight somewhat into the unlithified mixture of calcareous and argillaceous mud underneath. When growth of this bioherm with a micritic matrix ended, the domal top of it was approximately 3 m above the level surface of the coeval sediment around the bioherm. Both the domal elevation of the top and subsidence of the base of the structure can be read from the photograph in GYGI (1992, Fig. 19). Subsidence of the bioherm into unlithified mud below is evidence that it was an early cemented, wave-resistant structure from the beginning. The bioherm is consequently a reef.

Two coral reef knobs in the boundary beds between the Effingen Member and the Günsberg Formation on the opposite, western face of the Cluse de Rondchâtel near Péry are shown in Fig. 11 in GYGI et al. (1998). The domal top of the reef on the left side of the photograph is approximately 4 m above the flat surface of coeval sediment beside the reef. All of the reefs which began to grow in the uppermost Effingen Member were eventually choked by calcareous ooid sand which was shed from the prograding front of sand banks in the lowermost part of the Günsberg carbonate platform. The abundance of plate-like microsolenid corals in calcareous matrix of micrite in a bioherm in the lower, proximal part of the Günsberg Formation north of Grandval in Canton Bern (unpublished section RG 410, unit no. 27) is almost equal to that in the partially argillaceous Liesberg Member (compare Fig. 7 in ZIEGLER 1962 with Fig. 32 on p. 54 in GYGI 2000a). Much siliciclastic mud and some silt bypassed the bioherm above Grandval during its growth in the lower part of the carbonate platform of the Günsberg Formation.

Minor coral bioherms grew within the wide sand bank of calcareous ooids which is now the Tiergarten Member. One of the bioherms in the member is easily accessible 1 km south of the village of Dornach in Canton Solothurn at locality Stützli, beside the road leading from Dornach down to the farm Tüfleten. Several coral bioherms in the upper Tiergarten Member are well accessible in the forest above the left (western) flank of Chastelbach Gorge west of the village of Himmelried in Canton Solothurn. The largest bioherms in the Tiergarten Member with an elevation from base to top of as much as 10 m were found 2 km south of the village of Duggingen in Canton Basel-Landschaft, at the steep left (southern) flank of Seetel Valley called Eigenhollen. A forest road on the western flank of the small ravine 500 m northwest of the farm Eigen (south of Seetel Valley) intersects one of these bioherms, and another bioherm is above the same road where it ends above a small waterfall emptying into Seetel Valley. The bioherms in the Tiergarten Member probably grew in major, deep tidal channels like the channel between two of the Joulters Cays, which is shown in the aerial photograph of Fig. 4.8. Small coral bioherms are aligned in a row in the middle of this channel, which is about 4 m deep. The author dived to

these reef knobs as a participant in a field trip which was organized by the Third International Coral Reef Symposium at Miami, Florida, USA, in May 1977.

Both biostromes and bioherms of hermatypic corals are often closely associated with calcareous oolite. A bank of calcareous ooid sand accumulated in the lower part of the inner carbonate platform of the St-Ursanne Formation in the course of shallowing above a coral biostrome with locally abundant oncoids in what is now the Chestel Member. The oolitic sand bank of the upper Chestel Member then began to prograde basinward over the coral biostrome which was growing in somewhat deeper water in the distal part of the Grellingen Member (Fig. 1.5). This situation was similar to what is shown in the aerial view of some of the Joulters Cays north of Andros Island in the Bahamas, in the photograph shown in Fig. 4.8. The wide tidal channel between two of the Joulters Cays figured empties into a delta of calcareous ooid sand. The delta prograces over a coral biostrome growing on level bottom at a water depth of more than 5 m (gray shade in the lowermost part of the photograph). The ooid sand thereby progressively smothers the coral biostrome. It is evident from this, why corals are below and calcareous oolite is above in most of the shallowing-upward successions in sediments of the Late Jurassic in the region investigated. Exceptions from this normal, vertical facies succession are few like, for instance, in the uppermost Geissberg Member of Bözberg region west of the city of Brugg in Canton Aargau. Calcareous ooid sand was sedimented in that region directly upon calcareous mud with mainly bivalves of the Geissberg Member. The sand was probably shed from a tidal delta, without the advent of hermatypic corals in between (GYGI 1969a, Pl. 19, upper synthetic section).

The maximal water paleodepth where hermatypic corals of Late Jurassic age survived in the habitats investigated could be calculated semiquantitatively in section RG 307 in the quarry of La Charuque near Péry (GYGI 2000a, Pl. 22). Coral colonies appeared at this locality for the first time in bed no. 144 of section RG 307, in the uppermost part of the shallowing-upward Effingen Member. The Effingen Member was sedimented out of water with a varying mixture of calcareous and argillaceous mud in suspension. The upper surface of the calcareous ooid sand bank above, of bed no. 162 in the lower Günsberg Formation, eventually became permanently emergent (exposed) during an extended time span because of a sea level fall, which occurred before rapid rise no. 6 (in Fig. 5.1). Evidence of this prolonged exposure is lenses of lignite coal and gyrogonites of characean freshwater algae, which are associated with some carbonized tree branches above the oolite. It was stated above that the coral reefs which began to grow in the uppermost Effingen Member, and which were ultimately smothered by the lower part of the calcareous ooid sand shoal of unit no. 162 in the lower

Günsberg Formation, were early cemented and were, therefore, not compacted. Neither was the calcareous ooid sand above the bioherms compacted before lithification. The ooid sand of unit no. 162 is now cemented into a limestone with grainstone texture, according to the nomenclature proposed by Dunham (1962, Tab. 1). The sediment thickness from the base of bed no. 144 with the first hermatypic corals upward to the top of the calcareous oolite unit no. 162, which was for some time a permanently emergent surface, was measured to be 44 m. If compaction of the marly mud in the uppermost Effingen Member above bed no. 144 is taken into account, and when isostatic subsidence is assumed to have been at least half of the sediment thickness measured, then the greatest water depth where hermatypic corals could begin to grow and survive at this locality near Péry was approximately 20 m.

Colonies of hermatypic corals with a plate-like growth form are an adaptation to a low level of illumination. This is why such corals are the first to colonize a substrate in a shallowing-upward succession. The water depth of around 20 m, where plate-like corals first appeared in the region investigated, is much less than the 160 m, which is the maximal depth where plate-like corals can survive in particularly clear water in the Recent (see below). The comparatively shallow lower bathymetric limit to the habitat of Oxfordian hermatypic corals, which was semiquantitatively calculated above, could have been caused by turbidity of the water. This is unlikely, because there are no hermatypic corals, for instance, in the Mumienkalk Bed from a water depth of approximately 120 m. The sedimentation rate of this bed in normally quiet bottom water was minimal. Transparency of the entire water column in the pertinent part of the basin must, therefore, have been exceptionally high. Nevertheless, illumination was too dim for hermatypic corals on this part of the sea floor.

It was stated above that the thickness of the shallowing-upward succession in section RG 307 at Péry, in which the maximal water depth of about 20 m of the habitat of hermatypic corals was calculated, was measured to be 44 m. Sedimentation of the succession ended at the surface of a sand shoal of calcareous ooids which then became emergent for an extended time span. The older, but otherwise analogous succession in section RG 306 at Liesberg from the base of the Liesberg Member to the top of the oolitic Chestel Member above is 49 m thick. There is no evidence that corals at Liesberg first appeared at a water depth which was significantly different from the corresponding depth at Péry. The advent of prolific, mainly plate-like microsolenid corals at the base of the Liesberg Member at Liesberg was sudden. It occurred at the surface of mixed calcareous and argillaceous mud. It can be read from Fig. 32 on p. 54 and from Pl. 31 in Gygi (2000a) that plate-like coral colonies continue to prevail in the coral assemblage from the base of the Liesberg

Member upward through a thickness of 20 m to the top of bed no. 105 of section RG 306 at Liesberg. A measured thickness of 10 m of mud-grade, shallowing-upward sediment was calculated in section RG 306 to correspond to a decrease in water depth of 5 m. The bathymetric range of plate-like coral colonies in the Liesberg Member at Liesberg was consequently at a water depth of between 20 and 10 m. Massive coral colonies first appear in section RG 306 at the base of bed no. 106. Accretion of calcareous ooids began at the base of bed no. 109 when water was approximately 6 m deep.

Water depth was not critical of whether coral assemblages grew in the form of biostromes or as bioherms. The biostromes in the Liesberg Member and the individual bioherms growing close together in the boundary beds between the uppermost Effingen Member and the lowermost Günsberg Formation probably began to grow at approximately the same water depth of around 20 m in turbid water with much siliciclastic material in suspension. The cause of why corals in similar environments formed either biostromes like in the Liesberg Member, or individual bioherms close together like in the boundary beds between the Effingen Member and the Günsberg Formation, could not be found.

Primary production and corresponding supply of food animals to coral polyps were at a higher level in nutrient-rich water with some turbidity because of clay minerals and calcareous mud in suspension than in clear water. The Liesberg Member and the boundary beds between the Effingen Member and the Günsberg Formation have an elevated content of clay minerals. Dupraz and Strasser (2002, p. 449) concluded, based on the investigation of coral reefs in the St-Ursanne Formation and in the Günsberg Formation, that coral reefs growing in mixed carbonate-siliciclastic settings were "most diversified". Conversely, hermatypic corals, for instance, in the pure calcareous units of the Grellingen Member and the Buix Member in the St-Ursanne Formation, or in the Olten Member of the Balsthal Formation, lived in water with a low content of nutrients and animal food for corals. This is the probable reason why the average lateral interspace between lagoon reefs in the Buix Member is much greater than that between reef knobs in the boundary beds between the Effingen Member and the Günsberg Formation above. It is important to note that there is no simple relation between the amount of food supply and the density of hermatypic coral colonies in a given rock *at a small scale*. It was stated above that plate-like coral colonies are as much as 30% of the rock volume in the partly argillaceous Liesberg Member. Conversely, massive and thickly branching colonies are up to 50% of the rock volume in the lowermost part of the core in a reef in the Buix Member, in unit no. 18 of the unpublished section RG 336 at St-Ursanne. The pertinent lagoon reef grew in water with a very low content of nutrients.

In a greater context, an obvious relation existed between the content of nutrients in the water and the abundance of hermatypic corals in the pertinent environment. It can be read from Fig. 32 in Gygi (2000a), that plate-like coral colonies are approximately 30% of the volume of rock with a considerable content of clay minerals in the upper Liesberg Member in the type section RG 306 of the member at Liesberg (Gygi 2000a, Pl. 31). The coral bioherms which nucleated on the sediment surface of mixed calcareous and argillaceous mud in the uppermost Effingen Member at Péry were vigorously growing at a time, when water with much argillaceous mud, some siliciclastic silt, and with much lime mud in suspension bypassed their environment. These coral bioherms at the base of the Günsberg Formation are much closer together than similar bioherms in the Buix Member of the upper St-Ursanne Formation, which grew in a lagoon with sedimentation of particularly pure carbonate mud. The Olten Member at the distal margin of the carbonate platform of the Balsthal Formation had a primary matrix of pure calcareous mud. Plate-like corals are much less abundant in the biostrome of the Olten Member than in the Liesberg Member. This can be read from a comparison of Fig. 12 with Fig. 32 in Gygi (2000a).

Hermatypic corals in the Late Jurassic had to cope with rapid relative sea level rises of no more than around 10 m (Fig. 5.1). No evidence of a significant tidal range was found in the sediments investigated. The greatest paleodepth of approximately 20 m, where plate-like hermatypic corals could survive in the Oxfordian Age in northern Switzerland, is to be compared with the observation that living plate-like coral colonies were found in the Recent in the exceptionally clear water of the Red Sea at a depth of as much as 160 m (Kaise et al. 1993 in Insalaco 1996a, p. 186). Recent platy corals living in particularly deep water are part of a community which had to adapt to the numerous, great, and relatively short-term eustatic sea level fluctuations of as much as between 120 and 130 m (see above) during the Quaternary. This may be exceptional, and the much lesser paleobathymetric range in which hermatypic corals lived in Late Jurassic time was possibly the normal case through

much of geologic history since the Triassic, when scleractinian corals acquired symbiotic zooxanthellid representatives in their tissue (Stanley 1981). Neither high barrier reefs nor lagoon reefs with high and steep flanks similar to those that were figured in cross-section from the lagoon of the Bermuda Atoll by Gygi (1970, Fig. 17b–c) or by Garrett et al. (1971, Fig. 4) grew in the Late Jurassic in the region which is now northern Switzerland, because rapid, eustatically driven sea level changes were modest in the Oxfordian Age. No evidence was found that platy corals first appeared in the Liesberg Member at Liesberg at a water depth that was significantly different from the approximately 20 m which was calculated at Péry in the uppermost Effingen Member below the Günsberg Formation. But it must be noted that Hallock and Schlager (1986, Fig. 1) showed how different the maximal bathymetric range of hermatypic corals in the Recent is depending on the depth of the euphotic zone. Lathuilière et al. (2005, Fig. 7) thought to have discovered a mid-Oxfordian "reef tract" in the French part of the Jura Mountains near Bonnevaux-le-Prieuré 16 km southeast of Besançon, which allegedly rose steeply from a water depth of more than 40 m. Both the paleobathymetry and the sea floor topography concluded by these authors are untenable. Evidence of this was given at the end of Sect. 4.12 above.

Ahermatypic corals in the sediments investigated are solitary, small, and rare. One such specimen with a diameter of 4 mm was found in a bioherm of siliceous sponges from a water paleodepth of about 50 m in the lower Villigen Formation at Mellikon in Canton Aargau. The coral is visible in the right part of the polished slab from bed no. 17 of section RG 70, which is photographed in Fig. 2 of Pl. 6 in Gygi (1969a). An enlarged photograph of the same specimen is Fig. 24 in Gygi (1992). Several ahermatypic corals were found in the boundary beds between the Villigen and the Schwarzbach Formation from a water paleodepth of approximately 100 m (Fig. 4.9) at locality Summerhalde near the city of Schaffhausen (section RG 239 in Gygi 2003, Fig. 160 on p. 140).

7.1   Direct Measurement of Age Using Nuclear Decay

Pellets of mature, pure glauconite from the thin beds at the base of the Upper Jurassic Series in Canton Schaffhausen in northernmost Switzerland were dated with the K-Ar method for the first time by GYGI and MCDOWELL (1970) and again by FISCHER and GYGI (1989). The ages that were calculated by FISCHER in FISCHER and GYGI (1989, Fig. 3) were measured in glauconite pellets which were excavated from section RG 81b near Gächlingen. Section RG 81b is shown in Fig. 8.3 in this study. Pellets large enough to be separated from predominantly argillaceous sediment occur in this section in bed no. 11 called Glaukonitsandmergel Bed (GSM in the lower right of Fig. 1.5), in bed no. 13 called Mumienmergel Bed (MUM), and in the unnamed bed no. 15 of section RG 81b. The biochronologic ages of the radiometrically dated beds range from the *Cordatum* to the *Transversarium* Chron. These beds include an abundant and diverse assemblage of well-preserved ammonite casts. Glauconite pellets occur in section RG 81b both in the mainly argillaceous matrix of the sediments and within calcareous steinkerns (casts) of ammonites within the same beds. The dated glauconite pellets from the matrix of the sediment can be assumed to be practically coeval with the glauconite pellets within ammonite casts. Therefore, it was possible to assign radiometric ages with the best accuracy that could be achieved at the time directly to some particularly well-preserved casts of ammonites, which were identified and figured by GYGI in the paper by FISCHER and GYGI (1989, Figs. 4–6).

The color of the glauconite pellets which were separated from the matrix of the dated sediments varies between light green and very dark green. The intensity of the green color of glauconite depends on the content of ferrous iron Fe^{++} in the mineral. The color intensity grows with increasing content of ferrous iron in glauconite. Coupled with augmenting content of iron is growth of the content of potassium, which was used for dating. This was confirmed by MEUNIER

in EL ALBANI et al. (2005, Fig. 8). The specific weight, the magnetic susceptibility, and the resistance to abrasion of glauconite pellets increase with growing content of iron. All of these properties were used in mineral separation with the purpose to enrich the samples prepared for dating with glauconite pellets with a high content of potassium. Glauconites with the darkest green color sank together with heavy minerals in the heavy liquid used in mineral separation. When glauconite pellets with a very dark green color are powdered and x-rayed in a diffractometer, they give a high and narrow peak of the basal reflection. This is evidence of a high degree of crystal lattice order. The heaviest, very dark-green glauconites with a well-ordered crystal lattice were so rare, that mostly pellets with a lesser potassium content and thereby with a lower degree of crystal lattice order had to be used for dating. A diffractogram with the low and broadened basal reflection of the bulk of glauconites that were used for dating was shown in Fig. 3 in GYGI and MCDOWELL (1970).

The imperfect lattice order of the greatest part of the glauconite pellets dated is a probable cause of partial loss of argon, which is the product of nuclear decay of K^{40}, by diffusion of this gas out of the crystals. The glauconite ages that were calculated by FISCHER in FISCHER and GYGI (1989) are, therefore, possibly too young. The age of 149.2 million years of the *Cordatum* Chron, which is shown in Fig. 3 in FISCHER and GYGI (1989), is nearly 7% younger than the corresponding age given in the time scale which was published by GRADSTEIN et al. (2004, p. 310, and Table 18.2). The time scale in GRADSTEIN et al. (2004) is based on paleomagnetism which was measured in Europe in incomplete or even in stratigraphically condensed successions, or in successions with few ammonites. Therefore, the figures in GRADSTEIN et al. (2004) cannot be taken at face value. Nevertheless, their figures are presumed to be close enough to the real ages that they could be shown in Figs. 1.6 and 5.1 of this study. The pertinent figures in the allegedly "concise" geologic time scale by OGG et al. (2010) are the same as

R.A. Gygi, *Quantitative Geology of Late Jurassic Epicontinental Sediments in the Jura Mountains of Switzerland*,
DOI 10.1007/978-3-0348-0136-2_7, © Springer Basel 2012

those in GRADSTEIN et al. (2004). It is the opinion of VILLENEUVE in GRADSTEIN et al. (2004, p. 95) that radiometric K-Ar dating of glauconite cannot be precise. This author even regards the $^{40}Ar/^{39}Ar$ method in glauconite dating to be generally unsuitable, unless special analytical procedures are applied.

7.2 Relative Time Calibration According to Ammonite Biochronology

Taxonomy of ammonites which are essential in biochronology of the Late Jurassic was published in previous papers and in books by the author and by several coauthors. Ammonite taxonomy and biochronology conceived in these publications are based exclusively on the investigation of casts (steinkerns). Preservation of parts of the outer shell is very unusual in the casts investigated. The suture line of septa with the outer shell of the ammonite is more often visible on the surface of casts when the outer shell is removed by dissolution in the course of diagenesis. Traces of the soft body of the animals were never found or sought. Only *artificial* "species", either previously described or recently published to be new, can therefore be used in this study. These are taxa that were characterized according to morphologic features that can be measured on the exterior of the investigated casts. Assumption of the degree of morphologic variability within a distinct taxon is a matter of discretion. Pertinent judgment is consequently a source of never-ending controversy among taxonomists. A comparatively narrow variability was chosen by the author of this study in previous taxonomic work in order to arrive at a satisfactory biochronologic discrimination. For instance, he proposed the new taxon *Gregoryceras tenuisculptum* GYGI, although it was based on a single specimen in a mediocre state of preservation (GYGI 1977, Pl. 1, Figs. 5a, b) from the Mumienmergel Bed no. 7 in section RG 212 (Fig. 8.3 in this study). Nevertheless, the taxon of the *Densiplicatum* Chron was found to be justified by BERT (2004), and it was later found in Tunisia (BERT et al. 2009, Fig. 12). Because no organic matter of ammonites is preserved and because of inevitably arbitrary judgment in taxonomy of fossils, nothing can be said about the variability in the *biologic* ammonite species living at the time and that could produce fertile offspring, let alone about the possible existence of sibling species that are not uncommon in neozoology.

One reader of GYGI (2003) informed the author in a letter that he noted that the diameter of perisphinctid ammonites listed in some of the tables of dimensions differs from the maximal diameter that can be measured in photographs of the corresponding specimens. The reader, therefore, suspected, that either the scale of a distinct photograph indicated in the pertinent caption or the figures listed in the

corresponding table of dimensions were wrong. The reason for such discrepancies is that the author of this study measured the diameter and the corresponding dimensions which are stated and calculated in the tables of his taxonomic publications, where the section of the last whorl of a given specimen is best preserved and is presumably not deformed by early diagenetic compaction of the ammonite cast. This cannot be read from the photograph of an ammonite. It can only be perceived by someone who handles the pertinent cast. A special case is the ammonite cast with the individual number J24344 photographed in Fig. 136 on p. 117 in GYGI (2003). The pertinent shell was rapidly and entirely embedded into calcareous mud with the plane of coiling vertical compared to the sediment surface at the corresponding stage of deposition. After the ammonite shell was dissolved during early diagenesis, the unlithified calcareous mud which had filled in the shell was plastically compressed along with compaction of the encasing sediment. The slight compression of the ammonite cast occurred in a plane rectangular to the plane of coiling of the ammonite whorls. Because of this, the dimensions of the cast were measured in a plane inclined 45° against the plane of compression, in order to counteract or hopefully to eliminate the bias caused by minor deformation. All of this should have been explained, but it was not mentioned in the text about taxonomy on p. 118 in GYGI (2003).

The boundary beds between sediments of the Middle and the Late Jurassic in excavation RG 208 at Ueken near Herznach in Canton Aargau are thin and in iron-oolitic facies with a macrofossil assemblage of mainly ammonites. These beds in section RG 208 are not stratigraphically condensed in the strict sense. They are shown in Figs. 4.14 and 8.2. There is no evidence that a hiatus exists in excavation RG 208 between the *Paucicostatum* Zone, which is the last zone of the Middle Jurassic, and the *Scarburgense* Zone, which is the oldest zone of the Oxfordian. Conversely, there is a hiatus in excavation RG 208 encompassing the *Praecordatum, Bukowskii,* and the *Costicardia* Zone (GYGI 1981, Fig. 2, and Fig. 8.2 in this study). The corresponding boundary beds in section RG 210 at Gansingen in Canton Aargau are stratigraphically condensed (Fig. 8.3). The hiatus at the base of Oxfordian sediments in sections RG 81b and RG 212 in Canton Schaffhausen represents the first four Oxfordian ammonite chrons (Figs. 1.6 and 8.3). The beds directly above this hiatus are stratigraphically condensed. The sediments of Late Jurassic age which were deposited at a normal rate are represented in Figs. 1.5 and 4.15. Sediments laid down in water deeper than approximately 35 m include a complete succession of ammonite zones from the base of the Upper Jurassic Series to the top of the Kimmeridgian Stage (Fig. 1.6). The Kimmeridgian Stage is conceived here in the same sense as it is adopted by French and Russian earth scientists, or by GRADSTEIN et al. (2004, p. 310).

Problems with ammonite biochronology of the Late Jurassic in northern Switzerland are the following. The zonal indices of the *Cordatum*, *Densiplicatum*, and *Antecedens* Zones were only found in thin, stratigraphically condensed beds sedimented on the floor of the starved basin, but not in the proximal, thick, and normally deposited part of succession no. 1. This can be read from Fig. 1.5. The zonal indices of the *Hypselum* and the *Berrense* Zone are both represented by a single specimen collected from talus. The duration of a distinct ammonite chron could not be established. Only an *average* time span of approximately 370,000 years per chron could be calculated using the recently published radiometric time scales by GRADSTEIN et al. (2004) or by OGG et al. (2010). The ammonite chron as it is conceived in this study is the shortest time span that can be discriminated by means of morphologically defined ammonite taxa, when only specimens collected from *in situ* in sections measured bed by bed are used.

There is as yet no general agreement on how and where to establish a scheme of ammonite zones of the Oxfordian and the Kimmeridgian Stage which can be used as a biochronologic scale for worldwide correlations based on ammonites. One example of dissenting opinions about this is the history of how the *Transversarium* Zone of the middle Oxfordian was interpreted, and mostly misinterpreted, since the time when it was introduced by A. OPPEL in 1863. Initially, OPPEL (1863, p. 165) defined a *Transversarium* Zone *sensu lato* in northern Switzerland. This zone included both of what is now the Birmenstorf and the Effingen Member, and OPPEL mentioned "*Ammonites plicatilis* SOWERBY, *A. transversarius* QUENSTEDT and *A. Schilli* OPPEL" (sic) to occur in the zone. OPPEL (loc. cit.) stated that the taxa mentioned by him occur in the Birmensdorf Member of northern Switzerland (as the member was conceived and spelled at the time) and that this member was the "proper region" of the *Transversarium* Zone. Soon after, OPPEL restricted the *Transversarium* Zone to the Birmensdorf Member and designated the township in Canton Aargau, which was at that time spelled Birmensdorf, to be the type locality of the revised zone. The name of Birmensdorf in Canton Aargau was later changed to Birmenstorf in order to avoid confusion with the township of Birmensdorf in Canton Zürich, which is spelled like this to the present time and is 18 km southeast of Birmenstorf in Canton Aargau. W. WAAGEN (OPPEL and WAAGEN 1866, see mainly p. 244) published the pertinent study by OPPEL after OPPEL's early death. This lucid and comprehensive paper was ignored by almost all of subsequent authors. OPPEL defined the revised *Transversarium* Zone both lithologically and with ammonite biochronology.

The (bio-) zone in the sense of OPPEL is a stack of strata, which is defined at a type locality and that includes an assemblage of macrofossils named after a single taxon with

an eye-catching morphology, like, for instance, that of the ammonite taxon *Gregoryceras transversarium* (QUENSTEDT). The zone in the sense of OPPEL is, therefore, unequivocal and can be checked in the field. The author of this study documented with very numerous and well-preserved, figured ammonites mainly in the book by GYGI (2001, Fig. 1) and in other publications that the Birmenstorf Member in Canton Aargau and with it the revised *Transversarium* Zone of OPPEL and WAAGEN (1866) comprise what are in this study the *Densiplicatum*, *Antecedens*, and *Transversarium* Zones. OPPEL and WAAGEN (1866, p. 245) included their *Ammonites Schilli* OPPEL in the list of ammonites which occur in the *Transversarium* Zone. GYGI (1969a, p. 66) drew attention to the difficulty in delimiting the Birmenstorf Member lithologically from the Effingen Member above and proposed a solution. GYGI (2000c, Fig. 3) followed the delimitation of the lithostratigraphic members as proposed by GYGI (1969a, p. 66) and stated that *Larcheria schilli* (OPPEL) occurs in Canton Aargau in the lowermost Effingen Member directly above the Birmenstorf Member. *Larcheria schilli* (OPPEL) is consequently younger than *Gregoryceras transversarium* (QUENSTEDT). The holotypes of both *Gregoryceras transversarium* and *Larcheria schilli* were found in Canton Aargau in northern Switzerland.

7.3 The Ammonite Zones of the Oxfordian and Kimmeridgian Stages in Northern Switzerland

The ammonite *zone* in this section is the range zone of the International Stratigraphic Guide, Second Edition, p. 85, edited by SALVADOR (1994). It is equivalent to the ammonite *subzone* in earlier publications by the author of this study. The ammonite zones in this section are, as far as possible, those which were published as subzones by CARIOU and HANTZPERGUE (coord. 1997) in France. The book coordinated by CARIOU and HANTZPERGUE (1997) is a comprehensive summary of the biochronology of all macro- and microfossils of the Jurassic in western and southern Europe. It was stated above that the entire collection of the author of this study is now kept in the Museum of Natural History at Basel. Close to 10,000 ammonites are stored in the collection. The ammonites were found, with a few exceptions, *in situ* mostly in northern Switzerland. Among these, more than 6,000 specimens are prepared, and most of the perisphinctids are measured in great detail. In case the index of an ammonite zone or a subzone currently used in adjacent France or in Germany is not represented in the author's collection, then a new, time-equivalent zone is proposed here in order to be used in Switzerland. The three new ammonite zones that are defined in this study are based on well-preserved specimens, which were figured in previous publications by the author. All of the zonal indices of the Oxfordian and the Kimmeridgian in Switzerland are figured in stratigraphic order in Figs. 7.1–7.6.

Scarburgense Zone: The oldest specimen of the index taxon *Cardioceras (Scarburgiceras) scarburgense* (YOUNG and BIRD) which was found by the author *in situ* in northern Switzerland is from the iron-oolitic clayey marl of the uppermost Herznach Formation, bed no. 20 in section RG 307 in the quarry of La Charuque at Péry in Canton Bern. This ammonite was figured by GYGI (1990a, Pl. 3, Fig. 11). Section RG 307 is represented in GYGI (1990a, Fig. 4) and in GYGI (2000a, Pl. 22). Two specimens of *Cardioceras (Scarburgiceras) scarburgense* found in the iron mine of Herznach in Canton Aargau, probably in bed E 2 of JEANNET (1951, Fig. 2), were figured by MARCHAND et al. (2000, Pl. 1, Figs. 11–12). Well-preserved specimens of the taxon in the lowermost *Renggeri* Member of section RG 280 in the clay pit of Andil at Liesberg in Canton Basel-Landschaft first appear 1.2 m above the base of the member (Fig. 7.1A). The *Renggeri* Member was named after the ammonite *Creniceras renggeri* (OPPEL). The specific name *renggeri* was introduced by OPPEL (1863, p. 203) in a way which is not in agreement with modern rules of zoological nomenclature. Nevertheless, GYGI (1991a, p. 124) advocated validating the name, because it is now widely used and because it was given to an ammonite taxon which is abundant, easy to identify, and has a vast geographic range. A complete adult of *Creniceras renggeri* (OPPEL) from the lowermost *Renggeri* Member at Liesberg was figured by GYGI (1990a, Pl. 6, Fig. 2). Data presented in that paper are evidence that typical specimens of *Creniceras renggeri* coexisted near Liesberg with *Cardioceras (Scarburgiceras) scarburgense. Creniceras renggeri* occur in Europe in Switzerland, France, and southern England. *Creniceras renggeri* (OPPEL) was found in the lowermost part of the black shales of the Kidod Formation near the village of Majdal Shams 13 km southwest of the summit of Mt. Hermon in the borderland between Lebanon, Syria, and Israel. This can be read from the map in PICARD and HIRSCH (1987, Fig. 18).

Praecordatum Zone: The zone is named after the ammonite *Cardioceras (Scarburgiceras) praecordatum* R. DOUVILLÉ. A well-preserved specimen of the taxon was found in the lower *Renggeri* Member of section RG 280 in the clay pit of Andil near Liesberg in Canton Basel-Landschaft, 3.5 m above the base of the member. This ammonite is shown in Fig. 7.1B.

Bukowskii Zone: The zonal name is derived from *Cardioceras (Scarburgiceras) bukowskii* MAIRE. An incomplete specimen of the taxon which was found 47 m above the base of the *Renggeri* Member in section RG 280 in the clay pit of Andil near Liesberg was figured by GYGI (1990a, Pl. 6, Fig. 1). Several well-preserved specimens of the taxon were

collected by B. LANGE from an outcrop at locality RG 387 called La Cornée near Rebévelier in Canton Bern, where a small creek cuts into clayey marl of the upper part of the *Renggeri* Member. One of the specimens found by B. LANGE is shown in Fig. 7.1C.

Costicardia Zone: The zonal index is *Cardioceras costicardia* BUCKMAN. A very well-preserved *Cardioceras costicardia,* subspecies *vulgare* ARKELL, was found by B. HOSTETTLER in the uppermost *Renggeri* Member at locality RG 387, La Cornée near Rebévelier in Canton Bern. This specimen is shown in Fig. 7.1D$_{1-2}$.

Cordatum Zone: The index *Cardioceras (Cardioceras) cordatum* (SOWERBY) is uncommon in northern Switzerland. A single specimen was found to date by R. and S. GYGI in the region investigated, in Churz Tal near Siblingen in Canton Schaffhausen, in bed no. 14a of excavation RG 207. The ammonite is shown in Fig. 7.1E$_{1-2}$. Section RG 207 is figured in GYGI and MARCHAND (1982, Fig. 2). Specimens of the coeval taxon *Cardioceras (Cardioceras) persecans* (BUCKMAN) are much more common than the zonal index in the *Cordatum* Zone in northern Switzerland. Several *Cardioceras (Cardioceras) persecans* (BUCKMAN) which were collected in northern Switzerland from *in situ* were figured by GYGI and MARCHAND (1982).

Densiplicatum Zone: Two well-preserved specimens of the zonal index *Cardioceras (Vertebriceras) densiplicatum* BODEN were found in bed no. 7 of excavation RG 212 in Churz Tal near Siblingen in Canton Schaffhausen. One of them is shown in Fig. 7.1F$_{1-2}$. Section RG 212 is represented in Fig. 8.3. The geographic range of the coeval taxon *Cardioceras (Vertebriceras) vertebrale* (SOWERBY), which is the index of the *Vertebrale* Subzone in the *Plicatilis* Zone in southern England, is possibly restricted to that region. No representative of the taxon was probably ever found in Switzerland. Conversely, *Cardioceras (Vertebriceras) densiplicatum* BODEN is fairly abundant and has a by far greater geographic range. The *Plicatilis* Zone was conceived by CALLOMON (1960, p. 189, and Fig. 5 on p. 196). CALLOMON (1960) knew of only wholly septate nuclei of *Perisphinctes (Arisphinctes) plicatilis* in southern England and elsewhere in Europe when he declared the taxon to be a zonal index. Complete adults of *Arisphinctes* are very large perisphinctid ammonites which can only be unambiguously identified if at least part of the body chamber is preserved. One nucleus of *Perisphinctes (Arisphinctes) plicatilis* (SOWERBY) was collected from the Schellenbrücke Bed in the Herznach iron mine and was figured by GYGI (1998, Pl. 9, Fig. 4). The specimen is evidence that the vertical range of the taxon *plicatilis* begins in northern Switzerland in the *Cordatum* Zone like in southern England (oral communication by J. H. CALLOMON). It can be read from Fig. 230 on p. 163 in GYGI (2001) that the top of the vertical range of *Perisphinctes (Arisphinctes) plicatilis* is in the

Fig. 7.1 Index ammonites of *Scarburgense* to *Schilli* Zone. *Arrows* point to the position of the last septum of the phragmocone. (**A**) ***Scarburgense* Zone:** *Cardioceras (Scarburgiceras) scarburgense* (Young and Bird), section RG 280, bed no. 7, 1.2 m above base of bed, lowermost *Renggeri* Member, clay pit of Andil, Liesberg, Canton Basel-Landschaft. Refigured from Gygi (1990a, Pl. 3, Fig. 4). Scale ×1. (**B**) ***Praecordatum* Zone:** *Cardioceras (Scarburgiceras) praecordatum* R. Douvillé, section RG 280, bed no. 7, 3.5 m above base of bed, lower

upper *Densiplicatum* Zone. The author of this study made it plain in several of his publications, mainly in Gygi (1998, 2001), that the conception of the *Plicatilis* Zone in southern England by Callomon (1960) was inadequate. The zone is therefore omitted in this study.

The base of the *Transversarium* Zone, as it was revised by Oppel in Oppel and Waagen (1866), is for instance at Birmenstorf directly above iron oolite of the Bathonian. The pertinent boundary was excavated by the author at the eastern end of the vineyard Nettel in a trench on August 16, 1973. The position of the pertinent trench no. 1 is indicated in the geological map of Fig. 2 in Gygi (1977). The base of the zone in Oppel's sense in excavation RG 208 at Ueken is above iron oolite of the Schellenbrücke Bed of the *Cordatum* Zone. It is, therefore, coeval with the base of the *Plicatilis* Zone as it was conceived by Callomon (1960, p. 189, and Fig. 5 on p. 196). This is shown in Figs. 2 and 3 in Gygi (2000c). Time-equivalent with *Cardioceras densiplicatum* Boden of the Boreal Faunal Province is in the Tethyan Faunal Province *Gregoryceras tenuisculptum* Gygi (Gygi 1977, Fig. 4). The holotype of this taxon was found in bed no. 7 of excavation RG 212 at Siblingen in Canton Schaffhausen, and it was figured by Gygi (1977, Pl. 1, Fig. 5a–b). The taxon was found in the Monti Lessini, Province of Verona, Italy, by A. Benetti and N. Pezzoni. Two specimens from there were figured by Gygi (1990d, Fig. 2a–c). More specimens of the taxon were figured or mentioned to occur in southeastern France, Poland, and Spain (Bert 2004). *Gregoryceras tenuisculptum* was recently mentioned to occur in North Africa (Bert et al. 2009, Fig. 12). The taxon can therefore be used as a substitute of *Cardioceras densiplicatum* as zonal index in the Tethyan Faunal Province.

Antecedens Zone: The zonal index *Perisphinctes (Dichotomosphinctes) antecedens* Salfeld is comparatively common in northern Switzerland. The best-preserved specimen was found in bed no. 16a of excavation RG 207 in Churz Tal near Siblingen. This ammonite is shown in Fig. 7.1G. Section RG 207 is represented in Gygi and Marchand (1982, Fig. 2).

Transversarium Zone (revised): A *Transversarium* Zone *sensu lato* was introduced by Oppel (1863, p. 165). It was stated above that Oppel's revised and final definition of the zone was published by W. Waagen after Oppel's early death (Oppel and Waagen 1866). The base of Oppel's *Transversarium* Zone was at the base of the *Densiplicatum* Zone, and it included the *Schilli* Zone. Another revision of the *Transversarium* Zone is now needed in order to define the zone according to the rules of the International Stratigraphic Guide edited by Salvador (1994). This revision of the *Transversarium* Zone is made at the end of this section. The revised definition of the zone is the result of ammonite biochronology based on the several thousand ammonites, which were excavated from the Birmenstorf Member in Canton Aargau and from the coeval part of the lowermost Effingen Member in Canton Schaffhausen since 1962 (see Gygi 2000c, Fig. 3).

A. Oppel selected the ammonite taxon *Gregoryceras transversarium* (Quenstedt) as zonal index probably because of its extraordinary morphology, although ammonites of the taxon are uncommon. The holotype of the zonal index is refigured in Fig. 7.1 H$_{1-2}$. The holotype was found in the territory of the township in Canton Aargau, which was called Birmensdorf at the time of Quenstedt and Oppel. The name of this village is now spelled Birmenstorf (see above). Oppel (1863, p. 165) made plain that the "proper region" of *Gregoryceras transversarium* was the "geognostic horizon with (siliceous) sponges" that Moesch (1863, p. 160) had named Birmensdorf Beds. Oppel (1863, p. 167) stated that the base of his new *Transversarium* Zone was "above the iron ore with *Amm. Lamberti, cordatus* and *perarmatus*" (sic). The base of Oppel's *Transversarium* Zone is, therefore, unequivocal, even though Mangold and Gygi (1997) documented with detailed, measured sections

---

Fig. 7.1 (continued) *Renggeri* Member, clay pit of Andil, Liesberg, Canton Basel-Landschaft. Refigured from Gygi (1990a, Pl. 4, Fig. 8). Scale ×1. (**C**) *Bukowskii* **Zone:** *Cardioceras (Scarburgiceras) bukowskii* Maire, locality RG 387, from bed of a small creek cutting into the upper *Renggeri* Member northwest of locality La Cornée, Rebévelier, Canton Bern. Refigured from Gygi (1990a, Pl. 7, Fig. 1). Scale ×1. (**D**) 1–2. *Costicardia* **Zone:** *Cardioceras (Cardioceras) costicardia* Buckman, sub-species *vulgare* Arkell, locality RG 387, from bed of a small creek cutting into the uppermost *Renggeri* Member northwest of locality La Cornée, Rebévelier, Canton Bern. Refigured from Gygi (1995, Fig. 3/2a–b). Scale ×1. (**E**) 1–2. *Cordatum* **Zone:** *Cardioceras (Cardioceras) cordatum* (Sowerby), excavation RG 207 beside water conduit, bed no. 14a, Glaukonitsandmergel Bed, Churz Tal, Siblingen, Canton Schaffhausen. Refigured from Gygi and Marchand (1982, Pl. 10, Fig. 1a–b). Scale ×1. (**F**) 1–2. *Densiplicatum* **Zone:** *Cardioceras (Vertebriceras) densiplicatum* Boden, excavation RG 212 above shooting range, bed no. 7, Mumienmergel Bed, Churz Tal, Siblingen, Canton Schaffhausen. Refigured from Gygi and Marchand (1982, Pl. 11, Fig. 5a–b). Scale ×1. (**G**) *Antecedens* Zone: *Perisphinctes (Dichotomosphinctes) antecedens* Salfeld, excavation RG 207 beside water conduit, bed no. 16a, lower Mumienkalk Bed, Churz Tal, Siblingen, Canton Schaffhausen. Refigured from Gygi (2000a, Pl. 3, Fig. 4). Scale ×1. (**H**) 1–2. *Transversarium* Zone: *Gregoryceras transversarium* (Quenstedt), holotype, from soil above uncondensed part of Birmenstorf Member, Birmenstorf, Canton Aargau. Refigured from Gygi (1977, Pl. 7, Fig. 1a–b). Scale ×1. (**I**) *Schilli* Zone: *Larcheria schilli* (Oppel), not from *in situ*. Specimen fallen probably from bed no. 46 in lowermost Effingen Member in section RG 226, in road cut south of quarry of Unteregg, Veltheim, Canton Aargau. Refigured from Gygi (2000a, Pl. 8, Fig. 2). Scale ×1. Two specimens of the taxon of lesser preservation which were collected from *in situ* in the reference section RG 276 of the *Schilli* Zone in the quarry of Chalch at Holderbank in Canton Aargau are shown in Fig. 4a–b in Gygi (2001). Figs. E, F refigured with permission of Geobios

and with figured ammonites excavated from these sections that the iron oolite directly below the lowermost, stratigraphically condensed bed of the Birmenstorf Member is in a certain region in Canton Aargau of middle Bathonian age. This is the case, for instance, at Birmenstorf. OPPEL (1863) described and figured 13 new ammonite taxa from his *Transversarium* Zone at Birmenstorf alone. According to MOESCH (1867, p. 132), ten of the pertinent types of OPPEL's new taxa were found by C. MOESCH at the locality called Nettel. The vineyard of Nettel is approximately 700 m northeast of the village of Birmenstorf.

A geological map at the detailed scale of 1:5,000 of the vineyards called Bergreben and Nettel northeast of Birmenstorf was prepared by the author of this study in 1973. The aim was to find an outcrop of the Birmenstorf Member which could serve as reference section of the *Transversarium* Zone and to find topotypes of the new taxa described by QUENSTEDT (1847), OPPEL (1863), and MOESCH (1867). The geological map was printed as Fig. 2 in GYGI (1977). The result of the mapping was that an outcrop of the Birmenstorf Member near Birmenstorf did not exist, and probably never existed in the past. The ammonites from Birmenstorf figured by QUENSTEDT, OPPEL, and MOESCH must, therefore, all have been collected loose from the soil above the uncondensed part of the Birmenstorf Member in the vineyards northeast of the village of Birmenstorf.

GYGI (1969a, p. 64) proposed a lithostratigraphic definition of the Birmenstorf Member which is in agreement with the practice of how the member was previously conceived in the geological maps by F. MÜHLBERG. GYGI (1969a, Pl. 17) chose section RG 60 in the natural cleft of Eisengraben in the territory of the township of Gansingen in northeastern Canton Aargau to be the type section of the Birmenstorf Member, because he could not find a section of the member when doing detailed geological mapping at Birmenstorf. The type section RG 60 near Gansingen was later completed and replaced by sections RG 210 and RG 230. Excavation RG 210 was made on the northern rim of Eisengraben, and excavation RG 230 was located in a meadow about 30 m further north. These sections were initially labeled as "Grabung 1 and 3" in Pl. 11 in GYGI (1977). Then they were assembled in GYGI (2000c, Fig. 2) and in GYGI (2001, Fig. 1). The combined sections RG 210 and RG 230 encompass the entire Birmenstorf Member. GYGI (1977, p. 517) designated the assembled sections RG 210 and RG 230 to be the *reference section* of the *Transversarium* Zone in the sense of HEDBERG (1976, p. 58). The deep cleft of Eisengraben opened when an inclined, coherent block with a stack of early Callovian strata and the almost complete thickness of the Birmenstorf Member above slid about 10 m downslope to the south, after the foot of the stratified block was truncated by the incision of the small valley south of Eisengraben.

A *Gregoryceras transversarium* (QUENSTEDT) was found by R. and S. GYGI in 1973 in excavation RG 230 in the upper, particularly fossiliferous part of the uncondensed, normal facies of the Birmenstorf Member. This *Gregoryceras* is figured in GYGI (1977, Pl. 8, Fig. 1a, b). The excavations made by R. and S. GYGI in the Birmenstorf Member in Canton Aargau and in the thin time equivalents of the member in the lowermost Effingen Member in Canton Schaffhausen, mainly in the Mumienmergel Bed (abbreviated MUM in Fig. 1.5) and in the Mumienkalk Bed (MUK), yielded several thousand ammonites. The sections of eight of these excavations which were made since 1963 are represented in Pl. 11 in GYGI (1977). The ammonites best preserved from these excavations were published in the several papers which are mentioned in Sect. 1.2, beginning with GYGI (1966), and mainly in the book by GYGI (2001). The ammonite assemblage in the Birmenstorf Member in Canton Aargau, which is the type region of the *Transversarium* Zone, is consequently now documented in detail.

By far most of the ammonites that were collected in the Birmenstorf Member in northern Canton Aargau were found in the upper part of the stratigraphically uncondensed facies of the member. The uncondensed part of the member is on average approximately 5 m thick. The systematic, bed-by-bed excavations in the Birmenstorf Member revealed that the lowermost bed of the member with an average thickness of less than 10 cm is stratigraphically condensed in the strict sense (see below) in the whole of northern Canton Aargau, and that this bed represents both the *Densiplicatum* Zone and the *Antecedens* Zone. The condensed, lowermost bed of the Birmenstorf Member, which is no. 10 in the large excavation RG 208 at locality Brunnrain near Ueken in Canton Aargau, was uncovered and searched for macrofossils in the summer of 1971 on a surface of 119 m^2. The position of bed no. 10 in that excavation is shown in the photograph of Fig. 5 on p. 18 in GYGI (2000a). The section of excavation RG 208 is drawn as no. 2 in Pl. 11 in GYGI (1977). Details of section RG 208 are represented in this study in Fig. 4.14 (bed nos. 5, 6) and in Fig. 8.2 (bed nos. 7–9). Locality Brunnrain is 500 m southwest of the village of Ueken, about halfway between the village and the eastern part of the Herznach iron mine below Hübstel hill. The locality of excavation RG 208 is named Buech in the last edition of sheet no. 1069 at the scale of 1:25,000 of the Swiss National Map (Landeskarte). The *Densiplicatum* Zone in bed no. 10 of excavation RG 208 was documented with a *Cardioceras* which was identified by D. MARCHAND and figured in GYGI and MARCHAND (1982, Pl. 12, Fig. 3a, b), and with the *Perisphinctes (Arisphinctes) parvus* GYGI which is figured in GYGI (2001, Fig. 32). The *Antecedens* Zone is documented in bed no. 10 with the index taxon of the zone, a *Perisphinctes (Dichotomosphinctes) antecedens* SALFELD. This specimen was figured by GYGI (2001, Fig. 62c). There are few ammonites in the lowermost bed of the Birmenstorf Member, in spite of the minimal average rate of sedimentation which was the cause of stratigraphic condensation. No ammonites from the condensed bed were figured by OPPEL (1863).

GYGI (1977) published a revision of the ammonite genus *Gregoryceras* based on new material which was found in many of his excavations in northern Switzerland. *Gregoryceras* lived in low paleolatitudes. They were found in Central and mainly in Southern Europe, Central Asia, North Africa, Madagascar, India, and Chile in South America. Two specimens of *Gregoryceras transversarium* were found in the Cordillera de Domeyko in the Atacama Desert of northern Chile: one by A. VON HILLEBRANDT and a particularly well-preserved specimen by G. CHONG. These ammonites were identified by the author of this study to be *Gregoryceras transversarium* (QUENSTEDT). Both specimens were figured by GYGI and VON HILLEBRANDT (1991, Pl. 6, Fig. 1, and Pl. 8, Fig. 1a, b). They are evidence that the geographic range of the *Transversarium* Zone is intercontinental.

The question of what is the vertical range of the index *Gregoryceras transversarium* (QUENSTEDT) of the *Transversarium* Zone was raised because of the Ph. D. thesis of HAUERSTEIN (1966). The reason why GYGI (1977, Fig. 4) thought that *Gregoryceras transversarium* first appeared in the *Antecedens* Chron is that he trusted HAUERSTEIN (1966), who stated that he found a *Gregoryceras transversarium* in what he called Mumienschichtle, in the then opencut iron mine at the southern foot of Stoberg hill northeast of Blumberg in southern Germany. GYGI (1977, p. 455) renamed HAUERSTEIN's Mumienschichtle Mumienmergel Bed in order to distinguish it from the Mumienkalk Bed directly above. HAUERSTEIN's labeling of the bed, in which he stated to have found his *Gregoryceras,* was in the meantime recognized to be an error. The *Gregoryceras transversarium* that HAUERSTEIN had found near Blumberg was examined by the author of this study at its repository at München, and the ammonite was figured by GYGI (1977, Pl. 6, Fig. 3). It is now certain that HAUERSTEIN found this *Gregoryceras* in what GYGI (1977, p. 455) called Mumienkalk Bed (see this study, Fig. 8.3). The Mumienmergel Bed and the Mumienkalk Bed were excavated by the author on May 18, 1963, as section RG 88, 10 m east of the excavation which was made by HAUERSTEIN a short time before in the opencut, then disused iron mine near Blumberg. Section RG 88 was figured as no. 9 in Pl. 11 by GYGI (1977). It follows from this that the vertical range of *Gregoryceras transversarium* (QUENSTEDT 1847), which is a rare taxon in the area investigated, is about the same as that of *Perisphinctes (Dichotomosphinctes) elisabethae* DE RIAZ, which is the most common perisphinctid ammonite taxon in the type *Transversarium* Zone of northern

Switzerland. *Perisphinctes (Dichotomosphinctes) elisabethae* DE RIAZ is shown as no. 13 in Fig. 1.5, and it is listed as no. 13 in Fig. 1.6 of this study. *Perisphinctes (Dichotomosphinctes) luciaeformis* ÉNAY is coeval. The revised *Transversarium* Zone in this study replaces the *Luciaeformis* Subzone, which is represented in GYGI (2001, Figs. 1 and 230). The base of the revised *Transversarium* Zone is here defined to be at the base of bed no. 9 in section RG 210 in the natural cleft of Eisengraben in the territory of the township of Gansingen in northern Canton Aargau. Section RG 210 is shown in Fig. 1 on p. 7 in GYGI (2001).

Schilli Zone (revised): OPPEL (1863) as well as OPPEL and WAAGEN (1866, p. 245) included *Ammonites Schilli* OPPEL into their list of ammonites occurring in the *Transversarium* Zone. According to OPPEL (1863, p. 246), the holotype of *Larcheria schilli* (OPPEL) was found at the locality that OPPEL called Kreisacker. The locality is now spelled Cheisacher and is in the township of Sulz in northeastern Canton Aargau. The local name means plowed land with a lot of rock fragments. No outcrop now exists in the area of Cheisacher. The slightly corroded holotype of *Larcheria schilli* (OPPEL) is a cast, that most probably weathered out of rock below arable land, and which was turned up to the surface by a plowshare. The cast was then picked loose from the soil. *Larcheria schilli* (OPPEL) is the index taxon of the *Schilli* Zone in the sense of GYGI (2001). Data published by GYGI (2001, p. 8 and Figs. 2, 3) are evidence that *Gregoryceras transversarium* (QUENSTEDT) and *Larcheria schilli* (OPPEL) did not coexist. The taxa *succeeded* each other in time in what is their type region of Canton Aargau. GYGI (2001), therefore, proposed to keep the *Schilli* Zone separate from the *Transversarium* Zone below and from the *Bifurcatus* Zone above.

GYGI (2001, p. 8) chose section RG 276 in the now disused quarry of Chalch in the township of Holderbank in Canton Aargau to be the *reference section* of the *Schilli* Zone, because several specimens of the zonal index were found in the section *in situ*. Section RG 276 is shown as Fig. 3 in GYGI et al. (1979). *Larcheria schilli* first appears in bed no. 39 of section RG 276 according to the private collector D. KRÜGER, who found the pertinent specimen. The two specimens of *Larcheria schilli* which are shown in GYGI (2001, Fig. 4a, b) were collected by R. and S. GYGI from *in situ* in what is now bed no. 42 of section RG 276. According to GYGI (1977, p. 446), *Larcheria schilli* first appears in easternmost Canton Aargau in bed no. 7 of excavation RG 51b near Oberehrendingen. This section was figured as no. 5 in Pl. 11 by GYGI (1977). A *Subdiscosphinctes (Aureimontanites) holderbankensis* GYGI was found in bed no. 8 of excavation RG 51b. The specimen is shown in Fig. 215 in GYGI (2001). The holotype of *Subdiscosphinctes (Aureimontanites) holderbankensis* was found by R. and S. GYGI in bed no. 40 of section RG 276 at

Holderbank, directly above bed no. 39, in which *Larcheria schilli* (OPPEL) first appears. The thickness of the Birmenstorf Member and of the *Transversarium* Zone consequently diminishes in Canton Aargau between Holderbank and Oberehrendingen to 0.5 m.

Larcheria schilli was found to date neither in Canton Schaffhausen nor in adjacent southern Germany. *Subdiscosphinctes (Aureimontanites) holderbankensis* can be used in that region as a substitute of *Larcheria schilli* in order to document the *Schilli* Zone. One specimen of *Subdiscosphinctes (Aureimontanites) holderbankensis* was found in bed no. 11 of excavation RG 212 above the shooting range in Churz Tal near Siblingen in Canton Schaffhausen (this study, Fig. 8.3). GYGI (2003, p. 180) concluded from this that the top of the *Transversarium* Zone in Canton Schaffhausen is at the top of the thin, unnamed glauconitic marl above the Mumienkalk Bed. This zonal boundary is in section RG 212 within the lower part of bed no. 10. The base of the *Schilli* Zone in that region is, therefore, where the succession begins to be without glauconite. This means, according to Fig. 8.3, that the average thickness of the *Transversarium* Zone in the sense of GYGI (1977) is in Canton Schaffhausen 0.4 m, and that the entire thickness of the incompletely sedimented and partially stratigraphically condensed succession no. 1 is in Canton Schaffhausen on average 0.5 m. The thin sediments of succession no. 1 in this region, including the hiati in the sediments, represent a time span of 8 ammonite chrons (this study, Fig. 1.6) or of as much as approximately 3 million years. Sedimentary succession no. 1 is complete in northwestern Switzerland, and it is in that region on average 185 m thick. This is 370 times more than the thickness of succession no. 1 in Canton Schaffhausen. Only a detailed study of a great number of ammonites collected from *in situ,* and that were prepared and identified thereafter, as well as use of mineral stratigraphy in time correlation between sediments from deeper and from shallow water made this conclusion possible. This exemplifies the practical importance of ammonites in unraveling the history of sedimentation during the Late Jurassic in northern Switzerland.

Rotoides Zone: The first four specimens of *Perisphinctes (Dichotomoceras) rotoides* RONCHADZE, *non* ARKELL, which were ever collected from *in situ* in a measured section in the Jura Mountains, are the specimens that were found by R. and S. GYGI in bed no. 46 of section RG 276 in the now disused quarry of Chalch near Holderbank in Canton Aargau. GYGI in GYGI and PERSOZ (1986, p. 422) drew attention to this and added that the age of *Perisphinctes rotoides* of recent authors following ARKELL (1935–1948) is much older. According to CALLOMON (1960, p. 197), *Perisphinctes (Dichotomosphinctes) rotoides* ARKELL, *non* RONCHADZE, first appears in the Vertebrale Subzone. Section RG 276 is represented as Fig. 3 in GYGI et al. (1979). One specimen of the four

Perisphinctes (Dichotomoceras) rotoides RONCHADZE which were found *in situ* near Holderbank in Canton Aargau is shown in Fig. 7.2A. The holotype of the zonal index is from a locality in the high Jura range above the township of Gex in France near Geneva. According to GYGI (2000a, p. 85), the taxon *rotoides* RONCHADZE occurs as far afield as Madagascar.

Stenocycloides Zone: *Perisphinctes (Dichotomoceras) stenocycloides* SIEMIRADZKI is fairly abundant in the lower Effingen Member of Canton Aargau. One specimen from bed no. 50 in section RG 276 in the quarry of Chalch near Holderbank in Canton Aargau is shown in Fig. 7.2B. Typical *Perisphinctes (Dichotomoceras) bifurcatus* (QUENSTEDT) were not found in this zone.

Bifurcatus Zone (revised): The only *Perisphinctes (Dichotomoceras) bifurcatus* (QUENSTEDT) that was ever collected in northern Switzerland from *in situ* in a level which is precisely recorded is the specimen that B. HOSTETTLER found in section RG 307 near Péry in Canton Bern near the base of the unit with coral reefs, no. 160 of the section. Section RG 307 is shown in Pl. 22 in GYGI (2000a). The ammonite was in direct contact with a coral colony, about 2 m above the base of a reef. The base of the pertinent reef is in the uppermost Effingen Member. The ammonite is figured natural size in HOSTETTLER (2006, Fig. 4). Another specimen of *Perisphinctes (Dichotomoceras) bifurcatus* (QUENSTEDT) was found by R. ÉNAY in 1964 in the limestone succession of the lower Günsberg Formation in the landslide of Gschlief north of Günsberg in Canton Solothurn (section RG 14 in GYGI 1969a, Pl. 18). This ammonite was figured by GYGI (1995, Fig. 17/2). Ammonites are rare in the Günsberg Formation. It can be presumed that ÉNAY found his specimen in bed no. 137 of section RG 14, because the author of this study found an ammonite in bed no. 137 in 1963.

A complete adult of *Perisphinctes (Dichotomoceras) bifurcatus* (QUENSTEDT) is kept in the ROLLIER collection in the Geological Institute of the ETH Zürich. The ammonite was figured by ÉNAY and GYGI (2001, Pl. 1, Fig. 1), and it is shown here in Fig. 7.2C. The specimen is labeled to be from Müliberg. This is a hill southwest of the village of Bözen in Canton Aargau, where the Gerstenhübel Beds crop out. To judge from the material of the cast of the specimen from Müliberg, the ammonite was collected from the micritic limestone unit of the Gerstenhübel Beds. Many more specimens of *Perisphinctes (Dichotomoceras) bifurcatus* (QUENSTEDT) out of the ROLLIER collection at Zürich were figured by ÉNAY and GYGI (2001) from the quarry of Hinterstein in the Effingen Member near Oberehrendingen in Canton Aargau. The matrix of these specimens is micritic limestone. The ammonites figured from Oberehrendingen were, therefore, probably all collected from the Gerstenhübel Beds like the specimen from Bözen. Among the ammonites in the ROLLIER collection from the quarry of Hinterstein are three specimens of *Mirosphinctes frickensis* (MOESCH), which were figured by ÉNAY and GYGI (2001, Pl. 4, Figs. 6–8). The matrix

Fig. 7.2 Index ammonites of *Rotoides* to *Hypselum* Zone. *Arrows* point to the position of the last septum of the phragmocone. (**A**) ***Rotoides*** **Zone:** *Perisphinctes (Dichotomoceras) rotoides* Ronchadze, *non* Arkell, from bed no. 46 in the lower Effingen Member in reference section RG 276 of the *Rotoides* Zone in the quarry of Chalch at Holderbank in Canton Aargau. Refigured from Gygi (2000a, Pl. 9, Fig. 1). Scale ×1. (**B**) ***Stenocycloides*** **Zone:** *Perisphinctes (Dichotomoceras) stenocycloides* Siemiradzki, section RG 276, bed no. 50, lower Effingen Member, quarry of Chalch, Holderbank, Canton Aargau. Refigured from Gygi (2000a, Pl. 9,

of these *Mirosphinctes* is micritic limestone like that of the several *Perisphinctes* (*Dichotomoceras*) *bifurcatus* figured from the same quarry. The three *Mirosphinctes* figured were, therefore, probably collected as well from the succession of limestone beds in the Gerstenhübel unit. If this is correct, their age would be the *Bifurcatus* Chron. This is noteworthy, because the exact horizon of the holotype of *Mirosphinctes frickensis* (MOESCH) is unknown and because GYGI in GYGI and VON HILLEBRANDT (1991, Pl. 3, Fig. 3) identified and figured a well-preserved specimen of *Mirosphinctes frickensis* (MOESCH) from Quebrada San Pedro near Caracoles in Region Antofagasta, Chile.

The revised *Bifurcatus* Zone as it is conceived here is the time equivalent of the *Grossouvrei* Subzone in the sense of the volume coordinated by CARIOU and HANTZPERGUE (1997, p. 85, and Table 10). The index of the *Grossouvrei* Subzone, *Perisphinctes* (*Dichotomoceras*) *grossouvrei* SIEMIRADZKI, was found by A. de GROSSOUVRE near Raix in Département Charente in southwestern France. According to ÉNAY in ÉNAY and GYGI (2001, p. 474), the taxon is probably endemic in Département Charente. It is coeval with typical *Perisphinctes* (*Dichotomoceras*) *bifurcatus* (QUENSTEDT), which are common and have a much greater geographic range. For this reason, the *Grossouvrei* Subzone is replaced in this study by the *Bifurcatus* Zone in the strict sense.

Hypselum Zone: A near-complete adult of the zonal index *Euaspidoceras hypselum* (OPPEL) was found in the quarry of Jakobsberg east of Auenstein in Canton Aargau. D. KRÜGER picked the specimen from talus below the succession of limestone bed nos. 55–80 in the upper Effingen Member in section RG 37, which is represented in GYGI (1969a, Pl. 17). The cast was figured natural size by GYGI (2000a, Pl. 10, Fig. 1), and it is shown here in Fig. 7.2D.

Berrense Zone: The zonal index *Epipeltoceras berrense* (FAVRE) is from what FAVRE (1876, p. 12) called "gray limestone" in the units of the Ultrahelvetic facies realm, which crop out on Mt. Montsalvens east of Lake Gruyère in the external Pre-Alps of Canton Fribourg in Switzerland. The specific name of the zonal index is misleading, because it refers to Mt. La Berra. This mountain is located 8 km northeast of Mt. Montsalvens. Only Tertiary rocks occur in Mt. La Berra. The specimen figured by FAVRE (1876, Pl. 3, Fig. 11) was selected as lectotype of *Epipeltoceras berrense* (FAVRE) by SCHMIDT-KALER (1961, p. 195). The type is refigured here in Fig. 7.3A. A small specimen of this taxon was found by the author in talus below the lower part of

section RG 279 of the Hornbuck Member near Siblingen in Canton Schaffhausen. The cast labeled Gy 1974 is kept in the Museum of Natural History Basel. Section RG 279 is figured in GYGI (1991b, Fig. 3) as part of section RG 82. It was stated above that C. MOESCH found a well-preserved *Epipeltoceras* in the glauconitic *Crenularis* Member directly above the top of the Geissberg Member in the quarry west of locality Fahr (unpublished section RG 36) east of the village of Auenstein in Canton Aargau. The pertinent *Epipeltoceras* is a complete adult which is transitional between *Epipeltoceras berrense* (FAVRE) and the younger *Epipeltoceras bimammatum* (QUENSTEDT). The ammonite found by C. MOESCH was figured by GYGI (2000a, Pl. 10, Fig. 5). The *Berrense* Zone, therefore, encompasses in northern Switzerland probably the entire Geissberg Member and possibly the uppermost part of the Effingen Member below, as well as the lower part of the Hornbuck Member, which is the time equivalent of the Geissberg Member (Fig. 1.5).

Bimammatum Zone: An incomplete specimen of the zonal index *Epipeltoceras bimammatum* (QUENSTEDT) was found by the author *in situ* in the uppermost Hornbuck Member, in bed no. 11 of section RG 76 on the northern slope of Hornbuck hill north of Riedern am Sand in southern Germany. The specimen is shown in Fig. 7.3B. The section at Hornbuck is represented as Fig. 2 in GYGI (1969a). Another *Epipeltoceras bimammatum* was found by R. TRÜMPY *in situ* in the uppermost Hornbuck Member in the gully of Steiggraben near Geisslingen in southern Germany, in the uppermost part of bed no. 42 in section RG 74. This section is figured in GYGI (1991b, Fig. 4). Steiggraben with section RG 74 is the name of a locality indicated on sheet no. 1050, and Hornbuck with section RG 76 is indicated on sheet no. 1051 of the Landeskarte der Schweiz 1:25,000. C. MOESCH found his first specimen of *Epipeltoceras bimammatum* (QUENSTEDT) in a quarry at Lauffohr, which is now a fraction of the city of Brugg in Canton Aargau. According to MOESCH (1867, p. 153), A. OPPEL created his new *Bimammatum* Zone (OPPEL 1863, p. 175) based on this specimen. To judge from OPPEL (1863, p. 179), C. MOESCH found the specimen in the uppermost part of the *Crenularis* Member. This is most probably bed no. 26 with much glauconite in the unpublished section RG 61, which was measured by the author of this study in the quarry at Lauffohr in 1962. Section RG 61 at Lauffohr must, therefore, be regarded to be the reference section of the *Bimammatum* Zone.

Fig. 7.2 (continued) *Fig. 3). Scale ×1. (C) **Bifurcatus** Zone: Perisphinctes (Dichotomoceras) bifurcatus (*QUENSTEDT*), hill of Müliberg southwest of the village of Bözen in Canton Aargau, lower Effingen Member, to judge from matrix probably out of Gerstenhübel Beds. Refigured from* ÉNAY *and* GYGI *(2001, Pl. 1, Fig. 1). Scale ×1.*

(D) *Hypselum* **Zone:** *Euaspidoceras hypselum* (OPPEL), from the quarry of Jakobsberg east of the village of Auenstein in Canton Aargau. Not from *in situ,* specimen fallen from the limestone succession in the upper Effingen Member, bed nos. 55–80 in section RG 37. Refigured from GYGI (2000a, Pl. 10, Fig. 1). Scale ×0.5

Gredingensis Zone (new): *Wegelea gredingensis* (Wegele) is selected here to be the zonal index of a new zone. Gygi (2003, Fig. 170) called this zone *Hauffianum* Subzone, although he had never found *Taramelliceras hauffianum* (Oppel) to that date in northern Switzerland. Oppel (1863, p. 175) stated that "*Amm. Hauffianus*" occurs *below* "*Ammonites bimammatus*". The introduction of the new zone is deemed to be justified for these two reasons. A near-complete specimen of *Wegelea gredingensis* (Wegele) was found by the author near Beggingen in Canton Schaffhausen. A photograph of the ammonite is Fig. 7.3C. The specimen is from the upper Küssaburg Member, bed no. 55 in the unpublished section RG 85. The section is schematically represented in the upper synthetic section on Pl. 19 in Gygi (1969a). Bed no. 55 of section RG 85 is 7 m below the slightly glauconitic marker bed of the Knollen Bed.

A. von Hillebrandt found a small ammonite cast which is probably conspecific with *Wegelea gredingensis* (Wegele) in the Cordillera de Domeyko in northern Chile. Gygi in Gygi and von Hillebrandt (1991) assigned the specimen to the genus *Larcheria* and figured it on Pl. 2, Fig. 6 of that paper. Gygi followed with this generic identification Tintant (1959, p. 129) and Énay (1964, p. 493). Énay (1966, p. 518) assigned the taxon *gredingensis* to the genus *Orthosphinctes*. Gygi (2000a, p. 92) disagreed with this, because the umbilicus of *gredingensis* is narrower than that in unambiguous *Orthosphinctes*, and because the peristome of *gredingensis* is possibly simple. This can be judged from Fig. 7.3C of the specimen from Beggingen. Moreover, the whorl section of *gredingensis* is elongate-oval. This is incompatible with *Orthosphinctes* which have a rounded whorl section. Gygi (2000a, p. 92) assigned the taxon *gredingensis* to the *new* genus *Wegelea* because of the morphologic features mentioned and because according to Fig. 1.6 in this study, there is a time gap of six ammonite chrons between typical *Larcheria* and the taxon *gredingensis*. Provided that the specimen from northern Chile is indeed a *Wegelea gredingensis* (Wegele) in spite of its small size, the *Gredingensis* Zone would have an intercontinental geographic range.

Planula Zone: One specimen of the index *Subnebrodites planula* (Quenstedt), *non* Hehl in Zieten, was figured by Gygi (2000a, Pl. 11, Fig. 5) from bed no. 14 of the unpublished section RG 84 in the quarry owned by the township of Hemmental in Canton Schaffhausen. The same specimen is shown here in Fig. 7.3D. Section RG 84 is schematically represented in the upper synthetic section of Pl. 19 in Gygi (1969a). Another specimen of the taxon was figured by Gygi (2003, Fig. 60) from bed no. 139 in section RG 82c in the old quarry at locality Steimürlichopf in Churz Tal near Siblingen in Canton Schaffhausen. Section RG 82c is represented in Gygi (1969a, Pl. 16). Both specimens figured are from the

lowermost part of the Wangental Member in Canton Schaffhausen, directly above the glauconitic Knollen Bed.

Galar Zone: The index *Sutneria galar* (Oppel) was found *in situ* in bed no. 114 in the uppermost, glauconitic part of the Letzi Member in section RG 70 at Mellikon in Canton Aargau. Section RG 70 is represented in Gygi (1969a, Pl. 17). This incomplete specimen was figured by Gygi (2003, Fig. 64). A well-preserved, complete adult of *Sutneria galar* with lappet, which was figured by Gygi (2000a, Pl. 13, Fig. 3) at the scale ×2, was found in the uppermost Wangental Member, in bed no. 18 of excavation RG 239. The specimen is refigured here at the scale ×2 in Fig. 7.3E. The major excavation RG 239 was made in 1974 in a disused quarry at locality Summerhalde, above the road leading from the city of Schaffhausen to the village of Hemmental. The section of excavation RG 239 is represented as Fig. 160 in Gygi (2003, p. 140).

Platynota A Zone (revised): The zonal index is *Sutneria platynota* (Reinecke), morphotype A Schairer. One specimen of the index was figured by Gygi (2003, Fig. 159a). It was found in the lowermost part of the glauconitic lower Baden Member, in bed no. 120 in section RG 70 near Mellikon in Canton Aargau. This section is represented in Gygi (1969a, Pl. 17). Another specimen of the index, a well-preserved and complete adult with lappet, is shown at the scale ×2 in Fig. 7.3F. This ammonite was found in the lowermost Schwarzbach Formation, in bed no. 22 of excavation RG 239 at locality Summerhalde near the city of Schaffhausen. The section of excavation RG 239 is represented as Fig. 160 in Gygi (2003, p. 140).

Desmoides Zone: The zonal index, *Orthosphinctes desmoides* (Wegele), was found in the lower, glauconitic Baden Member, in bed no. 124 of section RG 70 near Mellikon in Canton Aargau (Gygi 1969a, Pl. 17). A representative of the taxon is shown in Fig. 7.4A.

Platynota C Zone (revised): *Sutneria platynota* (Reinecke), morphotype C Schairer, first appears in bed no. 26 of section RG 239 at locality Summerhalde near the city of Schaffhausen. A complete specimen from bed no. 30 in the same section is shown in Fig. 7.4B. Section RG 239 is represented in Gygi (2003, Fig. 160 on p. 140).

Lussasense Zone (new): *Ataxioceras (Schneidia) lussasense* Atrops is well represented in what was called Hippolytense Subzone by Gygi (2003, Fig. 170). Specimens of the taxon *hippolytense* Atrops were not found to date in northern Switzerland. A specimen of *A. lussasense* is shown in Fig. 7.4C. The ammonite is from bed no. 124 in section RG 70 near Mellikon in Canton Aargau. Section RG 70 is represented in Gygi (1969a, Pl. 17).

Evolutum Zone (new): The zonal index *Ataxioceras (Parataxioceras) evolutum* Atrops is relatively abundant in northern Switzerland in what Gygi (2003, Fig. 170) called *Lothari* Subzone. The holotype of *Ataxioceras*

Fig. 7.3 Index ammonites of *Berrense* to *Platynota* A Zone. *Arrows* point to the position of the last septum of the phragmocone. (A)

Berrense Zone: *Epipeltoceras berrense* (FAVRE), lectotype, "gray limestone", Mt. Montsalvens east of Lake Gruyère, Canton Fribourg.

(*Parataxioceras*) *lothari* (OPPEL) was found by C. MOESCH in the Baden Member at Baden in Canton Aargau. No unambiguous representative of the taxon could be found *in situ* by R. and S. GYGI in Canton Aargau (see GYGI 2000a, Pl. 14, Fig. 2). One ill-preserved specimen of *Ataxioceras* (*Parataxioceras*) *lothari* (OPPEL) from bed no. 57 in section RG 21 in the quarry on Mt. Born west of Olten in Canton Solothurn was figured by GYGI (1995, Fig. 22). Three specimens of *Ataxioceras* (*Parataxioceras*) *evolutum* ATROPS were figured in GYGI (2003, Fig. 122), and another is represented in Fig. 153 in GYGI (2003). The taxon occurs in the lower, glauconitic Baden Member at Mellikon in Canton Aargau, in the upper part of bed no. 124 of section RG 70 which is represented in Pl. 17 in GYGI (1969a). The specimen shown in Fig. 7.4D in this study is from Mellikon. The cast of *Ataxioceras* (*Parataxioceras*) *evolutum* ATROPS and that of *Balticeras pommerania* DOHM, which are both shown in Fig. 153 in GYGI (2003), almost touch each other. When Fig. 8.1 in this study is considered, it cannot be taken for granted that the *Ataxioceras* (*Parataxioceras*) *evolutum* ATROPS and the *Balticeras pommerania* DOHM, which are both represented close together in Fig. 153 in GYGI (2003), are coeval. Nevertheless, it is probable that the two taxa *are* coeval, because the very large specimen of *Balticeras pommerania* DOHM, which was figured by GYGI (2003, Fig. 80) from bed no. 59 of section RG 21 in the quarry on Mt. Born near Olten (GYGI 1969a, Pl. 18), was found above the beds, in which *Ataxioceras* are most abundant. GYGI (2003, p. 134) stated, that 13 specimens of *Balticeras pommerania* DOHM were found since 1962 in northern Switzerland. One of them is the near-complete adult of giant size, which was figured by GYGI (2003, Fig. 153). *Balticeras pommerania* DOHM was described for the first time from a quarry near Zarnglaff, which was then in the German region of Pommern. The pertinent township is now in northwestern Poland and is called Czarnoglowy. The *Lothari* Subzone in GYGI (2003, Fig. 170) is here replaced by the time-equivalent *Evolutum* Zone, because the index of the new zone is fairly common in northern Switzerland and because a time correlation can possibly be made with the zonal index *Ataxioceras* (*Parataxioceras*) *evolutum* ATROPS and with *Balticeras pommerania* DOHM between the

Subtethyan Faunal Province in northern Switzerland and the Subboreal Faunal Province in northwestern Poland.

Tenuicostatum Zone (revised): The *Tenuicostatum* Zone is introduced here in order to replace the *Tenuicostatum* Subzone of CARIOU and HANTZPERGUE (coord., 1997, Table 12 on p. 88). Two specimens of the zonal index *Crussoliceras tenuicostatum* (GEYER) were figured by GYGI (2003, Fig. 135a, b). The specimen selected for this study is shown in Fig. 7.4E. The taxon first appears in the uppermost part of bed no. 124 in section RG 70 at Mellikon in Canton Aargau (GYGI 1969a, Pl. 17). Bed no. 124 of this section is the lower, glauconitic part of the Baden Member. The vertical range of the zonal index is mainly in the lower *Divisum* Zone. The glauconitic facies of the lower Baden Member at Mellikon, therefore, probably extends into the lowermost *Divisum* Zone. In case that this can be confirmed, Fig. 170 in GYGI (2003) should be adapted accordingly. No *Crussoliceras divisum* (QUENSTEDT) were found to date in northern Switzerland.

Balderum Zone (revised): A *Balderum* Subzone was informally proposed by GYGI and PERSOZ (1986, Table 2). According to GYGI (2003, Fig. 170), the zonal index *Idoceras balderum* (OPPEL) first appears in Canton Aargau above the base of the bedded, micritic limestone of the lowermost Wettingen Member, where it coexists with *Orthaspidoceras uhlandi* (OPPEL). *Orthaspidoceras uhlandi* is a large taxon. A near-complete, adult specimen of *Orthaspidoceras uhlandi* was figured by GYGI (2000a, Pl. 15, Fig. 1) from Mellikon in Canton Aargau. *Idoceras balderum* is much smaller and is, therefore, easier to find. This is the reason why it is proposed here to replace the *Uhlandi* Subzone in GYGI (2003, Fig. 170) by the *Balderum* Zone. A small *Idoceras balderum* was found by the author *in situ* in the lowermost Wettingen Member, in bed no. 124 of section RG 62 near Villigen in Canton Aargau (GYGI 1969a, Pl. 17). The specimen was figured by GYGI (2000a, Pl. 14, Fig. 3). A large specimen of the taxon from Mellikon is shown in this study in Fig. 7.5A.

Acanthicum Zone: OPPEL (1863, p. 219) named and described the zonal index *Aspidoceras acanthicum*, but he did not figure specimens of his new taxon. OPPEL stated on p. 220 that these ammonites are common and occur, among

Fig. 7.3 (continued) *Photograph courtesy of Dr. Ursula MENKVELD, Curator of Paleontology, Museum of Natural History, Bern. Scale ×1.* (**B**) *Bimammatum* **Zone:** *Epipeltoceras bimammatum* (QUENSTEDT), section RG 76, bed no. 11, upper Hornbuck Member, hill of Hornbuck, Riedern am Sand, Baden-Württemberg, southern Germany. Refigured from GYGI (2000a, Pl. 10, Fig. 4). Scale ×1. (**C**) *Gredingensis* **Zone:** *Wegelea gredingensis* (WEGELE), section RG 85, bed no. 55, approximately 7 m below the Knollen Bed (marker bed), upper Küssaburg Member, im Wisse Rise, Beggingen, Canton Schaffhausen. Refigured from GYGI (2000a, Pl. 13, Fig. 1). Scale ×1. (**D**) *Planula* **Zone:** *Subnebrodites planula* (QUENSTEDT), *non* HEHL in ZIETEN, section RG

84, bed no. 14, lower Wangental Member, quarry owned by the township, Hemmental, Canton Schaffhausen. Refigured from GYGI (2000a, Pl. 11, Fig. 5). Scale ×1. (**E**) *Galar* **Zone:** *Sutneria galar* (OPPEL), excavation RG 239, bed no. 18, base of Schwarzbach Formation, southernmost quarry at Summerhalde, Schaffhausen. Refigured from GYGI (2000a, Pl. 13, Fig. 3). Scale ×2. (**F**) *Platynota* **A Zone:** *Sutneria platynota* (REINECKE), morphotype A SCHAIRER, excavation RG 239, bed no. 22, lower Schwarzbach Formation, southernmost quarry at Summerhalde, Schaffhausen. Refigured from GYGI (2000a, Pl. 13, Fig. 2). Scale ×2

Fig. 7.4 Index ammonites of *Desmoides* to *Tenuicostatum* Zone. *Arrows* point to the position of the last septum of the phragmocone. (**A**) *Desmoides* **Zone:** *Orthosphinctes (Ardescia) desmoides* (Wegele), sub-species *quenstedti* Atrops, section RG 70, bed no. 124, glauconitic lower Baden Member, large quarry, Mellikon, Canton Aargau. Refigured from Gygi (2003, Fig. 90). Scale ×1. (**B**) *Platynota* **C**

other localities, at Mt. Lägeren in northern Switzerland. The eastern end of the long ridge of Mt. Lägeren is west of the village of Dielsdorf in Canton Zürich. It was stated above that the author of this study found in 1958 after a blasting in the middle quarry west of the village of Dielsdorf a large and well-preserved, blue-gray cast of an *Aspidoceras acanthicum* (OPPEL) in blue-gray rubble of the marly limestone blasted. The blue-gray unit of marly limestone in the middle quarry at Dielsdorf has a thickness of about 4 m. The base of this unit with *Aspidoceras acanthicum* is stratigraphically 63 m above the base of the assembled section of the three quarries west of Dielsdorf which was published by NOTZ (1924). The author gave the pertinent ammonite to the Paleontological Institute of the University of Zürich. The specimen could not be located during a recent visit to that institute among several large *Aspidoceras acanthicum* from Dielsdorf, which match OPPEL's description of the taxon well. An adult *Aspidoceras acanthicum,* complete with the peristome, among specimens from Dielsdorf in the collection of the institute could be prepared and photographed by R. ROTH thanks to the permission of H. BUCHER, head of the Paleontological Institute of the University of Zürich. This specimen is shown in Fig. 7.5B. A nucleus of *Aspidoceras acanthicum,* which was found by A. and H. ZBINDEN in the slightly argillaceous, lower Banné Member in the quarry of L'Alombre aux Vaches, 2 km southwest of Courgenay in the region of Ajoie in Canton Jura (unpublished section RG 341, see also JANK et al. 2006, Fig. 7/1), was figured by GYGI (1995, Fig. 17/4).

Caletanum Zone: A well-preserved, near-complete specimen of the zonal index *Aspidoceras caletanum* (OPPEL) was found by A. and H. ZBINDEN in the excavation for the foundations of the administration building of the township of Alle in the region of Ajoie in Canton Jura. The ammonite was found stratigraphically a few meters above the slightly glauconitic marl of the Virgula Beds in the sediments from shallow water in the upper Reuchenette Formation. The specimen was figured by GYGI (1995, Fig. 26) and was then assigned to the *Eudoxus* Zone. *Aulacostephanus eudoxus* (D'ORBIGNY) is coeval with *Aspidoceras caletanum* (OPPEL). The position of the Virgula Beds in the Reuchenette Formation of Ajoie region can be read from Fig. 10 in JANK et al. (2006), or from Fig. 1.5 in this study. The *Aspidoceras caletanum* from Alle is shown in Fig. 7.5C.

Autissiodorensis Zone: The zonal index is *Aulacostephanus autissiodorensis* (COTTEAU). It was stated in Sect. 5.1 that a well-preserved specimen of this taxon was found in the quarry Bargetzi in the city of Solothurn. The ammonite is from the lower part of the thinly bedded, unnamed micritic limestone unit, which is directly above the Solothurn Turtle Limestone Member in the region around the city of Solothurn. A photograph of the ammonite from the quarry Bargetzi is shown in Fig. 7.6. The specimen is kept in the Naturmuseum Solothurn with the number 10,842. It is unknown whether the thinly bedded limestone unit near Solothurn represents the entire *Autissiodorensis* Zone or only part of it. The zone is the youngest in the Kimmeridgian Stage *sensu gallico,* both in the Subboreal Faunal Province of northwestern Europe and in the Subtethyan Faunal Province of northern Switzerland.

Concluding remark: The *average* duration of one of the ammonite chrons listed in Fig. 1.6 was calculated to be approximately 370,000 years. It is likely that the length of time represented by these chrons is unequal. For instance, the *Balderum* Chron was probably of rather short duration, whereas the *Caletanum* Chron, as it is conceived here, represents more time.

7.4 Time Correlation

One difficulty in time correlation of the sediments investigated is the occurrence of nondeposition (omission according to HEIM 1934, p. 375, see below), which could last millions of years in relatively shallow water without emersion. Omission in such a case left no traces in the sedimentary succession that are easy to recognize. The rate of sedimentation could be extremely low and lead to stratigraphic condensation. There was great variation in "normal" sedimentation rates. Detailed time correlation of the sediments investigated using a single method, therefore, proved to be impossible. For instance, as mentioned above, the measured and averaged thickness of the proximal, aggradational part of succession no. 1 in northwestern Switzerland is 185 m, whereas the same, but incomplete and partially condensed succession no. 1 in Canton Schaffhausen has an average thickness of 0.5 m, or 370 times less. Another instructive example is the lowermost, condensed bed of the Birmenstorf Member in Canton Aargau, which represents

---

Fig. 7.4 (continued) *Zone: Sutneria platynota* (REINECKE), morphotype C SCHAIRER, excavation RG 239, bed no. 30, lower Schwarzbach Formation, southernmost quarry at Summerhalde, Schaffhausen. Refigured from GYGI (2003, Fig. 159b). Scale ×2. (**C**) ***Lussasense*** **Zone**: *Ataxioceras (Schneidia) lussasense* ATROPS, section RG 70, bed no. 124, glauconitic lower Baden Member, large quarry, Mellikon, Canton Aargau. Refigured from GYGI (2003, Fig. 125c). Scale ×1. (**D**) ***Evolutum* Zone**: *Ataxioceras (Parataxioceras) evolutum*

ATROPS, section RG 70, bed no. 124, glauconitic lower Baden Member, large quarry, Mellikon, Canton Aargau. Refigured from GYGI (2003, Fig. 122c). Scale ×1. (**E**) ***Tenuicostatum* Zone**: *Crussoliceras tenuicostatum* (GEYER), section RG 70, uppermost part of bed no. 124, glauconitic lower Baden Member, large quarry, Mellikon, Canton Aargau. Refigured from GYGI (2003, Fig. 135a). Scale ×2/3. Figs. A–E refigured with permission of Schweizerische Paläontologische Abhandlungen

Fig. 7.5

Fig. 7.6 Index ammonite of *Autissiodorensis* **Zone:** *Aulacostephanus autissiodorensis* (COTTEAU), lowermost part of thinly-bedded, unnamed micritic limestone succession above the Solothurn Turtle Limestone

Member, upper Reuchenette Formation, quarry Bargetzi, Kreuzen, city of Solothurn. Refigured from GYGI (1995, Fig. 24). Scale ×2/3

both the *Densiplicatum* and the *Antecedens* Chron. The average thickness of this marker bed in the starved basin in Canton Aargau is between 6 and 8 cm. The coeval sediments that were laid down at a "normal" rate in proximal succession no.1 in northwestern Switzerland, are on average almost 1,600 times thicker (see below).

These results were obtained using a combination of detailed lithostratigraphy, of ammonite biochronology, and of mineral stratigraphy based mainly on the clay mineral kaolinite. In marine limestones, this mineral has apparently a greater stability than other clay minerals and is, therefore, least susceptible to environmental or diagenetic change in

such rocks. PERSOZ in GYGI and PERSOZ (1986, p. 438) concluded, in agreement with authors cited by him, that kaolinite in the limestone samples analyzed by him "is mainly an inherited mineral". The maxima and minima of kaolinite content in the measured sections were first calibrated with the ammonite succession in the uncondensed sediments of the epicontinental basin in Canton Aargau. The kaolinite maxima and minima in the basin were then correlated landward by means of the vertical variation in kaolinite content in sediments from shallow water in northwestern Switzerland, where ammonites are rare (GYGI and PERSOZ 1986, Fig. 10). P. R. VAIL and A. L. COE made the

Fig. 7.5 (See preceding page) Index ammonites of *Balderum* to *Caletanum* Zone. (**A**) *Balderum* **Zone:** *Idoceras balderum* (OPPEL), section RG 70, not from *in situ*, to judge from the matrix: from the bedded limestone succession of bed nos. 126–171, lower Wettingen Member, large quarry, Mellikon, Canton Aargau. Refigured from GYGI (2003, Fig. 167). Scale ×1. A smaller specimen of the taxon collected from *in situ* in bed no. 124 of section RG 62 at Villigen in Canton Aargau is shown in Fig. 3 on Pl. 14 in GYGI (2000a). (**B**) *Acanthicum* **Zone:** *Aspidoceras acanthicum* (OPPEL), Wettingen Member, middle

quarry west of Dielsdorf, Canton Zürich. Preparation and photograph by Rosemarie ROTH, courtesy of Hugo BUCHER, Professor of Paleontology, University of Zürich. Scale ×0.5. (**C**) *Caletanum* **Zone:** *Aspidoceras caletanum* (OPPEL), excavation for the foundations of the administration building of the township of Alle, Canton Jura, stratigraphically a few meters above the Virgula Beds, upper Reuchenette Formation. Refigured from GYGI (1995, Fig. 26). Scale ×2/3. Fig. A refigured with permission of Schweizerische Paläontologische Abhandlungen sequence stratigraphic interpretation of sections which were previously measured by the author of this study. The result was published by GYGI et al. (1998). This paper corroborated the correlations in the sediments from shallow water which were arrived at by GYGI and PERSOZ (1986) and GYGI (1995).

Regional time correlation with non-biochronologic methods can in some cases be more accurate than correlation using ammonite biochronology. Mineral stratigraphy, as it was made with kaolinite by GYGI and PERSOZ (1986, Pl. 1) within sediments from shallow water where ammonites are rare, gave the best results in correlation. The few ammonites that were found to date in strata from very shallow water, corroborated the mineral stratigraphic correlations which were made mainly with kaolinite by PERSOZ in GYGI and PERSOZ (1986). Some of the ammonites from very shallow water were figured by GYGI (1995). Their position in the stratigraphic context is represented in Fig. 2 of that paper. Provided that MILANKOVITCH cycles left an unequivocal imprint on sediments investigated, they would provide for the most accurate method of time correlation. It was shown in Sect. 2.3.4 that the elementary marl-limestone cycles in mud-grade sediments from deeper water represent widely different time spans. A MILANKOVITCH-type periodicity in massive limestone units from very shallow water, like, for instance, in the calcareous oolite of Chestel Member (Fig. 4.19B), or mainly in the Holzflue Member (GYGI 2000a, Fig. 37 on p. 64), could not be recognized. Feasibility of time correlation by means of cyclostratigraphy in the sense of STRASSER (2007) is, therefore, questioned by the author of this study. This is stated based on section RG 381 in Court Gorge at Moutier that was measured by the author in 1982, and which was published in GYGI (2000a, Pl. 28). A critical review and discussion of the same section as published by STRASSER et al. (2000) is given in Sect. 11.6.2 in this study.

The Knollen Bed is a thin marker bed with some glauconite in Canton Aargau and in Canton Schaffhausen. The approximate age of the bed could be documented in Canton Schaffhausen with a *Wegelea gredingensis* (WEGELE) from 7 m below, and with a *Subnebrodites planula* (QUENSTEDT) from 2.2 m above the Knollen Bed. Both of these ammonites are figured in this study, one in Fig. 7.3C and the other in Fig. 7.3D. There are few ammonites both in the Wangen Member and in the Letzi Member in Canton Aargau. It is by means of the Knollen Bed that the Wangen Member can be assigned to the *Gredingensis* Chron and the Letzi Member to the *Planula* Chron. The Knollen Bed and sequence boundary O8 at the base of the Letzi Member in section RG 62 on Mt. Geissberg near Villigen in Canton Aargau were figured by GYGI et al. (1998, Fig. 12). The Knollen Bed can only be discerned in good outcrops. It is an excellent regional marker bed, but it is not conspicuous enough for geologic mapping. This is why F. MÜHLBERG lumped the well-bedded, micritic

Wangen Member below with the similar Letzi Member of MOESCH (1867) above, and called the aggregate inappropriately "Wangen Member" in his geologic maps of Canton Aargau. For the same reason, F. SCHALCH reunited the Küssaburg and the Wangental Member of WÜRTENBERGER and WÜRTENBERGER (1866) in the "Wohlgeschichtete Kalke" (German for well-bedded limestones) in his geological maps.

Time correlation of the lagoonal Buix Member in the internal part of the upper St-Ursanne Formation in Canton Jura with the Birmenstorf Member on the level floor of the starved basin in Canton Aargau, and with the Mumienkalk Bed (MUK) in deeper water in Canton Schaffhausen, could be made with the two ammonites *Perisphinctes (Perisphinctes) alatus* ÉNAY (identified by R. ÉNAY) and *Perisphinctes (Dichotomosphinctes) dobrogensis* SIMIONESCU of late *Antecedens* age. Both of these ammonites were found by V. PÜMPIN in the Buix Member at St-Ursanne in sediment between coral reefs. The ammonites were figured by GYGI (1995, Figs. 4 and 14). U. PFIRTER found the *Perisphinctes (Perisphinctes)* cf. *andelotensis* ÉNAY in the upper Pichoux Formation near Nuglar in Canton Solothurn, which was figured by GYGI (1995, Fig. 8). The *Perisphinctes (Perisphinctes)* aff. *andelotensis* ÉNAY from the Pichoux Formation near Büren in Canton Solothurn was found by P. BORER and was figured by GYGI (1995, Fig. 9). GYGI (2001, Fig. 168) figured a *Perisphinctes (Dichotomosphinctes) dobrogensis* SIMIONESCU from the lowermost Birmenstorf Member at Herznach in Canton Aargau. Three specimens of *Perisphinctes (Perisphinctes) alatus* ÉNAY from the Mumienkalk Bed at Gächlingen and at Siblingen in Canton Schaffhausen were figured by GYGI (2001, Figs. 85–87).

The Laufen Member from shallow water in the lower Balsthal Formation can be correlated using two *Lithacosphinctes* sp. from the Laufen Member in Schachental near Dittingen in Canton Basel-Landschaft. One of these ammonite casts is figured in GYGI (1995, Fig. 20). The other specimen is in the exhibition of the Museum at Laufen in Canton Basel-Landschaft. This ammonite cast was found in the largest quarry in Schachental at coordinates 604,100/253,050, approximately 3.5 m above the quarry base (communication by P. BORER). The two specimens are probably conspecific with a *Lithacospinctes* cf. *gigantoplex* (QUENSTEDT) from the upper Küssaburg Member in Canton Schaffhausen. The pertinent cast of a giant, near-complete ammonite is figured in GYGI (2003, Fig. 37).

Ammonites continue to be the most accurate means in *intercontinental* time correlation, where they occur in sediments of Late Jurassic age. GYGI (1977) revised the ammonite genus *Gregoryceras,* which is as a rule rare in Central Europe. Another revision of the genus was made by BERT (2004). The holotype of the type species of the genus, *Gregoryceras transversarium* (QUENSTEDT), is from the uncondensed, normal facies of the Birmenstorf Member below one of the vineyards at Birmenstorf in Canton Aargau, Switzerland (GYGI 1977, p. 487). This holotype was refigured photographically by GYGI (1977, Pl. 7, Fig. 1a, b). Another specimen of the taxon was found by R. and S. GYGI *in situ* in the uncondensed facies of the upper Birmenstorf Member in section RG 230 near Gansingen in Canton Aargau. This ammonite was figured by GYGI (1977, Pl. 8, Fig. 1). The age of the uncondensed part of the Birmenstorf Member is now named *Transversarium* Chron (see above). GYGI and VON HILLEBRANDT (1991, Tab. 10) called the corresponding time slice of the latest *Transversarium* Chron in the sense of OPPEL and WAAGEN (1866) *Parandieri* Subchron. GYGI in GYGI and VON HILLEBRANDT (1991) identified three and figured two specimens of *Gregoryceras transversarium* (QUENSTEDT),

which were found in the Cordillera de Domeyko in the Atacama Desert of Region Antofagasta in northern Chile. One of the specimens figured is very well preserved. Time correlation between Switzerland and Chile could be made thanks to these unequivocal representatives of *Gregoryceras transversarium* (QUENSTEDT) from Chile on the level of what was called an ammonite subchron by GYGI and VON HILLEBRANDT (1991, Tab. 10). The *Transversarium* Chron as it is revised here is approximately 158 million years old (see Fig. 1.6), to judge from GRADSTEIN et al. (2004, p. 310). Time correlation between Switzerland and Chile could be made with the error of an ammonite chron, or of about ± 370,000 years. The error of the intercontinental time correlation with ammonites was in this case approximately ±0.2% of the radiometric age of the *Transversarium* Chron.

8.1   Net Sedimentation at a Normal Rate

The maximal rate of normal sedimentation of mud-grade calcareous or argillaceous particles out of suspension in the deeper part of the Rhodano-Swabian Basin was in the deposits investigated probably less than 5 m of now compacted sediment thickness per 10,000 years. The lower limit to a normal rate of sedimentation in a given deposit is where fossils of distinct biozones, like for instance ammonite casts, are still vertically separate, but not yet mixed within one and the same level because of stratigraphic condensation (see below). Rates of sedimentation in deeper water were calculated using biochronology of ammonites combined with mineral stratigraphy and radiochronology. The mixed calcareous and argillaceous mud, which includes biostromes of abundant colonies of hermatypic corals in the Liesberg Member (LIE in Fig. 1.5), was calculated to have accumulated at an average rate of at least 2.7 m of compacted sediment per 10,000 years in the following way. The thickness of the Liesberg Member was measured to be 25 m at the type locality in the clay pit of Hinter Chestel south of Liesberg, in section RG 306, which is represented in GYGI (2000a, Pl. 31). The Liesberg Member is part of the *Antecedens* Zone. The entire thickness of sediments of the *Antecedens* Zone in section RG 306 is at least 100 m. On the assumptions, that sedimentation proceeded at an even rate throughout the *Antecedens* Chron and that a chron lasted on average 370,000 years as it was calculated above, the total thickness of the Liesberg Member at Liesberg was sedimented during about 93,000 years at the average rate of now compacted, ca. 2.7 m per 10,000 years. This is one of the highest rates of net sedimentation which was calculated in the deposits investigated.

A low rate of normal sedimentation in deeper water is that by which the stratigraphically uncondensed part of the Birmenstorf Member (BIR in Fig. 1.5) was laid down. The average, measured thickness of this part of the member is around 5 m between Günsberg in Canton Solothurn (section

RG 14 in the landslide called Gschlief, see GYGI 1969a, Pl. 18) and Gansingen in Canton Aargau (sections RG 210 and RG 230 combined north of Eisengraben, which are represented in GYGI 2001, Fig. 1). The uncondensed part of the Birmenstorf Member represents the *Transversarium* Chron of approximately 370,000 years and was, therefore, laid down at the average rate of about 0.14 m (compacted) per 10,000 years. The sedimentation rate of the Birmenstorf Member is approximately 20 times less than that of the Liesberg Member.

The thickness of the St-Ursanne Formation is more than 90 m in section RG 306 at Liesberg (GYGI 2000a, Pl. 31). It is concluded from an ammonite that V. PÜMPIN found at St-Ursanne that only about the uppermost 20 m of the St-Ursanne Formation at Liesberg were sedimented during the *Transversarium* Chron. The lower 70 m of the formation at Liesberg are, therefore, part of the *Antecedens* Zone with a minimal thickness of 100 m at this locality. When an equal duration of ammonite chrons is assumed, the lower part of the St-Ursanne Formation belonging to the *Antecedens* Zone was deposited during about 260,000 years, and the upper part in around 370,000 years during the *Transversarium* Chron. The entire St-Ursanne Formation at Liesberg was then laid down in approximately 630,000 years at an average rate of 1.5 m in 10,000 years. The average rate of sedimentation in the upper part of the formation of *Transversarium* age can be calculated to have been 0.54 m in 10,000 years, and the average rate in the lower part 2.7 m in 10,000 years. The average sedimentation rate in the lower St-Ursanne Formation was then 5 times the average rate of sedimentation in the upper part of the formation at Liesberg. This corroborates the conclusion of SCHLAGER (1999, p. 186) that the rate of "growth" (he probably meant the rate of vertical accretion) of a carbonate platform diminishes during the time of its sedimentation.

The Buix Member in the internal part of the upper St-Ursanne Formation was sedimented into a lagoon with coral reefs. The thickness of the member as measured near St-Ursanne is almost exactly 30 m. It was stated above in the text on sea level rise no. 5 in Sect. 5.1 that the Buix Member with its early cemented lagoon reefs was not compacted to a significant degree. The member represents the uppermost part of the *Antecedens* Zone and the entire *Transversarium* Zone, or a time span which can be estimated at about 450,000 years. The Buix Member was, therefore, sedimented at an average rate of approximately 0.67 m per 10,000 years. The bank of calcareous ooid sand of the Chestel Member below was drowned by a rapid relative sea level rise of 10 m late in the *Antecedens* Chron. The rate of this rise was more than the maximal rate of carbonate sedimentation that was possible in the internal, lower part of the St-Ursanne Formation, or else it could not have cut short accretion of calcareous ooids at the top of Chestel Member and created a lagoon which was at an early stage deeper than 6 m. Calcareous ooids cannot be accreted in water deeper than 6 m.

8.2   Stratigraphic Condensation

Stratigraphic condensation of a sediment according to HEIM (1934, p. 376) implies that fossils of different age, like, for instance, ammonite casts of two or of several biozones, are embedded in a thin bed side by side, more or less in the same level. Such a bed including ammonite casts from several ammonite chrons is the iron-oolitic bed no. 7 in section RG 210 near Gansingen in Canton Aargau (Fig. 8.3). Ammonite casts with ages between the middle Callovian and the *Cordatum* Chron of the early Oxfordian are mixed in this bed with a thickness of 15–20 cm. HEIM's definition of stratigraphic condensation is preferred to the use of the term by Anglo-Saxon authors, because condensation in the sense of HEIM (1934) can be unequivocally delimited from starved, but otherwise normal sedimentation at a very low rate. Stratigraphic condensation was exemplified by GYGI and MARCHAND (1982, Pl. 3, Figs. 1 and 2) with two ammonites from the thin Schellenbrücke Bed in the Herznach iron mine. These ammonites are refigured here in Fig. 8.1. The age of the older specimen A in this figure is the *Scarburgense* Chron of the earliest Oxfordian. The age of the younger ammonite B in direct contact with the older one is the *Cordatum* Chron. Specimen B is according to Fig. 1.6 in this study approximately 1.5 million years younger than specimen A. Ammonites of the same two taxa were found together at the same level in bed no. 8 of excavation RG 208, which was made in 1971 in the territory of the township of Ueken, a few hundred meters north of the Herznach iron mine. A photograph made at an early stage of the excavation is in GYGI (2000a, Fig. 5 on p. 18). The pertinent ammonites were figured by GYGI and MARCHAND (1982, Pl. 3, Fig. 4, and Pl. 5, Fig. 8). Bed no. 8 in excavation RG 208 is the lower part of the Schellenbrücke Bed. Bed no. 8 was formed by an

Fig. 8.1 Stratigraphic condensation documented with two ammonite casts of different age which touch each other. The ages of the casts differ by more than one million years. (A) *Pavloviceras pavlowi* (R. DOUVILLÉ), *Scarburgense* Chron, age ca. 161 million years. (B) *Cardioceras persecans* (S. S. BUCKMAN), *Cordatum* Chron, age ca. 159.8 million years. The ammonite casts were found in a block fallen from the Schellenbrücke Bed in the roof of a tunnel in the iron mine west of Herznach in Canton Aargau. Scale bar is 1 cm. Refigured from GYGI and MARCHAND (1982, Pl. 3, Fig. 1–2) with permission of Geobios

alternation of at least five events of sedimentation and subsequent subaqueous erosion of fresh, loose sediment by bottom currents. These events caused *Pavloviceras pavlowi* (R. DOUVILLÉ) of the *Scarburgense* Zone and *Cardioceras persecans* (S. S. BUCKMAN) and more *Cardioceras* of the *Cordatum* Zone to be finally placed into the same level in bed no. 8 of excavation RG 208. The main stages A–E of the processes of stratigraphic condensation in that bed are represented in Fig. 8.2 of this study.

The lowermost bed no. 10 of the Birmenstorf Member in excavation RG 208 near Ueken is 10 cm thick (GYGI 1977, Pl. 11, section no. 2). The bed includes ammonites of both the *Densiplicatum* Zone and of the *Antecedens* Zone, which were mentioned above in the section about the *Transversarium* Zone. Bed no. 10 of excavation RG 208 is, therefore, stratigraphically condensed in the strict sense. The bed was sedimented during a time span of the order of 740,000 years. The average thickness of this condensed marker bed at the base of the Birmenstorf Member is in Canton Aargau about 7 cm. The total, compacted thickness of coeval sediments that were laid down near Liesberg in northwestern Switzerland is, to judge from Pl. 31 in GYGI (2000a), probably as much as 110 m. This is, as mentioned above, almost 1,600 times more than the average thickness of approximately 7 cm of the lowermost, stratigraphically condensed bed of the Birmenstorf Member in Canton Aargau.

8.3   Omission (Nondeposition)

HEIM (1934, p. 372) called subaqueous nondeposition without intervening emersion *omission*. The longest time span of omission that could be documented with ammonites to have occurred in the sediments investigated is around six million years, as read from GRADSTEIN et al. 2004, p. 310. This case of long-term omission was found in the starved basin between Auenstein and Oberehrendingen in Canton Aargau. Omission began in this region when sedimentation of iron oolite ceased in the later Bathonian Age. The time span of omission ended when deposition of the lowermost, condensed bed of the Birmenstorf Member began in the *Densiplicatum* Chron of the middle Oxfordian. The Bathonian ammonites documenting the beginning of long-term omission in Canton Aargau were collected by R. and S. GYGI from the iron oolite in section RG 226 near Auenstein, in section RG 276 near Holderbank, and in section RG 51b near Oberehrendingen, as well as by the author from the iron oolite in section RG 64 in the road cut on Mandacher Höhe between Mandach and Villigen. Diagnostic specimens among these Bathonian ammonites were figured by MANGOLD and GYGI (1997). Increasingly long-term omission occurred during deposition of the Bärschwil Formation in the distal belt, where the formation thins in the form of a wedge toward the basin (Fig. 1.5). In this case, the time represented by omission varies from section to section (see Sect. 4.8.2). The growing temporal extent of the hiatus, which evolved in the distal direction between the thinning Bärschwil Formation and the Pichoux Formation above, is schematically represented in Fig. 2 in GYGI (1995).

8.4   Subaqueous Erosion of Unlithified Sediment by Bottom Currents

HEIM (1958, p. 643) called subaqueous erosion by bottom currents *dereption*. This incorrectly latinized term was probably never adopted by subsequent authors. The effect of subaqueous erosion of fine-grained, unlithified sediment by bottom currents was particularly well visible in excavation RG 208 which was made in 1971 on a surface of 119 m^2 at locality Brunnrain directly north of the former Herznach iron mine. The upper part of the Herznach Formation and the

lower Birmenstorf Member were excavated at this locality, with special attention to the boundary beds between the Middle and the Upper Jurassic Series. An episodic, strong, and essentially unidirectional bottom current driven by an exceptionally violent cyclone (hurricane or typhoon) washed out pockets as deep as 15 cm between the uppermost layer of carbonate concretions with iron ooids in the earliest Oxfordian part of what is now the upper Herznach Formation (Fig. 8.2A, B). Erosion of loose iron-oolitic sediment with a matrix of mixed calcareous and argillaceous mud occurred on a sea floor, which was then at the approximate depth of 80 m (see above). The age of the concretions in bed no. 8 of excavation RG 208 (Fig. 8.2E) was identified by D. MARCHAND to be the *Scarburgense* Chron thanks to the ammonite *Pavloviceras pavlowi* (R. DOUVILLÉ). The pertinent specimen was figured by GYGI and MARCHAND (1982, Pl. 3, Fig. 4). All of the loose, mud-grade sediment including the iron ooids dispersed in it, which was eroded by a unidirectional bottom current at stage B in Fig. 8.2, was carried away by the current and was resedimented elsewhere.

There was neither sedimentation nor emersion after this event in a time span of more than one million years until the earliest part of the *Cordatum* Chron. At that time, a thin layer of calcareous mud with some iron ooids was laid down at stage C upon the corroded surface of concretions and at the bottom of eroded pockets in between. The first ammonite casts of *Cardioceras* of the *Cordatum* Chron were formed by early lithification of mud *within* ammonite shells, which were embedded into the sediment in pockets laid down at stage C. The sediment with the early lithified ammonite casts in it remained to be loose until stage D. One of the *Cardioceras* that was taken from bed no. 7b in excavation RG 208 (Fig. 8.2E) was figured by GYGI and MARCHAND (1982, Pl. 7, Fig. 6a, b). The cast of this *Cardioceras* of *Cordatum* age was formed *below* much older casts of *Pavloviceras* from the *Scarburgense* Chron (Fig. 8.2C) because of erosion that occurred at stage B. Then the thin layer of loose sediment laid down at stage C was removed entirely by another event of strong bottom currents at stage D. Water depth had in the meantime increased to more than 90 m. Bed nos. 8 and 9 in Fig. 8.2E are the Schellenbrücke Bed. This unit was sedimented at a water depth of about 95 m directly below the basal bed of the Birmenstorf Member with the first pellets of pure glauconite, which indicate a minimal water depth of approximately 100 m (Fig. 4.9). Excavation RG 208 near Ueken documented a stratigraphically condensed, deepening-upward succession in bed nos. 8 and 9 with an average thickness of around 20 cm. Water depth in Canton Aargau increased from approximately 80 m at the beginning of the Oxfordian Age to 100 m in the *Densiplicatum* Chron in the time span of 5 ammonite chrons that is equivalent to 1.85 million years.

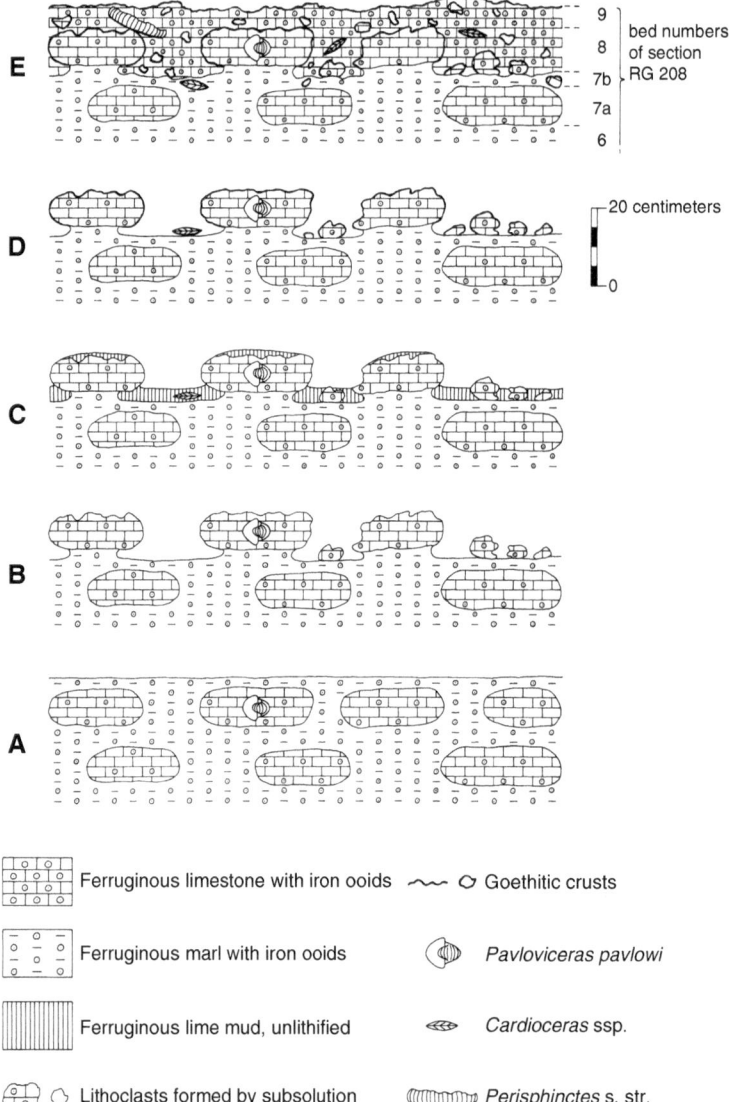

Fig. 8.2 (**A–E**) Main stages in combined processes of stratigraphic condensation of the Schellenbrücke Bed, the youngest unit of the Herznach Formation, beds no. 8 and 9 in excavation RG 208 at locality Brunnrain in the township of Ueken in Canton Aargau, directly north of the Herznach iron mine. The Schellenbrücke Bed at this locality is the product of growth of calcareous concretions in marl, of subaqueous erosion of loose sediment by unidirectional bottom currents, of subsolution of denudated concretions, of early lithification of calcareous mud within ammonite shells, and of burrowing. The figure shows the course of sedimentation from the beginning of the Late Jurassic. Previous stages of deposition during the last part of the Middle Jurassic are represented in Fig. 4.14. The age of the ammonite *Pavloviceras pavlowi* (R. DOUVILLÉ) which is shown in stages A–E of this figure is the *Scarburgense* Chron. The ammonites *Cardioceras* ssp. in stages C–E are of the *Cordatum* Chron. Stage (**A**) Soft marl with iron ooids of the early Oxfordian *Scarburgense* Chron with two layers of calcareous concretions. (**B**) Stage after an event of erosion of loose iron oolitic marl above and between the upper layer of concretions by a storm-driven, unidirectional bottom current, and subsequent subsolution (corrosion) of concretions at a varying degree in the upper layer. (**C**) A thin

layer of ferriferous, calcareous mud was sedimented during the *Cordatum* Chron after a period of omission that lasted more than one million years (see Fig. 1.6). Casts of *Cardioceras* of the *Cordatum* Chron were formed by early lithification of calcareous mud within shells. The mud the casts were embedded in was not lithified. (**D**) Calcareous mud sedimented during stage C remained to be loose until another storm-driven, unidirectional bottom current churned up the mud into suspension and removed it entirely. The early lithified casts of *Cardioceras* at the base of eroded pockets are the only remains of the mud which was sedimented during stage C. (**E**) Corroded calcareous relics of concretions, which were left over by subsolution at stage B, and casts of *Cardioceras* formed during stage C were embedded by burrowing organisms into the soft marl of bed 7b during a short time of omission. Then sedimentation of ferriferous, calcareous mud and accretion of iron ooids resumed and continued at a slow rate during the major part of the *Cordatum* Chron. Burrowing organisms brought corroded carbonate relics of subsolution formed during stage B, which were subsequently enveloped with a crust of goethite with a shining surface, into various positions within bed nos. 8 and 9. Refigured from GYGI and MARCHAND (1982, Fig. 6) with permission of Geobios

8.5   Subsolution: Subaqueous Corrosion of Early Lithified Limestone

HEIM (1958, p. 643) drew attention to "corrosive surfaces of limestone beds" in the marine environment and called the process of submarine corrosion *subsolution*. HEIM obviously meant corroded instead of corrosive. Several examples of subsolution are shown in Fig. 8.3 of the present study. German geologists call the iron-oolitic limestone beds of middle and late Callovian age in the region around Blumberg in southern Germany Grenzkalk. This unit also occurs in adjacent Canton Schaffhausen, in section RG 81b near Gächlingen and in section RG 212 near Siblingen (Fig. 8.3). The uppermost beds of the Grenzkalk unit in sections RG 81b and RG 212 were deeply corroded by subsolution during a long time span of omission in the late Callovian. The interspace between the remaining calcareous nodules was subsequently filled with mud of clayey marl, including some iron ooids. The uppermost layer of remnant iron-oolitic limestone nodules in bed no. 10 of section RG 81b was initially a continuous limestone bed. When sedimentation of bed no. 10 was completed, the thickness of the bed was at least as much as the vertical extension of the hummocky nodule in the upper part of bed no. 10 which now projects upward through the Glaukonitsandmergel Bed no. 11 (GSM in Fig. 1.5) and into the Mumienmergel Bed no. 13 above (MUM in Fig. 1.5).

The process of subsolution that was active in a long time span during the late Callovian created in section RG 81b near Gächlingen on the upper surface of iron-oolitic limestone bed no. 10, the uppermost layer of the Grenzkalk unit, a hummocky relief of as much as 20 cm. The successive, thin Oxfordian bed nos. 11–13 between hummocks of Callovian iron-oolitic limestone in this section were sedimented horizontally upon the corroded, upper surface of the Grenzkalk unit, as it is shown in Fig. 8.3. No particles of iron-oolitic limestone derived by subsolution from bed no. 10 were found in bed no. 11, or in bed no. 13 of section RG 81b. Only the very thin bed no. 12 in between includes some small particles of Callovian iron-oolitic limestone up to 1 cm across. The upper bedding plane of the Mumienkalk Bed no. 14 in section RG 81b (MUK in Fig. 1.5) is irregular, because subsolution reduced the thickness of the early lithified limestone bed unequally in a way similar to the process that corroded the upper bedding plane of the uppermost layer of bed no. 10 in the same section. Bed no. 15 of section RG 81b is in part of the section an unnamed, micritic limestone including cauliflower pellets of glauconite. The limestone bed is laterally replaced by glauconitic marl with corroded particles left over by subsolution of the limestone bed. The time equivalent of limestone bed no. 15 in section RG 81b is in section RG 212 the lower, glauconitic part of bed no. 10.

The remnant of a thin, corroded glauconitic limestone bed that is shown in lowermost unit no. 10 in section RG 212 in Fig. 8.3 is laterally replaced by glauconitic marl with limestone particles left over by subsolution, similar to bed no. 15 in section RG 81b.

Section RG 210 in the natural cleft of Eisengraben (iron ditch in English) in the territory of the township of Gansingen in northeastern Canton Aargau is shown in Fig. 8.3 is. Bed no. 6 in this section is micritic limestone with about 5% of fine-grained detrital quartz. The upper bedding plane of the limestone is uneven because of subsolution, and it is encrusted with goethite. Bed no. 6 is the uppermost part of an early Callovian limestone succession, which could be dated with the ammonite *Macrocephalites (Dolikephalites) gracilis* SPATH to be of the *Gracilis* Zone or of the early *Calloviense* Zone, respectively. The pertinent ammonite was found by B. PAGANINI in the corresponding limestone succession at locality Ischlegli north of the village of Elfingen, 2 km west-southwest of Eisengraben. The specimen with the individual number J27255 is kept in the Museum of Natural History at Basel. Bed no. 7 of section RG 210 is a stratigraphically condensed iron oolite. Flat nodules of silty limestone, like that in bed no. 6 below, are abundant in the iron oolite of bed no. 7. The silty limestone nodules in bed no. 7 are as much as 10 cm thick and up to 30 cm wide. They are encrusted with goethite all around. Ammonite casts are abundant in the iron oolite of bed no. 7, and they prevail in the macrofossil assemblage. This is evidence that the iron oolite was sedimented in water deeper than 35 m. It was stated above that the ages of the ammonite casts in the iron oolite of bed no. 7 range between the middle Callovian and the *Cordatum* Chron of the early Oxfordian. The upper surface of the iron-oolitic bed no. 7 with abundant, flat limestone nodules is uneven because of subsolution. The surface is encrusted with goethite. Bed no. 8 above is the lowermost, stratigraphically condensed limestone bed of the Birmenstorf Member. Some iron ooids and pellets of glauconite are dispersed in the micritic matrix of the bed. Nodules including exotic parts of early Callovian, silty limestone and younger, middle Callovian/early Oxfordian iron oolite in primary contact with each other are common in bed no. 8. The exotic particles derived from different, older beds within such nodules in bed no. 8 are in the contact in which they initially were in bed no. 7 below, from where the composite nodules were separated by subsolution. The nodules in bed no. 8 are encrusted with goethite all around.

The mode of formation of the large exotic nodules in bed nos. 7 and 8 of section RG 210 was most probably subsolution in the following way. Rapid relative sea level rise no. 3 (in Fig. 5.1) occurred a short time before the close of the early Callovian. The rise was documented both near Péry and near Liesberg. Rise no. 3 occurred near Péry after

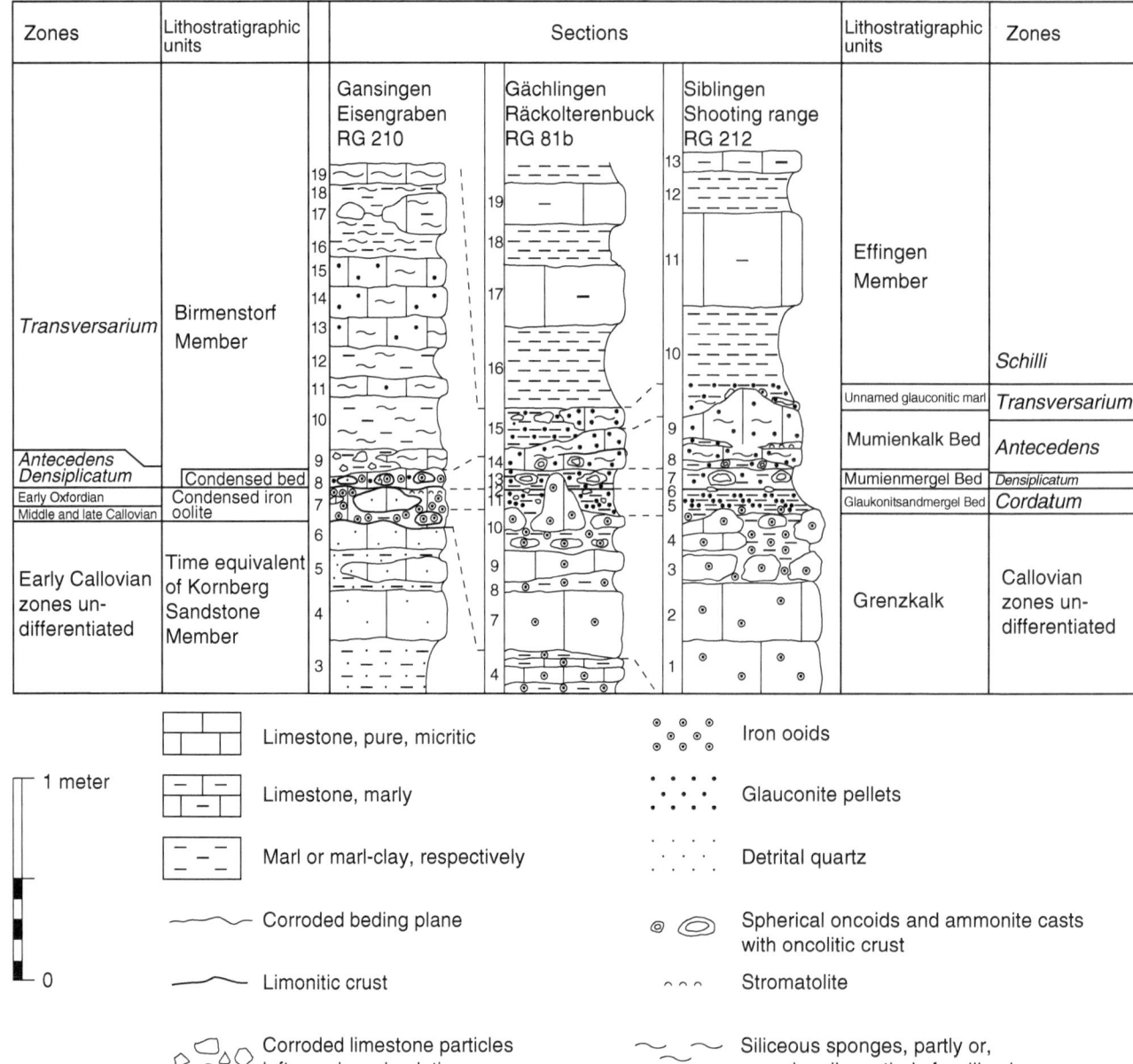

Fig. 8.3 Subsolution as an agent in stratigraphic condensation. Three sections with early lithified limestone beds which were subsequently corroded by subsolution. Explanation in the text. Section RG 210 is 2 km southwest of the village of Gansingen in northeastern Canton Aargau. Section RG 81b is 3 km north-northeast of the village of Gächlingen. Section RG 212 is in Churz Tal, 1 km northeast of the village of Siblingen. Sections RG 81b and RG 212 are in Canton Schaffhausen. For exact location with coordinates of the sections see Gygi (2000a, Tab. 1)

sedimentation of the marginal part of the early Callovian carbonate platform of the Dalle nacrée Member, and near Liesberg after deposition of the shallow subtidal time equivalent of the Dalle nacrée Member with hummocky cross-stratification (Sect. 5.1). It was stated above that the limestone succession with some detrital quartz at the base of section RG 210 in Eisengraben is dated with an ammonite to be of early Callovian age. The silty limestone succession near Gansingen is, therefore, the approximate time equivalent of the Kornberg Sandstone Member near Herznach in

northern Canton Aargau and of the Dalle nacrée Member in northwestern Switzerland. The Kornberg Member is represented in Fig. 4.10. Relative sea level rise no. 3, which was documented near Péry, near Liesberg, and near Herznach, was according to the work by Haq et al. (1988, Fig. 16) a eustatic rise. The rise is best documented in section RG 307 near Péry. The rise, because it was eustatic, increased water depth at the same time in section RG 210 near Gansingen, which is non-palinspastically 76 km east-northeast of Péry. The sea level rise began near Gansingen in

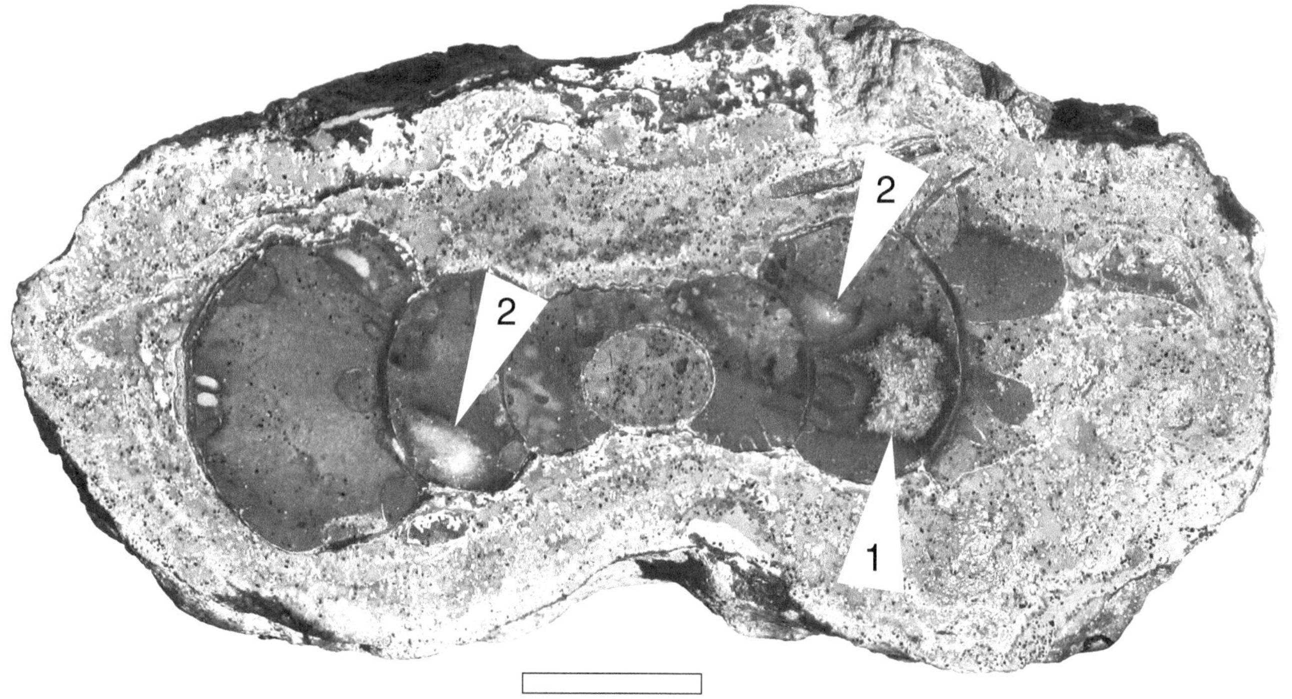

Fig. 8.4 Perisphinctid ammonite with a thick crust of micritic lime-stone, cross-cut and polished cast Gy 7449. The cast is figured in the position in which it was found with the plane of coiling nearly parallel to the Mumienmergel Bed, no. 7 in excavation RG 212 at Siblingen, Canton Schaffhausen (see Fig. 8.3). The mud-grade matrix of marl in the bed indicates that bottom water was quiet most of the time. Dark dots in the cast and mainly in the crust are cauliflower pellets of glauconite. The pellets are diagnostic of a depth of sedimentation greater than 100 m. The thickness of the oncolitic or rather stromatolitic crust around the cast is evidence that the cast was washed clear of embedding mud and that it was turned upside down several times. Upturning is corroborated by empty chambers in the phragmocone of the ammonite which were partially filled with lime mud. The chamber in the lower right (1) was partially filled in approximately the present position of the cast. The two chambers on the left (2) were partially filled with calcareous mud when the cast was upside down. The upper side of the crust around the cast is deeply corroded by subsolution and encrusted with goethite. Scale bar is 1 cm. Photograph by W. Etter

considerably deeper water than near Péry or near Liesberg, to judge from the ammonite in the early Callovian limestone succession at locality Ischlegli near Elfingen, 2 km west-southwest of section RG 210. The eustatic rise caused the source of both argillaceous and calcareous mud to recede landward (see also NORRIS and HALLAM 1995, p. 231). The rate of sedimentation thereby diminished in middle Callovian time in section RG 210 near Gansingen very much like at the same time in section RG 280 near Liesberg, or in section RG 307 near Péry. In this time span, slow sedimentation and early lithification of calcareous mud in section RG 210 near Gansingen alternated with subsolution of limestone. Nodules of iron oolite along with nodules including exotic particles were formed by the process. The nodules were periodically overturned and were encrusted with goethite all around (Fig. 8.3).

The large limestone nodules which are now within marl of bed no. 5 in section RG 210 near Gansingen can be regarded to represent an early stage of subsolution of a limestone bed which was initially continuous. But the possibility cannot be ruled out, that the early Callovian limestone nodules in bed no. 5 of that section are in fact concretions like the calcareous concretions that are characteristic of the Oxfordian Sornetan Member in northwestern Switzerland. An outcrop with layers of calcareous concretions within clayey marl of the Sornetan Member was photographed by GYGI (2000a, Fig. 31). The author of this study analyzed several thin sections of concretions in the Sornetan Member. The microsparitic mosaic of anhedral crystals of calcite which is visible in thin sections of these nodules is a second-ary texture. There can consequently be no doubt that the tough, calcareous nodules in the Sornetan Member are concretions. Their mode of formation was probably similar to that of nodules in the Blue Lias of England, which were studied by HALLAM (1964). The nodules in the Sornetan Member were alternatively called chailles or sphérites by French-speaking authors, like, for instance, GRESSLY (1838–1841). ÉNAY (1966) called them sphérites. Most of the French authors call time equivalents of the modern Swiss Sornetan Member in France "terrain à chailles" to the pres-ent time. The ferriferous limestone nodules including iron ooids within marl of the upper Herznach Formation represented in Fig. 8.2A, are concretions.

The large flat nodules of silty limestone in the iron-oolitic matrix of bed no. 7 in section RG 210 are most probably remnants of subsolution. The nodules were isolated by corrosion out of an initially continuous limestone bed of early Callovian age above bed no. 6 of the section. This youngest bed of early Callovian age in the section must have been dismantled into separate nodules in a process similar to that of subsolution, which isolated nodules from the uppermost part of the younger Grenzkalk unit in Canton Schaffhausen in sections RG 81b and RG 212. The nodules of silty limestone in bed no. 7 of section RG 210 at Gansingen became incorporated into iron oolite of this bed in the time span between the middle Callovian and the end of the *Cordatum* Chron of the early Oxfordian. Sedimentation and early lithification of parts of the iron oolite in bed no. 7 ended at the close of the *Cordatum* Chron. When sedimentation of bed no. 7 was completed, its thickness must have been much greater than it is now. Evidence that part of bed no. 7 was early lithified is subsolution following sedimentation. Subsolution of the upper part of bed no. 7 produced isolated nodules of iron oolite as well as nodules including both iron oolite and early Callovian silty limestone. The upper part of bed no. 7, which is not preserved, must have included nodules of early Callovian silty limestone like the lower, preserved part of the bed. Evidence of this is presence of nodules of Callovian silty limestone within bed no. 8. Subsolution of the upper part of bed no. 7 produced, depending on the site where the process was particularly active, nodules of pure iron oolite, nodules of early Callovian silty limestone, or nodules including parts of iron oolite and parts of early Callovian silty limestone in primary contact. Nodules including both iron oolite and silty limestone in primary contact are fairly common in bed no. 8. The rounded nodules of older rock which were incorporated into bed no. 8 of section RG 210 during the middle Oxfordian *Densiplicatum* and the *Antecedens* Chron are up to 10 cm wide. They are enveloped by a crust of goethite with a shining surface (Fig. 8.3).

Bottom water undersaturated with Ca(CO)$_3$ was the precondition of dissolution of early lithified calcareous sediment. This condition was temporary and must have existed at recurrent time intervals in bottom water, in which the stratigraphically condensed bed nos. 7 and 8 of section RG 210 were laid down. The macrofossil assemblage with mainly ammonites in bed nos. 7 and 8 of section RG 210 indicates, according to Fig. 4.9 in this study, that these beds were sedimented in water deeper than 35 m. Glauconite pellets in bed no. 8 of section RG 210 are evidence that water depth had increased at this location to approximately 100 m after the beginning of the *Densiplicatum* Chron, like at the same time in section RG 208 near Ueken. The crust of goethite enveloping the nodules in bed nos. 7 and 8 of section RG 210 near Gansingen is proof that the nodules were overturned from time to time, before they were finally embedded. The fact that this crust has a shining surface is evidence that the nodules were overturned without the

intervention of strong bottom currents. The shining crust of goethite around nodules in the stratigraphically condensed beds in section RG 210 must have been formed in a similar way as the thick oncolitic, calcareous crust including pellets of cauliflower glauconite which was primarily all around large ammonite casts in the stratigraphically condensed Mumienmergel Bed or in the Mumienkalk Bed in Canton Schaffhausen. Evidence was given in Sect. 4.3.2 that the stromatolitically encrusted ammonite casts in section RG 81b and RG 212 in Canton Schaffhausen (Figs. 8.3, 8.4) were occasionally overturned in water which was considerably deeper than 100 m. The external surface of the oncolitic crust around ammonite casts in the Mumienmergel Bed is often covered with a thin veneer of goethite with a shining surface.

Bioerosion is a special process of erosion of hard substrates of calcium carbonate in marine environments by boring fungi, algae, or sponges of the genus *Cliona*. Filaments of such soft organisms penetrate into hard substrate by dissolving calcium carbonate. Bioerosion by boring fungi must be considered to be a possible subsidiary process going on along with subsolution in the condensed, early lithified oolitic ironstones represented in Fig. 8.3 of this study. Fungal hyphae penetrated through a major calcareous fossil fragment which later became the core of a surficially coated iron ooid in an iron oolite of middle Bathonian age from a water paleodepth of between 20 and 30 m (see above). The pertinent bioclast with hyphae and sporangia of boring fungi in the core of this iron ooid was figured in a thin section by GYGI (1969a, Pl. 2, Fig. 5). The iron ooid was found in bed no. 3 of the unpublished section RG 55 near Schinznach-Dorf in Canton Aargau. Boring algae and boring sponges are very effective agents in bioerosion of Recent coral reefs and of limestone coasts. Hard surfaces of calcium carbonate bored by algae or by the boring sponge *Cliona lampa* DE LAUBENFELS in the shallow waters of Bermuda were found by the author of this study to be rough or harsh, whereas the eroded surfaces in the sections which are represented in Fig. 8.3 in this study are smooth. The smooth surfaces are, therefore, the result of subsolution rather than of bioerosion by boring organisms. Boring algae are unlikely to have contributed to possible bioerosion of limestone in the cases illustrated in Fig. 8.3, because the type of subsolution and possibly of bioerosion by fungi which is documented in the figured sections occurred at a water depth which was greater than approximately 35 m, according to the fossil assemblage with mainly ammonite casts in the eroded beds.

Ammonite casts prevail in the macrofossil assemblage in all of the beds mentioned in section RG 210 in Eisengraben near Gansingen. The situation is similar in section RG 81b near Gächlingen and in section RG 212 near Siblingen, where ammonite casts prevail both in the macrofossil assemblage in the iron-oolitic limestone of the Grenzkalk unit of middle and late Callovian age and in the Oxfordian sediments above. There are no indications of subaerial erosion in the three sections shown in Fig. 8.3. Ammonite casts prevail in the macrofossil assemblage in all of the beds corroded by subsolution which are represented in Fig. 8.3. This is evidence that subsolution occurred in the three sections shown in Fig. 8.3 at a water depth which was greater than about 35 m, according to Fig. 4.9 in this study. The relief produced by subsolution in the uppermost part of the Grenzkalk unit in section RG 81b near Gächlingen is the quantitatively most important example of subaqueous limestone corrosion which was found by the author in sediments in northern Switzerland. GYGI (1969a, p. 59 and 105, and Pl. 17) interpreted the large inclusions of older rocks in iron oolite corresponding to bed no. 7 of section RG 210 in Eisengraben in this study (Fig. 8.3) to be *in situ* relics of subaqueous limestone corrosion in the deep subtidal zone. Strong bottom currents were proved in Sect. 8.4 to have eroded unlithified, fine-grained sediment at a water depth of as much as more than 90 m (Fig. 8.2D). These currents could erode loose mud including iron ooids, but no evidence was found that they displaced or overturned the large concretions shown in Fig. 8.2B.

8.6   Subaerial Erosion of Limestone Interrupting Shallow-Marine Sedimentation

Subaerial erosion of shallow-marine, oolitic or pelletoidal limestone following local tectonic uplift in earliest Kimmeridgian time produced a supratidal relief of as much as 4 m in limestone of the uppermost Balsthal Formation. This uplift was confined to the region around Balsthal (see below). Erosion began after the beginning of the *Platynota* A Chron. This is concluded from the ammonite that B. MARTIN found 2.6 m below the paleosol on the eroded top of the Balsthal Formation near Balsthal. F. ATROPS (Lyon) identified the specimen. A photograph of it is shown in Fig. 19 in GYGI (1995). The greatest thickness of the paleosol was measured to be 0.6 m in bed no. 7 of the unpublished section RG 450. The section was measured on the rope at the perpendicular cliff below Point 702 on Chluser Roggen, 1 km to the south of Balsthal. The paleosol in this cliff is a gray-brown, saccharoidal limestone that weathers back as a deep notch. The sediment above the paleosol is again pelletoidal and oolitic limestone which was laid down in very shallow-marine water. The cliff of section RG 450 was measured and sampled in about 6 hours descending from Point 702 on a light steel rope. When the steel seat was fixed at the rope with a clamp, both hands could be used for measuring thicknesses and taking rock samples. A photograph of the pertinent device is shown in GYGI (2000a, Fig. 4). The erosion surface with its conspicuous relief and the notch of the paleosol in the cliff of Chluser Roggen can best be seen from a vantage point above the cliff of Wannenflue on the western flank of the cluse between Balsthal and Oensingen in Canton Solothurn, when looking north-northeast across the cluse. The photograph made from this vantage point is Fig. 37 on p. 64 in GYGI (2000a).

The same paleosol was accessible on a catwalk leading along the cliff above the right bank of Dünnern River at Innere Klus near Balsthal in 1986, when the unpublished section RG 439 was measured there. Calcrete nodules with radially oriented acicular calcite crystals occur in bed no. 15 of that section. The long axis of the calcite needles below the surface of the nodules is perpendicular to the surface. One of these nodules is shown in broken cross-section in Fig. 5.9. The acicular calcite crystals represented in Fig. 5.9 did not grow into an empty karst cavity like the "raggioni" (Italian for rays) which were figured by MUTTI (1994, Fig. 10). Evidence of this are some silt-grade detrital quartz grains that were found in two thin sections which were ground through radially arranged, elongate calcite crystals in another paleosol above the calcareous oolite of the Steinibach Member in section RG 4, bed no. 40B, at the cliff of Brocheni Flue. Section RG 4 was measured in 1962 on the rope 1 km west-southwest of Waldenburg in Canton Basel-Landschaft. The radiating needles of calcite in the paleosol from near Waldenburg are shown in Fig. 4.3. The detrital quartz grains were incorporated into the calcite needles when the crystals grew within and penetrated through preexisting material in a soil (GYGI 2003, p. 160). The only site where such a paleosol with acicular calcite crystals below the surface of calcrete nodules is now accessible on foot is where the upper surface of the Steinibach Member is intersected by a path leading through the uppermost part of Horngraben Gorge south of Aedermannsdorf in Canton Solothurn. This paleosol with a thickness of 30 cm is bed no. 3 of the unpublished section RG 15.

9.1 Endogenic Subsidence

GYGI (1986) called subsidence endogenic when it is caused by a non-sedimentary process going on deep in the interior of the earth. Endogenic subsidence of a given surface is the effect of processes other than compaction mainly of mud below this surface; of isostatic basement subsidence under the additional load of sediment, water, or ice; or of plastic, load-induced lateral flow of deeply buried rock salt. GYGI (1986) assumed that endogenic, regional subsidence was equable in Oxfordian time in all of the investigated parts of northern Switzerland. Later he recognized, that this was the case in Oxfordian time only between Canton Jura in the northwest and Balsthal in Canton Solothurn (GYGI 2003, p. 211). This situation thoroughly changed in Kimmeridgian time between the region around the city of Solothurn and the region of Ajoie in Canton Jura. Evidence of this is the well-preserved cast of the ammonite *Aulacostephanus autissiodorensis* (COTTEAU) shown in Fig. 7.6, which was found in the disused quarry Bargetzi at locality Steingruben in the northern outskirts of the city of Solothurn (see above). The base of the Reuchenette Formation is near Solothurn above the top of the Verena Member. The type locality of the Verena Member is at the hermitage of St. Verena in a small gorge 2 km north of the city of Solothurn. Only the upper part of the Verena Member crops out at the type locality. A photograph of the type locality of the Verena Member is Fig. 36 in GYGI (2000a). The bedding-plane parting, which is the boundary between the Verena Member below and the Reuchenette Formation above, is well visible in the upper right of that photograph. According to the ammonite from Solothurn mentioned above, the sediments of the entire Kimmeridgian Stage are approximately 52 m thick near the city of Solothurn.

The Kimmeridgian sediments from the base of the Reuchenette Formation upward were deposited in a shallow-marine and intermittently in a peritidal environment. According to the sections that were measured by M. JANK (see JANK et al. 2006, Fig. 10), the aggregate thickness of the Late Jurassic sediments above the base of the Reuchenette Formation in the region of Ajoie in Canton Jura is close to 140 m. Provided that the ammonite *Aspidoceras caletanum* (OPPEL) was properly identified, which JANK et al. (2006) reported in their Fig. 10 to have been found in the uppermost part of the Reuchenette Formation in Ajoie region, then the sedimentary succession of Late Jurassic age ends in Ajoie region in the *Caletanum* Zone (Fig. 1.6 in this study) and therefore short of the top of the Kimmeridgian Stage. The thickness of 140 m of the incomplete sediments of the Kimmeridgian Stage above the base of the Reuchenette Formation in Ajoie region is much greater than that of the complete Kimmeridgian Stage with a thickness of 52 m in the region around the city of Solothurn. The amount of endogenic subsidence was in later Kimmeridgian time in Ajoie region much greater than near the city of Solothurn. This can be concluded because sedimentation occurred in both regions mostly in a very shallow-marine and partially in a peritidal environment.

Endogenic subsidence going on at an equable rate in a vast area is the precondition of sedimentation with a more or less equal thickness in shallow-marine water or in a peritidal environment during an extended time span. The proximal part of succession no. 2, this is to say the Vorbourg Member, the Röschenz Member, and the Hauptmumienbank Member were sedimented in this way. Regional endogenic subsidence east of Olten grew early in the *Berrense* Chron greater than in the northwest (Fig. 4.15C, D). This was calculated by GYGI (2003, p. 210). The difference is not represented in Fig. 1.5 of this study, which only shows measured and averaged thicknesses. The thicknesses that are shown throughout the assembled transect of Fig. 1.5 are averaged and summed up above the sea floor as it was at the beginning of the Late Jurassic. Fig. 1.5 was drawn, as if the total amount of both endogenic and exogenic subsidence of the pre-Oxfordian basement was equal over the whole area covered by the section during deposition of succession nos. 1–3.

R.A. Gygi, *Quantitative Geology of Late Jurassic Epicontinental Sediments in the Jura Mountains of Switzerland*,
DOI 10.1007/978-3-0348-0136-2_9, © Springer Basel 2012

Endogenic subsidence going on at a low and equable rate over a long period of time in a vast region must be discerned from endogenic subsidence occurring in a small region at a much higher rate and in a short time. Such rapid, local endogenic subsidence could amount to a considerable total. One example of this is the great difference in thickness within the St-Ursanne Formation between Kleinlützel in Canton Solothurn and Liesberg in Canton Basel-Landschaft (east of Delémont). This is represented in the paleogeographic map of Fig. 5 in GYGI (1990c). The measured thickness of the St-Ursanne Formation is somewhat more than 90 m in section RG 306 near Liesberg (GYGI 2000a, Pl. 31), and 35 m in the unpublished section RG 397 near Kleinlützel. The two sections are palinspastically only 5 km apart. The difference in thickness of approximately 60 m within the St-Ursanne Formation over such a short horizontal distance evolved in the basement north of Liesberg during less than one ammonite chron. It is important to note that this rapid and substantial, endogenically driven event of synsedimentary faulting left no trace in the facies of the St-Ursanne Formation at the sediment surfaces in the pertinent time span. A synsedimentary fault moving within the Günsberg Formation produced the breccia near Vermes in Canton Jura, which was figured by GYGI (2000a, Fig. 62).

Another example of locally enhanced endogenic subsidence is the thickness of the Effingen Member near Riniken northwest of the city of Brugg in Canton Aargau. The entire thickness of the Effingen Member near Riniken was semiquantitatively assessed by the author to be locally enhanced to at least 260 m (GYGI 1990c, Fig. 6, to the right of the center of the paleogeographic map). The figure of 260 m was arrived at adding in the following way to the partial thickness of the Effingen Member, which was measured in the exploration well drilled by NAGRA 1 km north of the village of Riniken (NAGRA 1990). The strata of the Upper Jurassic are almost horizontal between the NAGRA borehole and the hill of Iberg west of the well, where the glauconitic marker bed of the *Crenularis* Member crops out. The approximate difference in elevation between the top of the NAGRA borehole and the base of the Geissberg Member in Iberg hill was added to the partial thickness of the Effingen Member, which is preserved in the borehole. The average thickness of the Effingen Member in Canton Aargau is around 200 m. The possibility is envisaged below that the local, synsedimentary tectonics recorded in the St-Ursanne Formation and in the Effingen Member was shallow-rooted and was caused by lateral, plastic flow of Triassic rock salt beneath the overburden of younger sediments.

9.2 Exogenic Subsidence Under Load: Isostatic Equilibration of the Lithosphere

STECKLER and WATTS (1978) investigated subsidence of the eastern margin of the North American continent off New York City in the time interval between 140 million years before present and the Recent. They found that the amount of isostatic basement subsidence under the load of sediments was greater than subsidence caused by processes in the earth interior. This can be read from Fig. 6 in STECKLER and WATTS (1978). GYGI (1986, p. 488), based on detailed, but nevertheless semiquantitative calculations of sedimentation at the much smaller scale of the Rhodano-Swabian Epicontinental Sea during the Late Jurassic in northern Switzerland, concluded that loading of the lithosphere mainly with sediments was quantitatively the most important factor in basement subsidence in the region investigated. He further concluded that shifting of the areas with maximal sedimentation in succession nos. 1 and 2, as represented in his Fig. 3 (which is an early version of Fig. 4.15 in the present study) led to differential basement subsidence in time and space.

The result of the calculations made by GYGI (1986) was that the greater part of the final accommodation space for the proximal, thick part of succession no. 1 mainly in Canton Jura, and for the subsequent, thick part of succession no. 2, which was shed into the adjacent, starved basin near Balsthal in Canton Solothurn, was created essentially by the weight that the sediments exerted on the basement. Compaction of fine-grained sediments was quantified using mainly the nomogram of Fig. 11 in PERRIER and QUIBLIER (1974). The amount of basement subsidence under the weight of additional water because of net sea level rise was of a minor order. The information combined from the 203 detailed sections which the author had measured in northern Switzerland until the end of 1985, when the manuscript of the paper of 1986 cited above was completed, from about 2,200 thin sections, from a great number of polished rock slabs, and from more than 10,000 specimens of macrofossils collected from *in situ* in the measured sections, was processed in the calculations. Proximal succession no. 1 and the thick, basinal part of succession no. 2 near Balsthal are both shallowing upward, and they filled in the basin up to close below sea level. The isostatic adjustment of the lithosphere, first under thick succession no. 1 in the northwest, and then under thick succession no. 2 in adjacent Canton Solothurn and in Canton Aargau led to differential subsidence of neighboring basement blocks underneath.

The regional and temporal difference in exogenic subsidence between adjacent basement blocks during sedimentation of succession nos. 1 and 2 evolved in a belt that is approximately congruent with the curved strip covered by the Pichoux Formation. This belt can be as narrow as less than 2 km (see Sect. 4.12).

Two minor normal faults that were obviously moving during sedimentation were photographed by the author and were published as Figs. 2 and 8 in GYGI and PERSOZ (1986). One with the displacement of several meters of pure limestone in the uppermost Pichoux Formation in section RG 411 north of Grandval in Canton Bern is shown in Fig. 11.1 of this study. The other in the upper Effingen Member was visible in a temporary outcrop in section RG 37 on the western wall of the quarry of Jakobsberg east of Auenstein in Canton Aargau. It is refigured here as Fig. 4.12. Subsidence along both of these faults was continuously compensated by sedimentation of mud supplied from the north. The maximal vertical displacement by the fault shown in Fig. 4.12 is approximately 30 cm. It is evident from the figure that movement along the clear-cut fault going on in unconsolidated mud was gradual and very slow. The maximal displacement that is visible at the base of the photograph was reached after an estimated, compacted sediment thickness of 5 m was laid down until the activity of the fault died down. The sedimentation rate of the Effingen Member probably never exceeded 5 m of deposit (now compacted) per 10,000 years (see above). The small synsedimentary fault shown in Fig. 4.12 was, therefore, gradually active during a time span of at least 10,000 years. A shallow-rooted growth fault could have been the cause of the figured synsedimentary displacement near Auenstein.

The fact that thicknesses of members in the carbonate Balsthal Formation vary from measured section to section was tentatively ascribed to the activity of growth faults by GYGI (1986, Fig. 3D). The much greater differences in thickness of approximately 60 m, which evolved in a relatively short time span and over short horizontal distances both in the St-Ursanne Formation north of Liesberg and in the Effingen Member near Riniken cannot be the effect of growth faults, because of the much greater order of magnitude of the differences in thickness. Plastic lateral flow of Triassic rock salt under the load of younger sediments must

be ruled out as well to have been the cause when the following facts are taken into account.

A narrow and deep rift filled with sediments of the Late Paleozoic was discovered in 1983, when NAGRA drilled an exploration well for a planned repository of nuclear waste near Weiach in northwestern Canton Zürich (NAGRA 1988). Subsidence of the rift began in later Carboniferous time and ended in the Permian (DIEBOLD 1988, Fig. 1). Subsequent deep drilling and seismic exploration by NAGRA revealed that the pre-Mesozoic basement of northern Switzerland is intersected by normal faults at close intervals. Movement along some of the faults continued in Mesozoic time (DIEBOLD et al. 1991, Beilage (appendix) 35, DIEBOLD and NOACK 1997, Figs. 7.3–7.5, and BIRKHÄUSER et al. 2001, Fig. 4.18). The trend of the normal faults in the pre-Mesozoic basement is *straight*. Conversely, the basinward margins of the carbonate platforms of Bathonian to Oxfordian age which are shown in Fig. 1 in GYGI (1986) or in Fig. 1.4 in this study are *curved*. The platform margins are consequently incongruent with the normal faults of mostly pre-Mesozoic age. Nevertheless the detailed, previously published information was not taken into account or disregarded by some recent authors (compare GYGI 1986, Fig. 1, with WETZEL et al. 2003, Fig. 1). The faults recorded by NAGRA were driven by *endogenic* processes which occurred deep in the interior of the Earth (GYGI 1986, p. 469). GYGI (2003, p. 171) called isostatic basement subsidence under the load of sediments, under additional water load during a net, probably eustatic sea level rise, or by formation of a continental ice shield *exogenic*.

The economic salt deposit, which is worked in Middle Triassic sediments near Pratteln east of Basel, is according to HAUBER (1971, Fig. 5) between 25 and 60 m thick. The thickness of the coeval deposit of rock salt, which is currently worked near Möhlin further east in Canton Aargau, is up to more than 90 m (HAUBER 1980, Fig. 7). A deposit of rock salt in sediments of Middle Triassic age in the exploration well which was drilled by NAGRA near Riniken in

R.A. Gygi, *Quantitative Geology of Late Jurassic Epicontinental Sediments in the Jura Mountains of Switzerland*, DOI 10.1007/978-3-0348-0136-2_10, © Springer Basel 2012

Canton Aargau (NAGRA 1990, Beilage-Band, Beilage 5.1a) is 2.4 m thick. The compacted thickness of sediments between the top of Middle Triassic rock salt and the base of the Pichoux Formation (middle Oxfordian) in northwestern Switzerland is approximately 520 m. The corresponding, compacted thickness of the sediments between the top of Middle Triassic rock salt and the base of the Birmenstorf Member in eastern Canton Aargau is estimated at 420 m. To judge from Figs. 7 and 8 in HAUBER (1971), these (compacted) thicknesses of overburden were insufficient to initiate plastic lateral flow in rock salt below because of areally differing thicknesses of overburden. But it must be noted that HAUBER (1980, Fig. 4) documented a synsedimentary normal fault to have been active in Middle Triassic time when rock salt was precipitated between Rheinfelden and Möhlin in northern Canton Aargau. This normal fault bounded the salt pan to the south by its vertical displacement that was in excess of 110 m.

Rock salt flowing like mountain glaciers of ice from diapirs in the Zagros Mountains in Iran is evidence that plasticity of salt is similar to that of water ice.

The Late Paleozoic rift that NAGRA discovered near Weiach in northwestern Canton Zürich has a west-eastern trend (DIEBOLD 1988, Fig. 1, USTASZEWSKI et al. 2005, Fig. 2). There is no evidence that subsidence in this trough continued in the Mesozoic Era. This can be concluded from thicknesses of the Effingen Member. This member is at Riniken west of Brugg in Canton Aargau at least 260 m thick above the presumed western extension of the Weiach rift (GYGI 1990c, Fig. 6). The thickness of the Effingen Member in the NAGRA exploration well near Benken in northern Canton Zürich is 13.7 m (NAGRA 2001). The well near Benken is over the northern rim of the probable northeastern continuation of the Late Paleozoic rift near Weiach. No normal fault corresponding to the northern boundary of this rift near Benken is shown in the map of tectonic structures at the base of Mesozoic sediments in BIRKHÄUSER et al. (2001, Fig. 4.18). According to MÜLLER et al. (1984, p.112), some synsedimentary faulting driven by *endogenic* processes must have occurred during the Mesozoic in northern Switzerland, but it probably had a different pattern in Late Jurassic time than in the Late Paleozoic. This is concluded from the course of the isopachs represented in the paleogeographic map no. 5 in GYGI (1990c). The thickness of the incomplete Reuchenette Formation in Ajoie region in Canton Jura was measured by JANK (2004, Chap. 1, Fig. 4) to be approximately 140 m. The Reuchenette Formation is complete around the city of Solothurn, but it is in that region only about 52 m thick. The greater thickness of the formation in Ajoie region can be presumed to have been caused by enhanced endogenic subsidence during the Late Jurassic in a possible western extension of the Late Paleozoic trough near Weiach in northwestern Canton Zürich (NAGRA 1988,

Beilage 6.2a, and USTASZEWSKI et al. 2005, Fig. 2). This is unlikely according to the following argumentation.

The thickness of the sediments between the top of the Wangental Member and the base of the Glaukonitbank of ALDINGER (1945) in the NAGRA well near Weiach is 40 m. This thickness was measured in 1984 by the author of this study on the drill cores. The thickness of the coeval strata cropping out in the three quarries west of the village of Dielsdorf in Canton Zürich, which is 9 km south of the NAGRA well near Weiach, is 35 m according to the rather schematic section which was published by NOTZ (1924). K. B. FÖLLMI, K. TSCHANZ, M. TUCHSCHMID, and K. WINZELER, then of the Paleontological Institute of the University of Zürich, measured in 1984 in the three quarries west of Dielsdorf the thickness of the sediments between the top of the Wangental Member and the base of the Echinidenschicht of NOTZ (1924), which is the time equivalent of the Glaukonitbank of ALDINGER (1945). According to their unpublished section, which is kept in the Paleontological Institute at Zürich, and that was kindly shown to the author of this study by H. FURRER, they arrived at a total thickness of 43 m. The location of the quarries at Dielsdorf is presumably south of the southern rim of the Late Paleozoic rift at Weiach. The thickness of 43 m of the pertinent sediments at Dielsdorf is not less, but somewhat greater than the thickness of the same strata between the top of the Wangental Member and the Glaukonitbank in the NAGRA exploration well near Weiach. This well was drilled probably close to the axis of the Late Paleozoic trough. The exact position of the three quarries west of Dielsdorf in relation to the southern boundary of the Weiach rift cannot be specified, because thrusting along oblique planes dissected and offset the southern boundary fault of the rift. Nevertheless, there is no evidence of further subsidence of the Late Paleozoic rift near Weiach in Late Jurassic time. The much greater basement subsidence during the Late Jurassic in Ajoie region in northwestern Switzerland compared with the region of the city of Solothurn was, therefore, probably unrelated to a possible western extension of the Late Paleozoic rift near Weiach.

The difference between isostatic equilibration of the basement under the load of thick sediments of proximal succession no. 1, and under the much lesser load of mainly the Birmenstorf Member as well as of some additional sea water, which was moderately deepening at the same time in the adjacent starved basin because of relatively rising sea level, finally amounted to approximately 130 m (Fig. 4.15B). The pertinent differential, *exogenic* movement of isostatic equilibration of the basement occurred below the curved belt of the Pichoux Formation. The subsequent regional, isostatic equilibration of the basement under the belt with maximal sedimentation of succession no. 2, which shifted into the starved basin of eastern Canton Solothurn and Canton Aargau, was reversed and nearly compensated below the belt of the Pichoux Formation in relation to thick, proximal succession no. 1 (compare Fig. 4.15B with 4.15C). The *sum of subsidence* of the pre-Oxfordian basement thereby became almost equal at the end of deposition of succession no. 2 under both succession nos. 1 and 2 between Canton Jura and Balsthal in Canton Solothurn. This is represented in Fig. 4.15C.

GYGI (1986, p. 471) presumed that the great difference in time and space in loading of adjacent basement blocks with sediments and with additional water in the course of probably eustatically rising sea level was equilibrated by exogenic reactivation of preexisting faults deep in the basement. This assumption was made because the pre-Mesozoic basement was found by NAGRA to be fractured by normal faults at close intervals, and because the belt of the Pichoux Formation can be very narrow compared with the thickness of the entire lithosphere below. It was stated above that the distal, basinward rims of the carbonate platform of the St-Ursanne Formation and of the subsequent, prograding platforms of the Günsberg and the Balsthal Formation are curved (GYGI 1986, Figs. 1 and 1.4 in this study). Conversely, the horizontal course of the normal faults which were documented by NAGRA to have been active in the pre-Mesozoic basement is straight. None of the faults recorded by NAGRA can consequently be assigned to exogenic, isostatic equilibration of the lithosphere first under the load of thick, proximal succession no. 1, and then, after subsequent shifting of the belt with heaviest sedimentation, of thick succession no. 2 in the starved basin of eastern Canton Solothurn and Canton Aargau. Differential, isostatic subsidence of the lithosphere first under thick, proximal succession no. 1, and then under the thick, more distal part of succession no. 2 in the epicontinental basin was, therefore, possibly equilibrated in the basement by flexure rather than by fracturing along normal faults. This is now supposed because basement deformation was probably slow and because deformation amounted to a minimal tilt. Flexural deformation of the basement is consequently shown in Fig. 4.15 of this study.

There was not only locally enhanced, endogenically driven *subsidence* of the basement in small areas like near Liesberg or near Riniken. Local *uplift* must be assumed to have caused emersion and concomitant subaerial erosion of the uppermost part of the Balsthal Formation in the region around Balsthal. An erosion surface below a paleosol was found in section RG 438 in Steinibach Gorge north of Balsthal, which is represented in GYGI (2000a, Pl. 44), as well as in the unpublished sections RG 439 at Innere Klus and RG 450 at Chluser Roggen near Balsthal. Subaerial erosion near Balsthal produced a supratidal relief of as much as 4 m, which is clearly visible at the cliff of Chluser Roggen (Fig. 37 in GYGI 2000a). A calcrete nodule from the paleosol above the erosion surface at Innere Klus near Balsthal is shown in Fig. 5.9 of this study. Another example of probably local uplift and subsequent subaerial erosion is the eroded upper surface of bed no. 232 in section RG 307 at Péry (GYGI 2000a, Pl. 22). Blackened lithoclasts in the lowermost part of bed no. 233 above are shown in Fig. 5.7.

11.1 Generalities

This study is an account of sedimentation during mainly the Oxfordian Age of the Late Jurassic. Sediments were investigated in the folded and in the tabular part of the Jura Range mainly in northern Switzerland. Sedimentary geology of these deposits is of general interest, because the term *Jurassic* that is used worldwide is derived from strata cropping out in the Jura Mountains. Facies is a term that became widely known after a study of Oxfordian strata in the Jura Mountains of northern Switzerland. The pertinent sediments are a shallow water-to-basin transition which is now assigned to the middle Oxfordian. Sediments of the modern Oxfordian Stage and the Kimmeridgian Stage in the sense of GRADSTEIN et al. (2004) include in northern Switzerland an uninterrupted succession of stratigraphically uncondensed strata with ammonites which were laid down in an epicontinental sea. The iron-oolitic boundary beds between the Callovian and the Oxfordian Stage in section RG 208 near Ueken in Canton Aargau are thin, but they are not stratigraphically condensed in this section. The beds are shown in Figs. 4.14 and 8.2. A diverse ammonite assemblage of almost 10,000 specimens was recently collected from *in situ* out of sections and excavations which were measured bed by bed. Every ammonite chron listed in Fig. 1.6 is documented with a figured specimen of the pertinent zonal index taxon, from the *Scarburgense* Zone at the base of the Oxfordian Stage upward to the *Autissiodorensis* Zone at the top of the Kimmeridgian Stage. Several of the zonal ammonite index taxa can be used in intercontinental time correlation. The course of sedimentation could be documented in detail thanks to the complete ammonite succession, which is represented in Fig. 1.6.

11.2 Endogenically Driven, Vertical Movements of the Basement

The lithosphere in the area investigated was in Late Jurassic time in an extensional regime. Regional subsidence caused by processes deep in the interior of the Earth was the result. Endogenic basement subsidence is independent of loading. Subsidence by endogenic processes was not equable in all of northern Switzerland. It varied in the Oxfordian and mainly in the Kimmeridgian Age. Averaged endogenic subsidence was normally equable in a restricted region and was moving at a low rate, to judge from the Vorbourg and the Röschenz Member in the lower Vellerat Formation. It can be read from Fig. 4.15C that a slight areal difference in endogenic subsidence occurred in the part of the basin where the Effingen Member is thickest. The Effingen Member was not sedimented into a tectonically evolving trough (compare Fig. 4.15B, C in this study with Fig. 15e–g in ALLENBACH 2002). There is no indication that the rate of endogenic subsidence in the area where the thickest part of the Effingen Member was laid down was significantly greater than the endogenic subsidence rate where the much thinner, coeval Röschenz Member was sedimented on a widespread, often peritidal surface between land in the north and the Günsberg carbonate platform. The same, areally equable endogenic subsidence as that during sedimentation of the Röschenz Member continued when the Hauptmumienbank Member was deposited directly above in a vast, shallow lagoon rimmed by the wide calcareous-oolitic sand bank of the Steinibach Member.

Endogenic *subsidence* is difficult to quantify, but some evidence was presented above that the amount during sedimentation of succession no. 1 was modest. The rate of endogenic subsidence could locally augment to a high degree, and the pertinent amount could increase to tens of meters in a short time. As mentioned above, an example of this is the St-Ursanne Formation with a thickness of approximately 95 m in section RG 306 at Liesberg (GYGI 2000a, Pl. 31), compared with 35 m in the unpublished section RG 397 at Kleinlützel, which is palinspastically only 5 km away (GYGI 1990c, Fig. 5). The locally enhanced thickness of the Effingen Member near Riniken, a village 2 km northwest of the city of Brugg in Canton Aargau (GYGI 1990c, Fig. 6), is another example. Subsidence along a normal, major synsedimentary fault is documented in Fig. 11.1. A minor, but particularly instructive synsedimentary fault is shown in Fig. 4.12.

Local *uplift* could be concluded in two cases in the Balsthal Formation. Endogenic uplift above sea level of fine-grained calcarenite, concomitant lithification and subsequent subaerial erosion of the uppermost part of bed no. 232 in the Verena Member in section RG 307 at Péry (GYGI 2000a, Pl. 22), are documented to have occurred in *Planula* time with black pebbles above an erosion surface with a relief of 20 cm (Fig. 5.7). The top of the oolitic sand bank in the uppermost Balsthal Formation around Balsthal in Canton Solothurn was raised in earliest Kimmeridgian time to at least 4 m above sea level, was lithified in the process, and was subsequently eroded (GYGI 2000a, Fig. 37 on p. 64, and calcrete nodule from a paleosol in Fig. 5.9 in this study).

The allegation by WETZEL et al. (2003, p. 154) that "there is no visible evidence for synsedimentary faults in outcrops" is consequently untenable. So is the assertion by ALLENBACH (2002, p. 323) in the abstract of his paper about the Swiss Jura, "that epicontinental areas were thought to have been tectonically stable during the Mesozoic". ALLENBACH (2002) ignored what MÜLLER et al. (1984) had previously published. He was unaware of isostatic equilibrium of the lithosphere by subsidence when being loaded with sediments. In addition, he disregarded the results of the calculations made by GYGI (1986), which were visualized in Fig. 3 of that paper and are refigured in Fig. 4.15 in this study. ALLENBACH (2002, title and p. 340) *did* recognize differential subsidence of the basement in Oxfordian time in northern Switzerland, but it is evident from his Fig. 15 that he confounded cause and effect of the greater part of subsidence. He ignored the conclusion by GYGI (1986, p. 488) that "the majority of basement subsidence was the effect, not the cause of sedimentation".

GYGI (1986, p. 472) estimated the initial water depth at the beginning of the Oxfordian in northwestern Switzerland at 80 m. This estimate was confirmed above in this study. The entire stack of Oxfordian sediments accommodated in that part of the basin was calculated on the same page in the paper by GYGI (1986) to have had an initial (decompacted)

thickness of 380 m at the end of the Oxfordian Age. The Oxfordian strata filled in the basin in northwestern Switzerland up to sea level. Total, *apparent* subsidence in Oxfordian time therefore amounted in that region to approximately 300 m. The measured (compacted) and averaged thickness of Oxfordian sediments in northwestern Switzerland is close to 310 m. The weight of these deposits entailed isostatic, *exogenic* basement subsidence of approximately 200 m, according to calculation step no. 15 on p. 472 in GYGI (1986). *Relative* sea level rise during the Oxfordian in the region of northwestern Switzerland was calculated in step no. 16 on the same page to have been around 100 m. This is 1/3 of the total, *apparent* subsidence. GYGI (1986, Fig. 4) further concluded that part of the relative rise of 100 m was *net eustatic sea level rise* that he estimated on p. 488 at between 25 and 30 m. When isostatic basement subsidence caused by the load of this additional water is rated at 40% of the rise following ZIEGLER (1982, p. 106), accommodation space created by eustatism would have amounted in the Oxfordian Age to approximately 40 m. Coeval *endogenic* subsidence in the area was consequently on average of the order of 60 m.

WETZEL et al. (2003, p. 166) recalculated the figures published by GYGI (1986) based on data from three exploration wells which were drilled by NAGRA in northern Switzerland. The authors did not mention GYGI's paper. The result of their work on p. 168 of the paper reads: "about 1/3 of the total subsidence was calculated to be of non-isostatic, *tectonic* origin" (italics by the author of this study). WETZEL et al. (2003) consequently disregarded both the carefully documented eustatic sea level rises in the Oxfordian Age previously concluded by several authors (reviewed and supplemented in GYGI 1986) and the sequence stratigraphy elaborated by A. L. COE and P. R. VAIL based on GYGI's sections in GYGI et al. (1998). The non-eustatic part of what WETZEL et al. (2003) unspecifically called "tectonic subsidence" is in this study termed *endogenic subsidence*. The amount of endogenic subsidence during the Oxfordian Age in the area investigated can only be estimated indirectly, because sea level falls could not be quantified. For the same reason, net eustatic sea level rise in the pertinent time span cannot be accurately assessed. Instead, only discrete, rapid relative sea level rises were quantified in Sect. 5.1 and in Fig. 5.1.

It is evident from Fig. 1.5 that the basinal rim of the thick stack of aggradational, shallowing-upward sediments in proximal succession no. 1 was stationary throughout sedimentation of the entire succession. This is probably the effect of eustatically, most of the time gradually rising sea level. The eustatic component of a few meters of the rapid part of sea level rise no. 7 in Fig. 5.1 in the *Hypselum* Chron caused the small transgression of the uppermost Effingen

Member over the basinward margin of the Günsberg Formation in section RG 14 at Günsberg (from bed no. 162 upward in the section, see GYGI 1969a, Pl. 18). Coeval backstepping of hermatypic corals from a biostrome in the uppermost Günsberg Formation in Court Gorge at Moutier (GYGI 2000a, Pl. 28, bed no. 39 in section RG 381) over peritidal sediments in the uppermost Röschenz Member proceeded to as far as near Porrentruy. The pronounced deepening-upward tendency during sedimentation of the Letzi Member in Canton Aargau and the coeval demise of the coral biostrome of the Olten Member, as well as subsequent backstepping of hermatypic corals from Olten to Balsthal, were obviously caused by a substantial amount of eustatic sea level rise. All of these vertical and lateral facies shifts are visualized in Fig. 1.5 in this study by a gray band. For instance, the eustatic component of the rapid relative sea level rise of 10 m in the *Antecedens* Chron, no. 5 in Fig 5.1, is *calculated* in this study to be approximately 6 m. The same eustatic rise was *estimated* by GYGI (1986, p. 475) at 7 m.

WETZEL et al. (2003) announced in the title of their paper that "reactivated basement structures" influenced the facies of sediments in the tectonically "quiescent" Jurassic epicontinental basin of northern Switzerland. The authors meant endogenic faulting. The author of this study inferred that the difference in thickness of the St-Ursanne Formation of approximately 95 m in section RG 306 near Liesberg (GYGI 2000a, Pl. 31) and 35.2 m in the unpublished section RG 397 near Kleinlützel, that is geographically 4 km to the northwest, evolved by the play of a normal, synsedimentary fault in a short time span late in the *Antecedens* Chron. The vertical displacement of about 60 m by the fault (or possibly a flexure) is incompatible with the tectonic quiescence alleged by WETZEL et al. (2003). Movement along the fault neither significantly affected depth of deposition nor facies of the sediments in the St-Ursanne Formation from shallow water directly above the fault. Vertical displacement by the fault was consequently continuously compensated by sedimentation on the downthrown side of the fault. The result of such a process at a much smaller scale in the proximal part of the Pichoux Formation can be read from Fig. 11.1.

This contradicts the assertions by WETZEL et al. (2003, p. 162) that "the facies boundaries of the lower to middle Oxfordian deposits coincide fairly well with the NE-SW-trending Late Paleozoic structures" and on p. 164 that "facies boundaries coincide spatially with tectonic structures in the basement". The authors continued on p. 168 that "tectonic movements such as synsedimentary faults could not develop". This is at variance with their assertion that facies boundaries in the Oxfordian coincide with Late Paleozoic tectonic structures in the basement. Their statement on p. 162 that "reefs nucleated in the vicinity of Late Paleozoic faults" is cryptic. The activity of endogenically driven normal faults varied in time and space. No such faults were found to have been active when the Herznach Formation was sedimented. Sedimentation of the Birmenstorf Member was not affected by proved faulting. Conversely, the play of normal faults produced widely varying thicknesses over a short horizontal distance in the coeval St-Ursanne Formation. Variable thicknesses in the Verena Member were probably caused by the activity of synsedimentary normal faults.

WETZEL et al. (2003) figured in the lower part of their Fig. 7 a simplified version of the assembled cross-section on Pl. l9 in GYGI (1969a). They labeled in the caption the sponge biostromes from the basin floor in the Birmenstorf Member and the lowermost Effingen Member between Holderbank and Schinznach-Dorf in Canton Aargau "sponge reef". These biostromes were shown in cross-section in GYGI et al. (1979, Fig. 3) and in a photograph of a bedding plane in GYGI (2000a, Fig. 17 on p. 27). The coral biostrome of the Olten Member is labeled "coral reef" in the same caption to Fig. 7 in WETZEL et al. (2003). The authors disregarded the improved assembled cross-section on Pl. 1A in GYGI and PERSOZ (1986). The isopach maps of Figs. 7 and 9 in WETZEL et al. (2003) of what the authors called Effingen Formation are a slightly revised version of Fig. 7 in GYGI (1969a) without reference to the source. GYGI (1990c, Fig. 6) drew a paleogeographic map of the *Hypselum* Chron with revised isopachs according to new data of ammonite biochronology that were not yet available to GYGI (1969a). This was ignored by WETZEL et al. (2003).

When the paleogeographic maps in GYGI (1990c) are compared with the trend of pre-Mesozoic faults in the basement, it becomes apparent that the pattern of curved margins of carbonate platforms and the progradation of successive platform margins is unrelated to normal faults underneath. The basinward rim of the Bajocian Hauptrogenstein carbonate platform runs according to BÜCHI et al. (1965, Fig. 12) north-south below the Aare River between Auenstein and Holderbank in Canton Aargau. The eastward progradation of the equally north-south running rim of the early Bathonian Spatkalk carbonate platform is shown in GYGI (1986, Fig. 1). The basinward margin of the Spatkalk platform marked the maximal extension of the Burgundy platform of PURSER (1979, Fig. 4) that extended like a broad tongue from northwest to southeast. It can be read from Fig. 1 in GYGI (1986) that the basinward margin of the early Callovian Dalle nacrée carbonate platform stepped back in comparison with the distal rim of the Spatkalk carbonate platform. This was the consequence of eustatic sea level rise. CONTINI (1976, Fig. 3) interpreted the marl of the Bärschwil Formation to have been filled into a trough. This shallowing-upward succession is in fact a physiographic high that eventually became the base of the St-Ursanne carbonate platform. Progradation of the subsequent Oxfordian carbonate platforms can be read from

GYGI (1986, Fig. 1). Progradation and backstepping of successive carbonate platform rims was evidently caused by the interplay of sediment supply and eustasy. Endogenically driven, synsedimentary faults moving in carbonate platforms could produce great local differences in thickness in a short time, but they did not noticeably influence facies nor progradation or backstepping.

11.3 Rates of Sedimentation compared with Rates of Exogenic Subsidence

It was stated by ZIEGLER (1982), calculated by GYGI (1986, p. 472, calculation step no. 15), and recalculated by WETZEL et al. (2003) that approximately 2/3 of the accommodation space in a basin is created by loading with sediments. The order of magnitude of the maximal possible rate of isostatic lithosphere equilibration when being *loaded* cannot be quantified directly. Only an indirect estimate can be made by comparison with a *calculated rate of lithosphere rebound* because of *unloading*. The rate at which the lithosphere reacted to unloading because of rapid melting of the thick ice shield over Fennoscandia after the end of the last Pleistocene glaciation could be assessed in the region around the Gulf of Bothnia between Sweden and Finland in the central part of Fennoscandia in a "natural experiment", as PRESS and SIEVER (1982, text and Fig. 18–30) put it. The rate of isostatic rebound, when the lithosphere is being unloaded, was found in that region to be less than the average rate at which ice melted away. Isostatic equilibration of the lithosphere by uplift of mainly central Fennoscandia since the last Pleistocene glaciation is, therefore, possibly not yet completed and continues at the present time. According to SAURAMO, as cited by PRESS and SIEVER (1982), the lithosphere under the Gulf of Bothnia rose during the past 5,000 years on average by 2 cm per year.

This rate of isostatic, *vertical* movement of the lithosphere caused by unloading is of the same order of magnitude as rates of *lateral* plate motion that were measured to occur in the Recent. The average rate of uplift, which was calculated to have occurred during the last 5,000 years in central Fennoscandia, is probably less than the possible maximal rate of isostatic lithosphere rebound that occurred during the time of most rapid melting of ice. The lithosphere can consequently react to unloading with a rate of isostatic uplift of at least 2 cm per year. When it is assumed, that the rate of isostatic uplift of the lithosphere because of *unloading* is similar to the rate of subsidence caused by *loading*, then the maximal sedimentation rates calculated or estimated, respectively, in northern Switzerland can be compared with the probably maximal possible rate of isostatic basement subsidence because of loading. The greatest rate of sedimentation that could be *calculated* is at least 2.7 m of compacted, mixed calcareous and argillaceous mud per 10,000 years in the Liesberg Member (Sect. 8.1). The maximal sedimentation rate of mixed calcareous and argillaceous mud in the Effingen Member was *presumed* above to have been no more than a compacted sediment thickness of 5 m per 10,000 years, or 0.5 mm per year. This *maximal* sedimentation rate is 40 times less than the calculated rate of isostatic uplift of the lithosphere under the Gulf of Bothnia of 2 cm per year. It can, therefore, be presumed that the lithosphere reacted promptly by isostatic subsidence when it was loaded in the Rhodano-Swabian Basin with sediment and water. Moreover it is to be expected that the basement reacts with subtle isostatic equilibration to even small increments of sediment or water load. Many of the sea level rises in the Late Jurassic were rapid, but they were of small scale. It is concluded based on all of the evidence mentioned above that the lithosphere in the Late Jurassic below what is now northern Switzerland was never off isostatic equilibrium (see main conclusion no. 5).

It can be read from Fig. 4.15A–C that there was great regional variability in basement subsidence. This was the consequence of lateral shifting of the belt with heaviest sedimentation. Water depth at the beginning of the Oxfordian Age between northwestern Switzerland and Canton Aargau was estimated above to have been approximately 80 m. The order of magnitude of this figure can be checked in the following way. Proximal, thick succession no. 1 filled the basin up to sea level. The measured and averaged thickness of this part of succession no. 1 is according to GYGI (1986, p. 470) 185 m. The initial thickness of this sediment stack, after it was partially decompacted using the nomogram of Fig. 11 in PERRIER and QUIBLIER (1974), was recalculated in this study to have been 240 m. The sediments filled the basin up to sea level. Evidence mentioned above indicates that only a small part of the pertinent accommodation space was provided by endogenic subsidence and by a net, probably eustatic sea level rise, which caused isostatic basement subsidence under the load of the additional water brought into the region by the rise. GYGI (1986, p. 471) concluded that exogenic basement subsidence under sediment load was approximately two-thirds of the sediment thickness that filled the basin entirely to sea level (Fig. 4.15B). The initial depth of the basin at the beginning of the Oxfordian in the northwest can therefore be concluded to have been of the order of 80 m, as it is represented in Fig. 4.15A. The same figure was calculated before by GYGI (1986, p. 472).

11.4 Interpretation of Sediments

11.4.1 Time Calibration and Correlation

Biochronology based on ammonites has its limitations, because collecting a sufficient number of specimens from *in situ* is time consuming. Time stratigraphy using ammonites requires a profound knowledge of taxonomy. An ammonite chron represents a time span of the order of several hundred thousand years. The accuracy of ammonite biochronology alone is, therefore, insufficient in a regional study of sedimentation. Lithologic marker beds in sediments from deeper water, like for instance the slightly glauconitic Knollen Bed in limestones of the Oxfordian, and the Glaukonitbank of ALDINGER (1945) in Kimmeridgian limestones, give excellent results in regional time correlation when the marker beds are used in combination with a detailed lithostratigraphic frame of reference of measured sections, and if it can be checked with ammonite biochronology that the marker beds are isochronous. Mineral stratigraphy was a successful method used in time correlation of sediments from deeper water with deposits from shallow water, because basinal sediments were time-calibrated by ammonite biochronology, and because deposits from very shallow water could be spot checked with the few ammonites that were found in such sediments. Sequence stratigraphy was used in order to double-check time correlations by the methods mentioned above.

11.4.2 Great Thickness Variation in Coeval Sediments

Ammonite biochronology was the only method by which the extreme thinning of succession no. 1 in the distal direction could be documented. The thickness of the succession diminishes from a measured and calculated average of 185 m in northwestern Switzerland to 0.5 m in Canton Schaffhausen. A hiatus representing the time span from the beginning of the *Scarburgense* to the end of the *Costicardia* Chron is at the base of succession no. 1 in Canton Schaffhausen. The Mumienmergel Bed and the Mumienkalk Bed of the middle Oxfordian are stratigraphically condensed in that region. The thickness of succession no. 2 can augment to a maximum of approximately 300 m in Canton Aargau, with a thickness of at least 260 m of the Effingen Member at Riniken, and a Geissberg Member with the abnormally great thickness of more than 30 m in the small valley of Chatzensteig, 2 km east of Effingen. Chatzensteig is 5 km west-southwest of the NAGRA exploration well near Riniken, and probably within the area of locally enhanced endogenic subsidence mentioned above. The entire Effingen Member in the NAGRA exploration well near Benken in Canton Zürich was measured to be 13.7 m thick (NAGRA 2001). This is about 5.3% of the probable compacted thickness near Riniken.

11.4.3 Paleobathymetry

Facies diagnostic of paleobathymetry are summarized and discussed in the following text in the order of their reliability. The order is from the supratidal down to the deep subtidal zone. Supratidal facies are paleosols with calcrete nodules (Fig. 5.9) or with rootlets (GYGI 2000a, Fig. 25 on p. 42). Limnic ostracods and gyrogonites of characean algae are evidence of freshwater ponds (OERTLI and ZIEGLER 1958, Pl. 1). Thin beds with lignite coal were found by several authors, beginning with HEER (1865, p. 125). Some layers of lignite coal include gyrogonites of characean algae and are, therefore, diagnostic of freshwater marshes. Facies of the upper intertidal zone are stromatolites with prism cracks (Figs. 4.4–4.6) or with bird's-eye pores (Figs. 4.6 and 5.8). Calcareous ooids are accreted on sand banks in the paleobathymetric range between the lowermost intertidal zone (Fig. 4.7) and the maximal water depth of 6 m. Care must be taken to distinguish environments where calcareous ooids are *accreted* from tidal deltas into which such ooids are *shed*. Sand banks on which calcareous ooids were accreted in the shallow subtidal zone down to a maximal water depth of 6 m can be identified, for instance, with cross-bedding and with large, parallel wave ripples. Very distinct ripples up to 15 cm high and about 70 cm apart were uncovered on a vast bedding plane in unit no. 39 of the upper Steinibach Member in the unpublished section RG 448 in the quarry of Mösliloch at Egerkingen in Canton Solothurn. A large pelletoidal sand wave grew in the uppermost Günsberg Formation to an elevation of more than 1 m, before it was truncated by a planed erosion surface. This is visible in the unpublished section RG 441 east of Liesberg, a photograph of which is shown in Fig. 5.5A. A tidal delta of calcareous ooid sand, which was shed near Péry in Canton Bern into water with a depth of approximately 6 m, is shown in the photograph of Fig. 5.2. Calcareous oolite in itself is undiagnostic of depth of sedimentation, unless there are criteria like large-scale sand ripples or sand waves.

The lower limit to the paleobathymetric range in which hermatypic corals could survive was calculated semiquantitatively in Chap. 6 in section RG 307 near Péry to be approximately 20 m in the uppermost Effingen Member. No evidence was found that the lower limit was significantly different at the base of the older Liesberg Member at Liesberg in Canton Basel-Landschaft, in the type section of the member (RG 306, Pl. 31 in GYGI 2000a). The lower boundary of 20 m is apparently independent of whether the corals lived on a sediment surface of pure calcareous mud or of mixed calcareous and argillaceous mud, respectively, and independent of the calculated rates of sedimentation. The lower limit to the habitat of hermatypic corals in the Oxfordian of northern Switzerland was in much shallower water than the maximal depth, where some Recent hermatypic corals can survive. There are great differences in the bathymetric lower limit to the habitat of Recent hermatypic corals (HALLOCK and SCHLAGER 1986, Fig. 1). It is, therefore, unfeasible to conclude bathymetric zones of Late Jurassic corals by comparison of the growth form of their skeletal parts with the colony shape of Recent corals, as it was done by LATHUILIÈRE et al. (2005). The fundamental difference in the paleobathymetry and in the growth form of hermatypic coral assemblages of Late Jurassic age as compared with the Recent has mainly the following cause. The amount of discrete, rapid relative sea level rises was in the Late Jurassic less than one-tenth of the sea level variations that occurred between individual glaciations of the Pleistocene. The sea level rise between the lowstand at the acme of the last Pleistocene glaciation and the highstand at the present time was calculated to be between 120 and 130 m. Conversely, the greatest rapid relative sea level rise that occurred during the Oxfordian amounted to approximately 10 m late in the *Antecedens* Chron (no. 5 in Fig. 5.1). The eustatic part of the rise was 6 m.

Ammonites of the genus *Cardioceras* were most abundant in the Boreal Faunal Province. The optimal water depth for *Cardioceras* in the environments investigated was during the early Oxfordian *Cordatum* Chron at the comparatively shallow paleodepth of approximately 35 m. *Cardioceras* were at this depth 77% of the ammonite assemblage in the fossil bed in the distal part of the Sornetan Member (Fig. 4.9). Marl with a high *content* of clay minerals was then laid down in that environment. The sedimentation rate of the fossil bed in the Sornetan Member was probably very low because of a rapid relative sea level rise (no. 4 in Fig. 5.1). Evidence of the low rate of sedimentation is the high concentration of macrofossils in the bed. A high sedimentation *rate* of argillaceous mud is unfavorable for ammonites (see main conclusion no. 27). The sedimentation rate of clay minerals was not detrimental to *Cardioceras* in the fossil bed, because the rate was low. Coeval *Cardioceras* are with 42% of the ammonite assemblage less abundant at the water depth of close to 100 m in the Schellenbrücke Bed with a low content of clay minerals at Ueken in Canton Aargau. *Cardioceras* are much less abundant in the thin, argillaceous Glaukonitsandmergel Bed of the *Cordatum* Chron in Canton Schaffhausen, which was sedimented at a very low rate at the water depth of approximately 115 m (Fig. 4.9).

Amoeboceras evolved from the genus *Cardioceras* at the turn between the *Antecedens* and the *Transversarium* Chron in the middle Oxfordian. No *Amoeboceras* at all were found in the Oxfordian/Kimmeridgian boundary beds from the water depth of 30–35 m in section RG 21 near Olten (GYGI 1969a, Pl. 18) and in coeval beds from somewhat deeper water in the unpublished section RG 28 at Schönenwerd west of Aarau, or in the glauconitic lower Baden Member of the same age in section RG 70 at Mellikon (GYGI 1969a, Pl. 17) from an estimated water depth of approximately 55 m. It is evident from the calculation by GYGI (1969a, p. 23) and from GYGI (2003) that ammonites are very abundant in the glauconitic lower Baden Member in Canton Aargau. *Amoeboceras* are fairly numerous, but of small or even dwarf size in the Oxfordian/Kimmeridgian boundary beds with a rich ammonite assemblage from the depth of sedimentation of approximately 100 m in section RG 239 at locality Summerhalde at Schaffhausen. It can be read from many figures in ATROPS et al. (1993, Pl. 1–2), that *Amoeboceras* at this locality and at this paleodepth are of smaller size than the much more abundant *Cardioceras* of the *Cordatum* Chron from nearly the same water depth in the Schellenbrücke Bed in Canton Aargau. *Amoeboceras* in the Birmenstorf Member from a water depth of close to 100 m, for instance, in section RG 276 at Holderbank in Canton Aargau are of about the same size as *Cardioceras* in the Schellenbrücke Bed. This becomes evident when figures in ATROPS et al. (1993) are compared with figures in GYGI and MARCHAND (1982). The Oxfordian/Kimmeridgian boundary beds at Schaffhausen were, therefore, sedimented in an environment which was possibly marginal to *Amoeboceras*. Both *Cardioceras* and *Amoeboceras* are most abundant in ammonite assemblages in the Boreal Faunal Province. There is consequently no reason to assume that *Amoeboceras* in northern Switzerland preferred cooler water than *Cardioceras*. This is an indication that bottom water at the depth of 100 m grew substantially warmer between the early Oxfordian *Cordatum* Chron and the end of the Oxfordian Age in the region investigated because of warming climate (see main conclusion no. 7).

The *vertical* facies succession in proximal, thick and aggradational succession no. 1 in northwestern Switzerland, from the base upward to the top, is diagnostic of diminishing water depth. Ammonites are uncommon, but they are the main element of the macrofossil assemblage in the thin iron oolite with few macrofossils of the earliest Oxfordian Age in the uppermost bed of the Herznach Formation at the base of succession no. 1. Several, independent lines of evidence indicate that this iron oolite was sedimented at a water depth of approximately 80 m. Ammonites are common and prevail by far in the macrofossil assemblage in the lower *Renggeri* Member above. Many of these ammonites are figured in GYGI (1990a, 1995). Ammonites are numerous and prevail in the macrofossil assemblage in the fossil bed sedimented at the water depth of close to 35 m in the middle of the distal part of the Sornetan Member (Fig. 4.9). Ammonites from the bed were figured by GYGI and MARCHAND (1993, Pl. 1–3). Large bivalves of the genus *Pholadomya* are typical of the macrofossil assemblage in the upper part of the Sornetan Member (Pholadomyen of ÉTALLON 1862) from the paleobathymetric interval of between 20 and 30 m. The lower limit to the bathymetric range of the habitat of hermatypic corals was calculated in section RG 307 at Péry to have been in the uppermost Effingen Member at the water depth of approximately 20 m. As stated above, the base of the Liesberg Member with biostromes of mainly platy corals (GYGI 2000a, Fig. 32

on op. 54) was probably at about the same water depth. There are small coral bioherms, but no ammonites in the calcareous oolite of the Tiergarten Member, which is indicative of a water depth of between low tide level (Fig. 4.7) and 6 m. The position of the prism-cracked, intertidal stromatolite in the Vorbourg Member above the St-Ursanne Formation at Liesberg is indicated in Fig. 1.5, and the photograph of a polished cross-section of the stromatolite is shown in Fig. 4.6.

Four principal marks of marine water paleodepth can be distinguished, with reliability decreasing with growing depth of deposition, in sediments of Late Jurassic age in northern Switzerland. *First,* particular types of stromatolites from the upper intertidal zone, which were mainly figured by GYGI (1992) and that are refigured in this study in Figs. 4.4–4.6, indicate approximately mean sea level, because no evidence of a substantial tidal range was found. The *second* paleobathymetric mark is at about 20 m below mean sea level. This is the greatest water depth where hermatypic corals could survive in the region investigated. No evidence was found of a significant variability of this depth. The *third* paleobathymetric mark was calculated in Sect. 4.2 to be at the water depth of around 35 m, from where downward ammonites *suddenly* prevail in the macrofossil assemblage. The *fourth* paleobathymetric mark is approximately 100 m below sea level. This is the lower limit of the depth range in which iron ooids could be accreted *in situ* on the basin floor. This bathymetric boundary touches and probably does not overlap with the upper limit of the depth range from where downward *cauliflower* pellets of pure glauconite were growing at a low rate, at or slightly below the surface of sediments of calcareous or argillaceous mud. The facies boundary between sediments with iron ooids and sediments with cauliflower glauconite pellets, both vertical and lateral, is an important mark of water paleodepth in the environments *of the region investigated.* This facies boundary is to be expected to be at a lesser water depth in a higher paleolatitude and at greater depth in a lower paleolatitude (Fig. 4.2).

Quantification of water paleodepth in the deep subtidal zone of the Late Jurassic is difficult, because ammonites are extinct. *Octopus,* the closest relatives of ammonites in the Recent, are undiagnostic, because they live, for instance, in coral reefs in very shallow water. Iron ooids are nowhere accreted in a Recent *marine* environment. Glauconite was identified mainly in thin section from the peritidal Vorbourg Member down to the thin, unnamed marl and limestone above the Mumienkalk Bed in Canton Schaffhausen (section RG 81b in Fig. 8.3). Some of the glauconites were identified in x-ray diffractograms by GYGI and MCDOWELL (1970, Fig. 3) or in the electron microprobe and were found to be pure. The depth of deposition of the unnamed bed above the Mumienkalk Bed is indicated in Fig. 4.9 to be around 125 m.

This depth was concluded departing from the 100 m at the beginning of the Oxfordian near Blumberg north of Canton Schaffhausen. 100 m was the probable water depth at which iron ooid accretion was replaced by formation of *cauliflower* pellets of pure glauconite when water depth increased from less to more than 100 m.

The 100 m can only be an approximate figure, because it was arrived at based on indirect evidence. So are the 80 m which were assumed to be the water depth at the beginning of the Oxfordian in Canton Aargau. The difference of 20 m is the probable amount of net relative sea level rise between the beginning of the Oxfordian and the end of the *Densiplicatum* Chron when water depth had increased to 100 m in Canton Aargau and when glauconite pellets first appeared at the base of the Birmenstorf Member. Isostatic basement subsidence under the additional weight of water brought into the basin by the rise of 20 m was 8 m. Endogenic subsidence was consequently modest when succession no. 1 was laid down in Canton Aargau. It was argued above that sedimentation of the Birmenstorf Member ended in Canton Aargau at the water depth of about 105 m. Provided that endogenic subsidence was of the same amount in Canton Aargau and in Canton Schaffhausen, then the sea floor in Canton Schaffhausen must have been at the depth of approximately 125 m at the end of the *Transversarium* Chron. Pellets of pure glauconite were found at the base of the Pichoux Formation close to the toe of the formation, in the lower Birmenstorf Member in Canton Aargau, and in Canton Schaffhausen in the thin beds from the Glaukonitsandmergel Bed upward to the top of the beds of *Transversarium* age.

11.4.4 Sea Floor Topography

The *lateral* facies succession in coeval sediments from the Tiergarten Member in the marginal, upper part of the carbonate platform of the St-Ursanne Formation over the depositional slope of the Pichoux Formation down to the Birmenstorf Member on the level basin floor includes the following environments. Small coral reefs grew probably in deep tidal channels in the sand bank of calcareous ooids of the Tiergarten Member. Some bivalves of the genus *Pholadomya* and large perisphinctid ammonites occur about halfway downslope in micritic limestone of the Pichoux Formation. Two of the pertinent ammonites from the Pichoux Formation are figured in GYGI (1995, Figs. 8 and 9). Siliceous sponges, mainly of the plate-like genus *Discophyma,* are very abundant in the Birmenstorf Member. One specimen of *Discophyma* is figured in polished cross-section in GYGI (1992, Fig. 32), and two specimens of the genus entirely preserved and uncovered on a bedding plane are shown in GYGI (2000a, Fig. 17 on p. 27). Ammonites out of the rich macrofossil assemblage in the Birmenstorf Member were figured in several papers by the author and coauthors, and mainly in GYGI (2001).

Abundant pellets of pure glauconite and few iron ooids were found in the stratigraphically condensed, lowermost bed of the Birmenstorf Member along with ammonites of the *Densiplicatum* and the *Antecedens* Chron. The vertical facies transition from the iron oolitic Herznach Formation below to the glauconitic facies in the lowermost bed of the Birmenstorf Member above occurred everywhere in Canton Aargau and in eastern Canton Solothurn at the same time. The facies change was brought about by eustatically rising sea level *before* rapid rise no. 5 (in Fig. 5.1) in the *Antecedens* Chron. The time of the facies change is documented with figured ammonites, which were found in excavation RG 208 near Ueken. The pertinent figures are specified below. This is compelling evidence that both the uppermost part of the iron-oolitic Herznach Formation below and the glauconitic part of the lowermost Birmenstorf Member above were sedimented upon a level sea floor at a water depth of approximately 100 m (see main conclusion no. 18). A possible exception to this was a minor step on the sea floor above the basinward rim of the early Callovian carbonate platform of the Dalle nacrée Member with its thin cover of Callovian iron oolite of the Herznach Formation, as it was assumed to exist by GYGI (2003, Figs. 173 and 187). This step could not be quantified, and it had no noticeable influence on the facies and the thickness of the Herznach Formation above. No step is, therefore, shown at the pertinent location in Figs. 1.5 and 4.15A of this study. Both the litho- and the biofacies of the Herznach Formation are invariable in the entire area between Herznach in Canton Aargau and Canton Jura in northwestern Switzerland.

The above conclusion that the Birmenstorf Member was sedimented in comparatively deep water on the level floor of a starved epicontinental basin depends on evidence that the stratigraphically condensed, lowermost marker bed of the member is isochronous everywhere in Canton Aargau. First of all, the age of the bed must be documented. The bed is a marly limestone with a thickness of about 10 cm in the following sections: bed no. 5 in section RG 32 near Küttigen (GYGI 1969a, Pl. 17), bed no. 10 in section RG 208 near Ueken (GYGI 1977, section no. 2 in Pl. 11), and bed no. 8 in section RG 210 near Gansingen (this study, Fig. 8.3). Abundant pellets of pure glauconite and sporadic iron ooids occur in the condensed bed in all of the three sections. It was stated above that bed no. 10 of section RG 208 near Ueken includes ammonites of both the *Densiplicatum* and of the *Antecedens* Chron. One *Cardioceras* of the *Densiplicatum* Chron from bed no. 10 was identified by D. MARCHAND and figured by GYGI and MARCHAND (1982, Pl. 12, Fig. 3). A coeval *Perisphinctes (Arisphinctes) parvus* GYGI from the same bed is shown in GYGI (2001, Fig. 32 on p. 27). A

Perisphinctes (Dichotomosphinctes) antecedens SALFELD, the index taxon of the *Antecedens* Chron from bed no. 10 in excavation RG 208, was figured by GYGI (2001, Fig. 62c on p. 46). Bed no. 10 of excavation RG 208 was broken up and searched for macrofossils on a surface of 119 m^2. The three ammonites mentioned were the only specimens diagnostic of an ammonite chron that could be found in the bed. In spite of stratigraphic condensation, ammonites in the marker bed in lowermost Birmenstorf Member are everywhere about as uncommon as they are in bed no. 10 in section RG 208. Proof that the lowermost, condensed bed of the Birmenstorf Member is isochronous in a vast region can, therefore, only be provided by indirect evidence, which is given in the following text.

The Schellenbrücke Bed of the *Cordatum* Chron directly below the lowermost bed of the Birmenstorf Member includes a wealth of well-preserved ammonites. The Schellenbrücke Bed is the uppermost unit of the Herznach Formation, where the bed occurs in Canton Aargau. Most of the ammonites of the genus *Cardioceras* which were collected from the Schellenbrücke Bed were found in excavation RG 208 near Ueken in bed nos. 8 and 9 (Fig. 8.2E). More *Cardioceras* were collected from bed no. 3 in section RG 32 near Küttigen in Canton Aargau (GYGI 1969a, Pl. 17). This bed is the type of the Schellenbrücke Bed in the former outcrop beside the cantonal road above the old bridge called Schellenbrücke. Some *Cardioceras* are from bed no. 7 in section RG 210 near Gansingen (Fig. 8.3). Numerous cardioceratid ammonites of the *Cordatum* Chron from the Schellenbrücke Bed were identified by D. MARCHAND and figured in GYGI and MARCHAND (1982). Perisphinctid ammonites out of the Schellenbrücke Bed were figured by GYGI (1998).

The abundant and well-preserved ammonites of the *Transversarium* Chron in the uncondensed, normal facies of the Birmenstorf Member in Canton Aargau were found mainly in the large excavations in and north of Eisengraben near Gansingen, in excavation RG 210 shown in Fig. 8.3 in this study and in neighboring excavation RG 230, that is represented in Fig. 1 in GYGI (2001). Most of the perisphinctid ammonites from the Birmenstorf Member were published by GYGI (2001). The several papers on other ammonite groups in the uncondensed, normal facies of the Birmenstorf Member are mentioned in Sect. 1.2. The Schellenbrücke Bed is documented to be isochronous mainly on the strength of a great number of figured cardioceratid ammonites. The normal facies of the Birmenstorf Member is concluded to be isochronous based on very numerous perisphinctid ammonites. The lowermost, condensed bed of the Birmenstorf Member between the Schellenbrücke Bed below and the uncondensed part of the Birmenstorf Member directly above must consequently be isochronous as well.

This and the fact that the vertical boundary between iron-oolitic facies below and facies with pellets of pure glauconite above is everywhere within the condensed bed in the lowermost Birmenstorf Member is proof that both the Herznach Formation and the Birmenstorf Member were sedimented on the level basin floor.

The top of thick, shallowing-upward succession no. 1 in northwestern Switzerland was probably *flat* at most of the stages of aggradational sedimentation. The top of the St-Ursanne Formation had little relief. This is documented by the widespread, peritidal facies of the Vorbourg Member directly above in the lowermost part of succession no. 2. The top of the Pichoux Formation off the basinward rim of the St-Ursanne Formation was a depositional *slope* with a primary inclination of between less than 0.5° and a maximum of approximately 4°. There are bioherms of hermatypic corals in the most proximal, upper part of the wedge-like Pichoux Formation (Fig. 1.5). It was stated above that the composition of the macrofossil assemblage in the formation changes from there, about halfway downslope, to some large perisphinctid ammonites along with few bivalves of the genus *Pholadomya*. The macrofossil assemblage grades at the toe of the slope, in the adjacent Birmenstorf Member on the *level* basin floor, into biostromes of siliceous sponges with abundant ammonites. The units mentioned are coeval according to ammonites which were figured by GYGI (1995, 2001). The lateral facies variation between the coeval units provides for the possibility to distinguish the carbonate *platform* of the St-Ursanne Formation from the very gently sloping *ramp* of the coeval Pichoux Formation, and to delimitate this ramp from the Birmenstorf Member on the *level basin floor* in eastern Canton Solothurn and in Canton Aargau.

The thin calcareous oolite in the middle of the Röschenz Member in succession no. 2 (Fig. 1.5), the coeval oncolite with hermatypic corals in the Grüne Mumienbank Bed at Péry (figured in GYGI 1992, Fig. 14), and the coeval Gerstenhübel Beds from deeper water in Canton Solothurn and in Canton Aargau are a lateral facies transition from very shallow water without ammonites over a slope (not exposed) to deeper water with ammonites. Ammonites from the Gerstenhübel limestone succession were figured by GYGI (2000a, Pl. 8, Fig. 4) and by ÉNAY and GYGI (2001, Pl. 1, Fig. 1). A subaqueous truncation surface and a debris flow directly above in the upper Effingen Member in section RG 37 at Auenstein in Canton Aargau are shown in Figs. 4.12 and 4.13.

Sedimentation of succession no. 2 ended in the central Jura Mountains with deposition of the very widespread, lagoonal oncolite of the Hauptmumienbank Member. The coeval calcareous oolite of the Steinibach Member rimmed the lagoon against the deeper part of the epicontinental basin. Both these members were laid down in very shallow water. Photographs of polished slabs of the Hauptmumienbank Member are shown in Fig. 14 on p. 24

in GYGI (2000a) and in Fig. 5.6 of this study. The top of the Hauptmumienbank Member at Roches north of Moutier is shown in Fig. 5.6 to be a planed and bored erosion surface. Coeval with this surface is a paleosol at the top of Steinibach Member near Waldenburg in Canton Basel-Landschaft, which is shown in Fig. 4.3. A similar paleosol at the top of the Steinibach Member crops out beside a path in the unpublished section RG 15 in the uppermost part of Horngraben Gorge south of Aedermannsdorf in Canton Solothurn. Both the Hauptmumienbank and the Steinibach Member were sedimented upon a level surface in very shallow water. After the end of deposition their top was at several localities emergent following a minor sea level fall, which occurred before rapid relative sea level rise no. 9 in Fig. 5.1. It is evident from Fig. 1.5 that succession no. 2 was shallowing upward in eastern Canton Solothurn. Sedimentation of the succession ended at Balsthal at the top of the sand bank of calcareous ooids in the Steinibach Member of the lower Balsthal Formation. The level sea floor at the beginning of the Oxfordian Age at the water depth of approximately 80 m between Canton Jura and Canton Aargau; the proximal, thick and shallowing-upward part of succession no. 1 in northwestern Switzerland; the depositional slope in the Pichoux Formation as well as such slopes in the proximal part of the Effingen Member; and the flat top of proximal succession no. 2 approximately at sea level are summarized in Fig. 1.5.

An almost perfectly flat upper surface of fresh sediment must have existed during the early *Berrense* Chron in the uppermost part of the Günsberg carbonate platform and landward in the adjacent, coeval part of the upper Röschenz Member. This is concluded, because the eustatic sea level rise no. 8 of only a few meters (in Fig. 5.1) was sufficient to allow for hermatypic corals to advance (or retrograde, respectively) from the upper Günsberg carbonate platform over a palinspastically restored distance of at least 30 km to Bressaucourt near Porrentruy in the interior of the shallow-water realm. Both the local coral biostrome of bed nos. 39–41 in section RG 381 in Court Gorge at Moutier (GYGI 2000a, Pl. 28) and the equally local coral biostrome near Bressaucourt southwest of Porrentruy are indicated by a single cross in Fig. 1.5. A depositional slope, which existed at the same time in the upper Effingen Member between Auenstein and Rekingen near Mellikon in northeastern Canton Aargau, is documented by the subaqueous truncation surface at Auenstein shown in Fig. 4.12, and by another such surface cropping out in the quarry in Musital southwest of Rekingen. A debris flow and a turbidite in the upper Effingen Member in Musital were figured by GYGI (1969a, Pl. 4, Figs. 12 and 13).

A detailed description and an interpretation of the mode of formation of the boundary beds between the Middle and the Upper Jurassic Series in section RG 210 in Eisengraben near Gansingen (Fig. 8.3) were given above in Sect. 8.5.

NORRIS and HALLAM (1995, p. 231) were of the opinion that what they thought to be "negligible sedimentation rates, growth of concretions, and reworking by bottom currents" was sufficient evidence to reinterpret the level floor of a starved epicontinental basin at a water depth of approximately 80 m, as it was concluded by GYGI and PERSOZ (1986, Pl. 1A), to be a "swell" which was allegedly influenced by bottom currents strong enough to overturn flat, cobble-size nodules. NORRIS and HALLAM (1995) did not figure the section in Eisengraben. Therefore, it can only be presumed that their concretions are in fact the large nodules mainly within the iron oolite of bed no. 7 in section RG 210. According to the description in Sect. 8.5, these nodules are most probably no concretions, but rather the remains of subsolution. The nodules were certainly not reworked by strong bottom currents, as it was presumed by NORRIS and HALLAM.

GYGI (1969a, p. 59) wrote that the nodules of limestone with some silt-grade detrital quartz, which are now incorporated in iron oolite of bed no. 7 in section RG 210, are relics of a limestone bed like that below the iron oolite, and that the nodules were formed *in situ by corrosion* (subsolution in this study). The process of subsolution going on in section RG 210 was probably similar to the one, which was active in the uppermost Grenzkalk unit in sections RG 81b and 212 in Canton Schaffhausen (Fig. 8.3). The large nodules of early Callovian limestone in iron oolite of bed no. 7 in section RG 210 were initially produced by subsolution, without intervening emersion. The nodules became from the middle Callovian on and in early Oxfordian time incorporated into the iron oolite which was sedimented in this time span in between them (Fig. 8.3). The abundant ammonite casts that prevail in the macrofossil assemblage within the iron oolite of bed no. 7 are evidence of water depth greater than 35 m. Reworking of flat, ellipsoidal nodules the size of several decimeters by bottom currents at a water depth greater than 35 m is highly improbable. The nodules are encrusted with goethite all around. Provided that they were transported, the crust would have been removed by abrasion.

NORRIS and HALLAM (1995, p. 231) disagreed with the sea floor topography which was concluded mainly by GYGI and PERSOZ (1986, Pl. 1A). They suggested what they thought to be "a more logical explanation". They postulated the existence of "an extensive swell in northern Switzerland", but they failed to indicate exactly when and where this swell evolved. The authors declared the existence of their swell based on what they had seen "in a trench in the woods near Mönthal" (p. 229). This rather vague localization of what they called a trench can only refer to the outcrop in the *natural cleft* of Eisengraben, which is in the territory of the township of Gansingen in northeastern Canton Aargau, not of Mönthal. The section in Eisengraben was figured for

the first time as RG 60 in GYGI (1966). It was measured again in 1971 and was then relabeled RG 210. NORRIS and HALLAM (1995) were unaware of the importance of this section, which was selected to be the reference section of the *Transversarium* Zone by GYGI (1977, p. 517). ALLENBACH (2002) agreed with the existence of the swell postulated by NORRIS and HALLAM (1995). He imagined that the nodules of early Callovian limestone encrusted with goethite in iron-oolitic bed no. 7 in section RG 210 in Eisengraben were clasts that were eroded by currents on this alleged swell, and that the clasts were reworked by the same strong currents on the sea floor (p. 332). No explanation was given of the envisaged processes of erosion and abrasion of the nodules when being reworked and of subsequent encrustation with goethite.

ALLENBACH (2002) was unaware of isostatic equilibration of the lithosphere by subsidence when being loaded with sediment. He consequently assumed on p. 328 that accommodation space for the thick Effingen Member was created by endogenic subsidence alone. ALLENBACH disregarded that isostatic subsidence of the basement under the load of the Effingen Member was calculated and visualized before by GYGI (1986, Fig. 3C). The process of isostatic basement subsidence caused by loading with sediment was discovered by WETZEL et al. (2003, p. 165), with ALLENBACH being a coauthor. Moreover, ALLENBACH (2002) did not pay attention to the ample paleobathymetric information which can be read from the composition of macrofossil assemblages, as was published mainly by SCOTT (1940), ZIEGLER (1967), and GYGI (1986, Fig. 6). The views on sea floor topography visualized by ALLENBACH (2002) in his Fig. 15 are, therefore, at variance with what was previously published based on an extensive set of dependable data.

ALLENBACH (2002, p. 338 and Fig. 15b) thought that the Bärschwil Formation was sedimented into a trough which evolved exclusively by endogenic subsidence. This assertion could only be made by ignoring all of the previously published information about paleobathymetry and sea floor topography (compare, for instance, GYGI and PERSOZ 1986, Pl. 1A, and GYGI 1986, Fig. 6, with ALLENBACH 2002, Fig. 15b). WETZEL et al. (2003) alleged that Oxfordian sea floor topography in northern Switzerland was predetermined by tectonic activity in the pre-Mesozoic basement. They asserted on p. 164 that the "facies boundaries coincide spatially with tectonic structures in the basement". When the pattern of *known* tectonic structures in the pre-Mesozoic basement is compared with *documented*, undisputed topographic highs of the sea floor in the Oxfordian Age like carbonate platforms and their curved basinward rims, it becomes obvious that the assertions by WETZEL et al. (2003) are untenable.

Progradation of the carbonate platform of the Günsberg Formation over the coeval, proximal part of the Effingen

Member cannot possibly have been controlled by synsedimentary faulting. Instead, the interrelation between sediment supply and sea level variation controlled the rate of progradation, and then halted progradation of the Günsberg Formation in the *Bifurcatus* Chron (see above). WETZEL et al. (2003) ignored or were unaware of the substantial tectonic activity in the pre-Oxfordian basement, which was previously *documented* to have occurred during the Oxfordian Age. Probably a deep-rooted normal fault driven by an endogenic process was active during sedimentation of the carbonate platform of the St-Ursanne Formation between Liesberg and Kleinlützel (GYGI 1990c, Fig. 5). The southern flank of the fault with a west-eastern trend is downthrown by 60 m. The order of magnitude of this fault excludes an exogenically driven growth fault to have been the cause. The downthrown block is to the *south* of a possible western extension of the Late Paleozoic rift which was discovered in the exploration well drilled by NAGRA in 1983 near the village of Weiach in northwestern Canton Zürich. The thickness of the Effingen Member in a small area near Riniken in Canton Aargau is 60 m greater than normal (GYGI 1990c, Fig. 6). In both of these cases, endogenic faulting in the pre-Mesozoic basement created substantial differences in the thickness of Oxfordian sediments, but *this synsedimentary tectonism left no traces in the facies at the surface* of the sediments which were laid down above the faults in motion. Sediment supply continuously compensated differential basement subsidence at the surface on the downthrown side of a fault (see Fig. 11.1).

Deep drilling and seismic exploration in northern Switzerland revealed that the difference in the amount of isostatic subsidence of adjacent basement blocks was not equilibrated by the reactivation of preexisting faults within the pre-Mesozoic basement during the Oxfordian Age. Local isostatic adjustment of the basement in the sense of STECKLER and WATTS (1978, p. 1) along normal faults at close intervals in the area of northern Switzerland was presumed for the first time to have occurred in Oxfordian time by GYGI (1986, p. 471). ALLENBACH (2002, p. 340) alleged the same to be a new conclusion, and it was announced as such in the title of the paper by WETZEL et al. (2003). The most probable assumption that can be made at the present time is that the difference in the amount of subsidence between adjacent basement blocks, which was caused by shifting of belts with maximal sedimentation, was equilibrated essentially by slight deformation of the basement in a *flexural tilt*. This was a *tectonic* process that was caused mainly by the *exogenic* process of sedimentation. The term "*tectonic subsidence*" which is currently used for basement subsidence caused by processes going on deep in the interior of the Earth is, therefore, equivocal. It should be replaced in sedimentary geology by the unequivocal term "*endogenic subsidence*".

11.5 Paleoclimate and Sediment Supply

11.5.1 Subtropical Climate Alternating Between Humid and Semiarid

Densely colonized coral patch reefs in the Buix Member and in the Tiergarten Member of the upper St-Ursanne Formation are evidence that climate during the late *Antecedens* Chron and in the *Transversarium* Chron of the middle Oxfordian was in the region investigated either tropical or subtropical. The thickly branching hermatypic coral *Cryptocoenia limbata* GOLDFUSS shown in Fig. 3.1 was collected out of a small coral reef in the upper St-Ursanne Formation, from unit no. 11 in the unpublished section RG 370, 3 km south of the village of Courtételle in Canton Jura. Close to 50% of the volume of this reef are skeletal parts of coral colonies. The distinctly delimited accretion bands with a thickness of between 4.3 and 5.7 mm which can be discerned in Fig. 3.1 could be proved in Chap. 3 to be annual by comparison with Recent corals collected by the author in Bermuda. According to INSALACO (1996b, p. 416), the thin dark part of the bands are "regions of thicker septa with more numerous dissepiments (high density band)". The regular alternation of high- and low-density bands in Fig. 3.1 is evidence that the temperature of shallow water the coral grew in varied between warm in the summer months and substantially cooler during the winter. The banding in the Late Jurassic coral skeleton from near Courtételle is diagnostic of seasonal climate like the very similar banding measured in the skeletal part of a Recent coral from Bermuda. The coral reefs of Bermuda grow in an environment with pronounced seasonal variation in temperature of both air and shallow water. Bermuda is near or at the northern boundary where coral reef growth is now possible in the North Atlantic Ocean. Climate in the region investigated in northern Switzerland was consequently *subtropical* in the middle Oxfordian *Transversarium* Chron. Weathering was most effective in times of humid climate with much rainfall on land in the northwest and north. Runoff and supply of terrigenous sediment to the adjacent shallow-marine realm in the south was at a maximum in such time spans. Conversely, the rate of terrigenous sediment supply dropped to a minimum in periods of drier climate.

11.5.2 Long-Term Climatic Change Throughout the Oxfordian Age

Oxfordian climate grew increasingly dry throughout the climatic variations during deposition of succession nos.

1–3. This becomes obvious, when the thickness and the carbonate content of the marly Bärschwil Formation in lower succession no. 1 is compared with the much thinner marl of the Bure Member in lowermost succession no. 3. Not only fine-grained siliciclastic material in the sediments laid down in deeper water of the epicontinental basin is allochthonous. Most of the calcareous mud in the same sediments from deeper water is allochthonous as well. The *Renggeri* Member in lowermost succession no. 1 is a homogenous clayey marl with a carbonate content of 32%, according to an analysis by PFRUNDER and WICKERT (1970) and to the classification by PETTIJOHN (1975, Fig. 10–41). There are no intercalations with enhanced carbonate content in the *Renggeri* Member. The lower part of the Effingen Member in lowermost succession no. 2 has an average carbonate content of at least 60%, to judge from Pl. 17 in GYGI (1969a). The marl of the Effingen Member includes intercalations with enhanced carbonate content. The calcareous layers in the succession of the Gerstenhübel Beds in the lower Effingen Member are micritic limestone.

No carbonate analyses are known from the marl of the Bure Member in the lowermost part of proximal succession no. 3. The thickness of the member is 9 m in its type section RG 454 in an exploration well at Bure in Canton Jura which is represented in Pl. 16 in GYGI (2000a). The marl of the Bure Member in section RG 454 includes some intercalations of marly limestone in the lowermost part and two limestone beds in the upper part. The Bure Member in an exploration well near St-Ursanne (section RG 443 on Pl. 20 in GYGI 2000a) is 10.5 m thick and has apparently a higher average carbonate content. The upper part of the Bure Member cropped out in 1981 along Chemin paulin road near Courgenay in Canton Jura when section RG 350 shown in Pl. 19 in GYGI (2000a) was measured. The limestones of the Courgenay Formation directly above the Bure Member in the upper part of succession no. 3 are in section RG 443 near St-Ursanne 50.7 m thick and are complete. These limestones are the bulk of succession no. 3. Both the volume and the time represented by the Bure Member are modest in comparison with the limestones of the Courgenay Formation above. Dolomitization in the Oxfordian sediments investigated occurred for the first time sporadically in the upper Günsberg Formation. A euhedral crystal of dolomite in a coral biostrome, bed no. 165 in the Günsberg Formation in section RG 14 at Günsberg, was figured by GYGI (1969a, Pl. 7, Fig. 30). Much dolomite was found in the uppermost Günsberg Formation in the gorge of Lochbach north of Selzach in Canton Solothurn (GYGI 2000a, Pl. 29). Dolomitization became pervasive in the Verena Member. GYGI (2000a, Fig. 16 on p. 26) showed a thin-section photograph of dolosparite from the Verena Member. Calcite pseudomorphs after sulfate minerals (Fig. 3.3) and some gypsum in the Verena Member are evidence that the climate became driest toward the end of the Oxfordian Age.

Ammonites of the genera *Cardioceras* and *Amoeboceras* were most abundant in the Boreal Faunal Province. The *Cardioceras* and *Amoeboceras* fossilized in the sediments investigated can be assumed to have lived in water with a similar range in temperature. It is probable that ammonite casts of these genera indicate comparatively cool bottom water in the rocks investigated. *Cardioceras* were in the early Oxfordian *Cordatum* Chron most abundant at a water depth of between 30 and 35 m in the fossil bed in the distal part of the middle Sornetan Member. The abundance of *Cardioceras* diminished with increasing water paleodepth in the coeval Schellenbrücke Bed and in the Glaukonitsandmergel Bed (Fig. 4.9). This is the contrary of what is to be expected, because bottom water most probably became cooler with growing water depth. No simple relation consequently existed between water temperature in the habitat of *Cardioceras* and the percentage of representatives of this genus in an ammonite assemblage.

Cardioceras evolved to *Amoeboceras*. The oldest, uncommon *Amoeboceras* were found in the uncondensed part of the Birmenstorf Member in Canton Aargau from a water depth of approximately 100 m. They are in this member of similar size like *Cardioceras* in the older Schellenbrücke Bed from a water depth of close to 100 m. Very few and smaller *Amoeboceras* from a lesser water depth occur in the Gerstenhübel Beds of the *Bifurcatus* Chron and in the Hornbuck Member of the late Oxfordian *Bimammatum* Chron (figured in ATROPS et al. 1993). *Amoeboceras* are fairly numerous in some of the Oxfordian/Kimmeridgian boundary beds from an estimated water depth of 100 m in section RG 239 at Summerhalde near Schaffhausen. All of the specimens in that section are of small and even of dwarf size, according to the classification of ammonite size in GYGI (2001, p. 12). As stated in Sect. 11.4.3 above, this is an indication that the habitat was marginal to *Amoeboceras,* when the Oxfordian/Kimmeridgian boundary beds in section RG 239 were sedimented. Based on these observations, the assumption can be made that the temperature of bottom water in the deep subtidal zone increased in the time span between the early Oxfordian *Cordatum* Chron and the Oxfordian/Kimmeridgian time boundary. This was possibly the effect of globally warming climate. ABBINK et al. (2001, Fig. 12) concluded warming of the climate in the course of the Oxfordian based on sporomorph data from the southern North Sea. It follows from this discussion that climate grew certainly drier and probably warmer throughout the Oxfordian Age.

The paleotemperatures as published by DROMART et al. (2003, Tab. 2) do not show a distinct rise in the latest part of the early Callovian, when the carbonate platform of the Dalle nacrée Member and its time equivalents in eastern France drowned. The authors found a low temperature at the turn from the *Mariae* to the *Cordatum* Chron in the early Oxfordian. This was not confirmed by RAIS (2007, Chap. 4,

Tab. 1) based mainly on belemnites out of the collection of the author of this study. It must be noted that several ages of GYGI's samples listed by RAIS (2007) are at variance with data that were previously published by GYGI. The age of samples RG 280/5a and b from Liesberg shown in Tab. 1 by RAIS (2007) is, according to Fig. 2 in GYGI (1990a), the late Callovian *Athleta* Chron, instead of the early Oxfordian *Mariae* Chron as was alleged by RAIS. Moreover, the age of sample RG 51/8 from Oberehrendingen is the *Schilli* Chron (GYGI 1977, p. 446) instead of what RAIS called *Plicatilis-Transversarium* Chron. The age of sample RG 356/202 from Grandval is the *Bifurcatus* Chron (GYGI 2000a, Pl. 38), instead of the *Plicatilis* Chron as was stated by RAIS. Apart from the untenable statements of age which were made by RAIS, her results do not show a significant rise in paleotemperatures from *Athleta* time in the late Callovian to *Bimammatum* time in the late Oxfordian.

11.5.3 Sediment Supply and Mode of Transport

The three principal marl-limestone successions in Oxfordian sediments reflect a long-term climatic alternation between humid and substantially drier time spans. The supply of terrigenous sediment to the Rhodano-Swabian Epicontinental Basin varied at the source, according to climatic conditions on neighboring land (GYGI 1986, p. 485). A climate with much rainfall lasted from the beginning of the *Schilli* Chron until the *Hypselum* Chron, when the lower part of succession no. 2 with a great amount of clay minerals and with some siliciclastic silt was sedimented. *Direct* evidence of a humid climate in this time span are land plants which were found in the Günsberg Formation by A. and H. ZBINDEN in section RG 307 near Péry, and by R. ALLENBACH at a locality in Court Gorge southwest of Moutier. The conifer plant remains from Moutier were published by ALLENBACH and KONIJNENBURG-VAN CITTERT (1997). Further evidence of times with much rainfall are mainly thin seams of lignite coal, which were found in the Günsberg Formation by HEER (1865, p. 125) near Pfeffingen in Canton Basel-Landschaft, by KEMMERLING (1911, p. 22) in Moutier Gorge, and by GYGI (2000a, Pl. 27, at the top of bed no. 80 in section RG 390 in Moutier Gorge, and Pl. 40, in the upper part of bed no. 25 in section RG 430 at Gänsbrunnen in Canton Solothurn). Limnic ostracods and characean algae lived at the same time in shallow freshwater ponds. Corresponding fossils are ostracod shells and gyrogonites of characean algae which were shown in Pl. 1 in OERTLI and ZIEGLER (1958). ZIEGLER (1962) found characean gyrogonites in many of his sections of the Röschenz Member. GYGI (2000a) isolated such gyrogonites from the Günsberg Formation, for instance, in section RG 315 near Sornetan in bed nos. 58 and 61 (Pl. 21), and in section RG 307 near Péry at the top of

bed no. 162 in Pl. 22. Characean gyrogonites were also found within cracks between prisms of an intertidal stromatolite in the Günsberg Formation, in bed no. 21 of the unpublished section RG 414 north of Grandval (east of Moutier). One of the prisms out of bed no. 21 in section RG 414 is shown in Fig. 4.5.

The high percentage of terrigenous sediment in the Effingen Member is an *indirect* indication, that there was a massive increase in rainfall and of concomitant weathering on land in the northwest and north from the beginning of the *Schilli* Chron onward. ZIEGLER (1962, p. 42) presumed that detrital silt of quartz, which he found to be abundant in particular beds in the Röschenz Member, may have been blown by wind from land into the marginal marine environment of the member. His assumption was based on the observation that grains of detrital quartz in the Röschenz Member are well sorted and mainly of silt size. This is confirmed in this study by the thin section photograph in Fig. 3.2. The assumption by ZIEGLER (1962) that siliciclastic silt in the Röschenz (alias *Natica-*) Member was transported by wind into the sediment was adopted and taken for granted by BOLLIGER and BURRI (1967). These authors observed that detrital quartz was associated with some feldspar of the same grain size. GYGI (1969a, p. 20) identified different types of detrital feldspar in thin sections, where grain size was sufficient to distinguish potassium-rich feldspar from plagioclase. He concluded on the same page from some of his particularly thinly ground and polished thin sections that the partially cavernous surface of detrital, cross-cut quartz grains was the product of corrosion by interstitial water with a high pH during diagenesis. BOLLIGER and BURRI (1970) thought that the surface morphology of the detrital grains of quartz and feldspar, which they had isolated from the Röschenz Member and from the coeval Effingen Member, was diagnostic of eolian transport.

Provided that it was indeed wind that blew much siliciclastic silt from land into the adjacent marginal marine and partially peritidal realm to the south and into marine sediments beyond like the Effingen Member, one would expect that siliciclastic silt was most abundant in pure limestones from shallow-marine water, which were sedimented in periods of relatively dry climate. The contrary is the case. The purest carbonate rock in the Oxfordian shallow-water realm of northern Switzerland is the Buix Member in the upper St-Ursanne Formation. The amount of detrital quartz recorded by F. PERSOZ by x-ray diffractometry in Buix Member in a section along the road north of the farm Foradrai, 2 km west-southwest of the village of Glovelier in Canton Jura, is minimal. The Buix Member is labeled KKS in Fig. 12 by PERSOZ in GYGI and PERSOZ (1986). No detrital quartz at all is indicated to occur, according to x-ray diffractometry, in the Verena Member in the same section near Foradrai shown in Fig. 12 by PERSOZ. GYGI

North   South

A

Fig. 11.1 Section RG 411 in the erosional cirque of Combe des Geais on the south slope of Mt. Raimeux north of the village of Grandval east of Moutier in Canton Bern. Close-up view of the north-south trending eastern wall of the cirque, looking eastward from Point 1165.3. The width of the outcrop is 100 m. PIC: Thickly bedded, micritic limestone of the Pichoux Formation. SUF: Tongue of the uppermost St-Ursanne Formation, massive, micritic limestone with hermatypic corals. GÜN: Marly limestone with some detrital quartz of the lowermost Günsberg Formation. The tongue of the St-Ursanne Formation and mainly the lowermost part of the Günsberg Formation are thinning toward the south. Southeastward is basinward according to Fig. 1.2. A: Normal fault. The block south (right) of the fault is downthrown by several meters. Motion along the fault occurred during sedimentation of the Pichoux Formation. Photograph by C. Lanz. Two coral patch reefs in the Günsberg Formation above the outcrop are now almost completely hidden in the forest. They were well visible in 1983 when the author photographed the corresponding Fig. 8 in GYGI and PERSOZ (1986)

arrived at the same result by scanning numerous of his thin sections from the Verena Member. Pervasive dolomitization (thin section photograph of dolosparite in Fig. 16 on p. 26 in GYGI 2000a) and pseudomorphs of calcite after sulfate minerals in the initially pure calcareous oolite of the Verena Member (thin section photograph of Fig. 3.3 in this study) are evidence that the Verena Member was sedimented in the driest time span of the Oxfordian.

Siliciclastic silt is in fact most abundant in Oxfordian sediments of northern Switzerland in the partially argillaceous deposits of the Röschenz Member and of the coeval Effingen Member. A simple and dependable field-geologic method in order to find much siliciclastic silt in marl or limestone is to rub a rusty geologic hammer against the pertinent rock. If there is much such silt in the rock, it will readily remove rust on the hammer. ROLLIER (1898, p. 58) stated that siliciclastic silt can occur in calcareous intercalations in the Séquanien (in the part of his Séquanien which is now the Röschenz Member) in a concentration high enough that good whetstones could be manufactured from pertinent beds in the Röschenz Member near Schelten in Canton Bern and at Damvant in Canton Jura. Figure 3.2 in this study is the photograph of a thin section through such a siltstone with calcareous matrix in the Röschenz Member, bed no. 50 of section RG 400 near Corban in Canton Jura. Section RG 400 is represented in Pl. 35 in GYGI (2000a).

Running water rather than wind transported most or all of the weathering products, mainly clay minerals and siliciclastic silt, in times of humid climate from land in the north into the adjacent, widespread marginal marine and peritidal environment of the Röschenz Member. Siliciclastic silt could be concentrated in the course of sediment transport by elutriation. Endogenic basement subsidence in the marine, proximal realm of what is now northwestern Switzerland was from the *Schilli* Chron on such that peritidal sedimentation of the Vorbourg and of the Röschenz Member could proceed for an extended time. Water currents driven by tides and occasional storms carried most of the siliciclastic material, which temporally resided in the Röschenz Member, along with the great quantity of calcareous mud that was precipitated in that environment (see conclusion in Sect. 4.14.2), across the carbonate platform of the Günsberg Formation and beyond into deeper water, where the Effingen Member was laid down. Transport of fine-grained sediment across a wide belt of very shallow water must have been of a large scale. Thinning of a tongue of the uppermost St-Ursanne Formation and mainly of the lowermost part of the Günsberg Formation directly above toward the deeper part of the epicontinental basin is visible in the north-south trending eastern wall of the erosional cirque of Combe des Geais north of Grandval (Fig. 11.1). The direction of thinning is diagnostic of the direction from where fine-grained sediment was transported from the peritidal area in the northwest into deeper water of the Rhodano-Swabian Basin in the southeast according to Fig. 1.2.

Sediment transport down the depositional slope of the marginal part of the deeper epicontinental, Rhodano-Swabian Basin is well documented in the Effingen Member. A turbidite of pure mud in a channel in the lowermost Effingen Member in section RG 226 in the deep road cut at Veltheim is figured in GYGI et al. (1998, Fig. 10). The thickness of intercalations with much siliciclastic silt in the Effingen Member in the deeper part of the basin is never more than a few centimeters. A photograph of a polished slab with two thin layers with much siliciclastic silt from the Effingen Member at Rekingen in northeastern Canton Aargau is shown in Fig. 12 on Pl. 4 in GYGI (1969a). Such thin, silty layers laid down in deeper water in the Effingen Member are probably the elutriated, lowermost part of small turbidites. Many of such silty layers have a matrix of mainly carbonate. Some semi-consolidated, plastic layers with much siliciclastic silt slid down on inclined substrate. They were dismantled and deformed in the process. One such isolated part of a silty layer is shown in Fig. 4.13. Thin, laminated beds with much siliciclastic silt crop out in the Effingen Member in section RG 87 at the head of the landslide of Bleiche on the western slope of Mt. Eichberg northwest of Blumberg in southern Germany (GYGI 1969a, Pl. 16). The beds were laid down at an estimated water depth of approximately 100 m on the basin floor, which was probably level at this locality. The laminated beds in the Effingen Member in section RG 87 could, therefore, be tempestites rather than turbidites.

Supply of weathering products from land in the northwest and north was minimal during sedimentation of the carbonate platform of the St-Ursanne Formation, the adjacent Pichoux Formation on the slope, and the coeval, thin Birmenstorf Member on the level basin floor (Fig. 1.5). A large part of a leaf of a land plant in lagoonal sediment of the Buix Member was found and figured by PÜMPIN (1965, Fig. 21). Evaporite minerals were not found to date in the St-Ursanne Formation. This is evidence that moderate rainfall on land was sufficient to support plant growth from late in the *Antecedens* Chron onward when

sedimentation of pure limestone of the St-Ursanne Formation began, and that climate was not arid until the end of sedimentation of the formation at the close of the *Transversarium* Chron.

The width of carbonate platforms depended on climate. The platforms of the St-Ursanne Formation and of the Balsthal Formation grew wide in times of relatively dry climate, when a minimum of terrigenous mud and silt was supplied to their environment. Rainfall on land in the northwest substantially increased at the turn from the *Transversarium* Chron to the *Schilli* Chron. As a consequence, a great amount of fine-grained terrigenous material bypassed the Günsberg carbonate platform in suspension in water over the platform from the early *Schilli* Chron on. The ample argillaceous material that water currents transported across the platform did not inhibit carbonate production on the platform altogether. Only the *rate* of production of calcium carbonate was thoroughly reduced by the high concentration of clay minerals in suspension in the water passing over the platform. This is reflected by the much lesser width of the carbonate platform of the Günsberg Formation as compared with the width of the preceding carbonate platform of the St-Ursanne Formation or of the succeeding carbonate platform of the Balsthal Formation.

11.5.4 Storms and Storm Wave Base

Storms in the region investigated could be violent, probably like Recent hurricanes are during summer months over the Caribbean Sea, the Gulf of Mexico, or over the Bahamas. A storm which occurred in the early Oxfordian *Scarburgense* Chron drove bottom currents that churned up mixed argillaceous and calcareous mud including iron ooids from the sediment surface at a water depth of approximately 80 m in section RG 208 near Ueken in Canton Aargau. The currents washed out deep pockets between calcareous concretions (Fig. 8.2B). The eroded sediment was redeposited elsewhere. Such an event probably mixed the entire water column thoroughly down to the sea floor. Bottom currents driven by another storm at least 1.2 million years later in the early *Cordatum* Chron washed some calcareous mud with iron ooids in section RG 208 out of the same pockets (Fig. 8.2C, D) and resedimented the mud elsewhere. Water depth had in the meantime increased at this locality to nearly 100 m.

The Mumienmergel Bed and the Mumienkalk Bed of middle Oxfordian age in Canton Schaffhausen (sections RG 81b and RG 212 in Fig. 8.3) were laid down in water approximately 120 m deep. The matrix of the Mumienmergel Bed was initially mixed argillaceous and calcareous mud, and the primary matrix of the Mumienkalk Bed was calcareous mud. Large, early lithified calcareous casts of cardioceratid and mainly of perisphinctid ammonites were formed in the unlithified, mud-grade matrix of both of these beds. Many of the casts were later winnowed clear of loose mud around them by episodic, storm-driven bottom currents. Some of the casts were broken soon after early lithification by a process unknown, when the matrix of the sediment was loose. None of the broken edges of the fragments are abraded. The fragment of a large cardioceratid ammonite cast with a thick oncolitic crust is shown in Fig. 4.1 G. The crust all around the fragment is evidence that the ammonite cast was broken at an early stage, before the encasing sediment was compacted and lithified. The fragment was winnowed clear of sediment before being encrusted. Storm-driven bottom currents strong enough to churn up mud from the sediment surface at the depth of 120 m are evidence of particularly violent cyclones, but it was probably not bottom currents which broke up ammonite casts. The fact that the broken edges of fragments of ammonite casts are not abraded is proof that the fragments were not transported by bottom currents.

The conclusion to be drawn from the Mumienmergel and the Mumienkalk Bed is that storm wave base *in the part of the Rhodano-Swabian Basin investigated* was during the *Antecedens* Chron of the *middle Oxfordian* considerably deeper than 120 m. It was proved in Sect. 4.3.2 that storm-driven bottom currents churned up unlithified mud from the sediment surface at this water depth. A storm-driven current of bottom water at the depth of around 80 m eroded in the *Scarburgense* Chron of the *earliest Oxfordian* unlithified, iron-oolitic mud from above concretions and out of pockets in between in the upper Herznach Formation in section RG 208 at Ueken in Canton Aargau (Fig. 8.2A, B). Sedimentation of homogenous mud of the marl-clay of the *Renggeri* Member began in the *Scarburgense* Chron of the *early Oxfordian* at a water depth of approximately 80 m and ended at a depth which can be estimated to have been about 50 m. The iron sulfide substance of the casts of stunted or dwarfed ammonites *living* in the environment and of thin burrows in the sediment is evidence that bottom water was dysaerobic and normally near-stagnant in the entire time span when the *Renggeri* Member was sedimented. Bottom currents driven by particularly violent storms very probably churned up mud from the sediment surface in the *Renggeri* Member of northwestern Switzerland like in the earliest Oxfordian sediment of the upper Herznach Formation not far away near Ueken (Fig. 8.2B). Nevertheless, no silty storm layers or tempestites occurred to the author in the perfectly fresh outcrop of the *Renggeri* Member when he measured section RG 280 in July 1978 in the clay pit of Andil near Liesberg at the time when the pit was worked (section drawn in Pl. 30 in Gygi 2000a). This is possibly because there was so little siliciclastic silt in the fresh sediment of mud that the silt did not become concentrated enough by elutriation during storms to form silt-rich tempestites that can be recognized as

such in a fresh outcrop. Bottom water was repeatedly quasi-stagnant and dysaerobic for a short time when the younger Effingen Member was laid down. These episodes occurred when the middle part of the member was sedimented at a water depth of the order of 60–70 m.

11.5.5 Kaolinite as a Means in Time Correlation

The clay mineral kaolinite is formed on land by weathering in a warm climate with adequate rainfall, according to Carroll (1958, Tab. 2), or to Griffin (1962) as cited by Persoz in Gygi and Persoz (1987, p. 59). It was documented above that marl-cay and marl in the sediments investigated were laid down when there was more rainfall on adjacent land in the northwest than when limestone was deposited in the area of northern Switzerland. It is, therefore, to be expected, that kaolinite in the shallow-marine sediments of the Late Jurassic in northern Switzerland would be more abundant in marl than in limestone. Persoz in Gygi and Persoz (1986, Fig. 10) showed that the contrary is the case. The most instructive example of this is the limestone succession of the Gerstenhübel Beds in the lower Effingen Member in Canton Aargau. Kaolinite is more abundant in micritic limestone of the Gerstenhübel Beds than in marl below and above. The highest percentages of kaolinite were found in the limestones of the Villigen Formation of the late Oxfordian. No direct link consequently exists between humidity of the climate and ample rainfall in the source area on land, and the kaolinite content of coeval sediments in an adjacent epicontinental sea. The Buix Member with up to more than 99% carbonate and with much kaolinite (Persoz in Gygi and Persoz 1986, Fig. 10) is the purest limestone in Oxfordian sediments. The part of a large leaf of a land plant in the Buix Member mentioned above indicates that rainfall on nearby land was sufficient at the time to support growth of large plants. It can, therefore, be presumed, that rainfall was sufficient on land northwest and north of where Switzerland is now, that *weathering* could produce the elevated amount of kaolinite which was measured by F. Persoz in the Buix Member. Kaolinite could consequently be formed on land not only when climate was humid, but as well in time spans with a moderately dry climate.

Micritic limestones from deeper water with maxima of kaolinite content are those of the Birmenstorf Member, of the Gerstenhübel Beds, and of the Villigen Formation (see Persoz in Gygi and Persoz 1986, Fig. 10, assembled section "Aargau"). All of these units are dated with figured ammonites. Ammonites in the lagoonal, shallow-water Buix Member are coeval with corresponding taxa in the Birmenstorf Member. The figured *Perisphinctes*

(Dichotomoceras) bifurcatus (QUENSTEDT), no. 19 in Figs. 1.5 and 1.6, which B. HOSTETTLER found at the base of a coral bioherm in section RG 307 at Péry (in lower unit no. 160 in Pl. 22 in GYGI 2000a), indicates that the lagoonal oncolite of the Grüne Mumienbank Bed with much kaolinite above the bioherm is coeval with the Gerstenhübel Beds from deeper water with a specimen of the same ammonite taxon, which is shown in Fig. 7.2C. The figured *Lithacosphinctes evolutus* (QUENSTEDT) out of the uppermost Balsthal Formation at Balsthal, no. 29 in Fig. 1.5, is evidence that the top of the shallow-water Balsthal Formation is coeval with the top of the Villigen Formation from deeper water in the epicontinental basin. The sharp drop in kaolinite content at the base of Kimmeridgian sediments, both from deeper and from shallowest water, which is represented in GYGI and PERSOZ (1986, Fig. 10) is consequently coeval. The vertical variation in kaolinite content is similar in successions from deeper water and from shallowest water. PERSOZ in GYGI and PERSOZ (1986, p. 439) concluded from this that kaolinite in the sediments investigated is of "detrital provenance", and that the mineral could, therefore, be used in time correlation. The ammonites from shallow-water sediments which were figured in the paper by GYGI (1995) corroborated the conclusion by PERSOZ.

11.6 Sea Level Variation

11.6.1 Eustasy in the Quaternary

Deep drilling through the Antarctic ice shield which was recently completed at the site called Dome C in East Antarctica revealed a regular succession of 8 climatic cycles of air temperature varying between high and low. The cycles are asymmetric. The first part of such a cycle was a long time of on average falling air temperature until a minimum was reached. In the second, short part of the cycle, temperature rose rapidly and returned to the initial level. The time which elapsed between the last four principal peaks of warm temperature before the warm period at the present time was on average about 100,000 years (MCMANUS 2004, Fig. 1). It can be read from the figure cited that the maxima and minima of temperature as well as the period of the last four glacial cycles in the Quaternary were almost equal. Sea level rise since the peak of the last glaciation of the Pleistocene 24,000 years before the present time (KELLER and KRAYSS 2005, Fig. 10) was calculated by BARD et al. (1990) based on corals from Barbados Island in the Antilles to have been about 120 m. LAMBECK and CHAPPELL (2001, Fig. 1B) rated the same rise off Huon Peninsula in Papua New Guinea at between 120 and 130 m. It can be concluded from this that accumulation of the thick ice shield over Fennoscandia

during the last Pleistocene glaciation took a long time, whereas the ice melted rapidly away at the end of the glaciation, as mentioned above. Eustatic sea level rise caused by melting of ice on land since the end of the last Pleistocene glaciation must, therefore, have proceeded at times at a high rate.

11.6.2 Eustasy in the Late Jurassic

The principal difference between eustasy in the Quaternary and in the Late Jurassic is not so much the pattern, but mainly the scale. Conformable evidence from cores drilled from Quaternary ice in Antarctica and from cores drilled from coeval deep sea sediment (MCMANUS 2004, Fig. 1) is that temperature fell on average slowly until the peak of a glacial cycle and then rose rapidly. The rapid rise in temperature after the last glacial cycle of the Pleistocene caused sea level to rise globally by at least 120 m at a high rate. This is to be compared with the rapid relative sea level rises in the Late Jurassic with a eustatic component, which amounted to at most 6 m. All of the individual sea level rises that are represented in Fig. 5.1 were rapid, and the falls indicated in between can be assumed to have been slow. Eustatic sea level falls moved in the same direction as endogenic basement subsidence. Endogenic subsidence was as a rule equable in a vast region and moving at a low rate. No incisions produced by erosion were found at the basinward rim of carbonate platforms. This is one of the reasons why the assumption is made that eustatic sea level falls were moving at a low rate. The question can be asked, whether sea level variation in the Late Jurassic with a similar pattern, but at a by far lesser scale than that in the Quaternary, was caused by a minor quantity of ice somewhere *on land,* which was accumulating slowly and then melted away rapidly. According to FIRSTBROOK et al. (1979), there was land neither around the North Pole nor around the South Pole during the Late Jurassic (see also ROWLEY and LOTTES 1988). HALLAM (1985, p. 433) stated that there were no polar ice caps in the Mesozoic. MOORE et al. (1992, p. 144) wrote that no glacial deposits were reported in the Kimmeridgian or in the Tithonian on land.

A minor quantity of ice may have existed *on land* in the Oxfordian and in the Kimmeridgian Age in northeastern Siberia, which was sufficient to cause sea level fluctuations of the order of 5 m like rise nos. 6–9 in the Günsberg Formation and in the lowermost Balsthal Formation above. Possible glendonites were reported to occur in the marine Koster Formation in northeastern Siberia by CHUMAKOV and FRAKES (1997). The type section of the formation beside the Artyk River is about 600 km north-northwest of the city of Magadan on the north shore of the

Sea of Okhotsk. The youngest possible calcite pseudomorphs after glendonite that CHUMAKOV and FRAKES (1997, Fig. 3) found in the uppermost Koster Formation are probably of Oxfordian age. According to KEMPER and SCHMITZ (1981, Pl. 1), glendonites are aggregates of large crystals mainly of calcite, which are probably pseudomorphs after the mineral Thenardite Na_2SO_4. Thenardite occurs on the sea floor in the Recent Arctic, and it is stable only in cold sea water. The mineral is highly soluble in warmer waters south of the Arctic. This is an indication that the Jurassic Koster Formation was sedimented in sea water which was about as cold as Arctic waters are in the Recent. The quantity of ice in the Late Jurassic on high mountain ranges in lower latitudes was probably insufficient to cause sea level variation of the scale which is represented in Fig. 5.1.

The more or less regular period of Oxfordian sea level fluctuation nos. 6–9 in Fig. 5.1 was close to 400,000 years. This is much longer than the period of approximately 100,000 years of the last four glaciations and interglacials in the Quaternary. The time intervals between rapid relative sea level rise nos. 11–15 in the Kimmeridgian Age are of widely varying duration. All of this is incompatible with glacio-eustasy. Some of the sea level rises which are documented in this study were slow and gradual (see Fig. 4 in GYGI 1986, lower part of succession no. 1 and upper part of succession no. 3). Most of the rises recorded were rapid. The amount of rapid rises was unequal, and these rises occurred after time spans of widely varying length. Provided that global sea level variation during the Late Jurassic was caused by MILANKOVITCH-type variation in insolation of the Earth, the sea level rises would all have been rapid and of a similar amount, and the period of variations would be regular. The question of what process drove the numerous rapid relative sea level rises from the Bathonian on through the Kimmeridgian Age consequently remains to be unanswered.

SCHULZ and SCHÄFER-NETH (1997) drew attention to the fact that the volume of water changes with variation in its temperature. They hypothesized that sea level fluctuations of the order of a few meters could have been caused by MILANKOVITCH-type variation in the temperature and thereby of the volume of ocean water. STRASSER (2007) claimed to be able to recognize an "astronomical time scale" in Oxfordian sediments in the Swiss Jura Mountains. He called on p. 427 of that paper the MILANKOVITCH-type cycle of 20,000 years, which is approximately the period of precession of the Earth's rotating *axis*, and the cycles of 100,000 and 400,000 years that are the effect of eccentricity of the Earth's *orbit* around the Sun, indiscriminately "orbital (Milankovitch) cycles." The pattern of sea level variation in cyclostratigraphy as conceived by STRASSER (2007) was imaged by VÉDRINE (2007, Fig. 7.2 on p. 108).

STRASSER et al. (2000, Fig. 3) interpreted the section in Court Gorge in the territory of Moutier, which is equivalent to section RG 381 on Pl. 28 in GYGI (2000a), in terms of lithostratigraphy, ammonite biochronology, sequence stratigraphy, and cyclostratigraphy. STRASSER (2007, Fig. 7) stated that he had seen in this section the Steinibach Member, Bure Member, La May Member, and (in Fig. 8) the Porrentruy Member. A definition and a short description of these members were published in the dictionary of Late Jurassic lithostratigraphic units in northern Switzerland by GYGI (2000b). Synonymy of the members was listed in GYGI (2000a). It can be read from Pl. 28 in GYGI (2000a) that none of the four members mentioned by STRASSER (2007) crop out in section RG 381 in Court Gorge. According to the ammonite zonation in Fig. 1.6 in this study, the base of the *Hypselum* Zone in the section of STRASSER et al. (2000, Fig. 3) is the transgressive surface 9 m above the base of this section. The base of the *Berrense* Zone is the transgressive surface 18 m above the base of that section, and the base of the *Bimammatum* Zone is 37 m above the section's base. STRASSER et al. (2000) thought they could delimit in the sediments in their section in Court Gorge time intervals of 400,000 years which they numbered 1–4. Their interval no. 1 approximately coincides with the *Hypselum* Chron, and interval no. 2 represents more or less the *Berrense* Chron. Interval no. 3 includes both the *Bimammatum* and the *Gredingensis* Chron and, therefore, represents a time span of certainly more than 400,000 years. The person who measured the section drawn in STRASSER et al. (2000) did not perceive the conspicuous transgressive surfaces 9 m and 18 m above the base of their section. These surfaces are distinctly indicated to be eroded and encrusted at the top of bed nos. 28 and 38 of section RG 381 on Pl. 28 in GYGI (2000a). Both the upper surfaces of bed nos. 28 and 38 are encrusted with goethite and with ostreid bivalves. The upper bedding plane of bed no. 28 in this section is planed, bored, and well exposed and accessible on the south side of the road. This was indicated for the first time by ZIEGLER (1956, p. 92), who drew attention to the particular property of both of the pertinent bedding planes. ZIEGLER gave a short and appropriate description of the bedding planes.

A rapid relative sea level rise of more than 10 m was necessary to drown an *entire* carbonate platform in the region investigated. This is exemplified by the early Bathonian carbonate platform of the Spatkalk Member, which drowned during a rapid relative sea level rise of more than 20 m (no. 1 in Fig. 5.1). A major, rapid relative sea level rise of an unknown amount drowned the late Bathonian carbonate platform of the Calcaire roux sableux Member just before the end of the Bathonian (no. 2 in Fig. 5.1). The early Callovian carbonate platform of the Dalle nacrée Member was drowned late in the early Callovian by a rapid relative sea level rise of approximately 15 m.

Only a rapid sea level rise could stop accretion of calcareous ooids at the top of the sand bank which is now the Chestel Member near St-Ursanne (Fig. 4.19B) in the interior of the carbonate platform of the St-Ursanne Formation. This relative rise, no. 5 in Fig. 5.1, must have been essentially eustatic, because sea level rises of the same order of magnitude were recorded in coeval sediments in eastern France and in southern England. The rapid relative sea level rise, which cut short production of calcareous ooid sand late in the *Antecedens* Chron on the shoal of what is now the Chestel Member, was estimated by GYGI (1986, p. 473) to have been at least 10 m. The eustatic component of the rise was assumed by GYGI to be 7 m following ZIEGLER (1982, p. 106). This estimate was made by comparison with the lagoon in the Bermuda Atoll. The sea level rise late in the Antecedens Chron could be corroborated, as rise no. 5 in Fig. 5.1 in this study, to have been 10 m based on direct evidence provided by the thickness of the Buix Member near St-Ursanne (see above). The rapid relative sea level rise of 10 m with a corrected eustatic component of 6 m late in *Antecedens* time was sufficient to drown the sand bank where calcareous ooids of the Chestel Member were accreted in the interior of the carbonate platform, but the amount of the rise was too small to drown the entire carbonate platform of the St-Ursanne Formation.

Production of calcareous ooid sand continued and could keep up with sea level rise no. 5 in the Tiergarten Member near the basinward rim of the carbonate platform of the St-Ursanne Formation. Conversely, inter-reef sedimentation of very pure, fine-grained calcareous detritus and possibly of some chemically precipitated calcareous mud in the lagoon in the platform interior (now Buix Member) could not keep up with the rapid sea level rise late in the *Antecedens* Chron. Sedimentation in the lagoon therefore lagged behind the rate of production of calcareous ooid sand on the shoal of the marginal Tiergarten Member. As a consequence, water depth must have rapidly increased in the lagoon at the initial stage of its evolution. The accommodation space for the tidal delta of calcareous ooid sand with a thickness of 5.7 m in the uppermost carbonate platform of the Günsberg Formation near Péry (Fig. 5.2) could only be provided by a rapid relative sea level rise during the *Hypselum* Chron (no. 7 in Fig. 5.1). The delta near Péry probably evolved not far from the basinward margin of the platform (Fig. 1.4), where there was the highest rate of carbonate production.

The fact that the basinward margin of proximal, thick succession no. 1 was stationary from the beginning of the Late Jurassic until the end of the *Transversarium* Chron is evidence that a major, net eustatic sea level rise occurred during sedimentation of succession no. 1. It becomes plausible from this that water depth considerably increased in the pertinent time span in the epicontinental basin, where

sedimentation of the entire succession no. 1 amounted to a measured average thickness of approximately 5 m in Canton Aargau or 0.5 m in Canton Schaffhausen (compare Fig. 4.15A with B). Most of the eustatic sea level rise from the beginning of the Oxfordian Age until late in the *Antecedens* Chron was probably gradual except in the *Cordatum* Chron. Water depth in northwestern Switzerland was at the beginning of the Late Jurassic approximately 80 m according to the calculations by GYGI (1986, p. 472). This figure could be confirmed in the present study, because Oxfordian sediments in northwestern Switzerland filled the Rhodano-Swabian Epicontinental Basin up to sea level. The calculations by GYGI (1986, p. 472) departed from this well-documented reference level.

Evidence was given above that water depth at the time boundary between the Middle and the Late Jurassic was probably about 80 m as well in Canton Aargau. The water depth of 100 m which was concluded above for the vertical and lateral facies boundary between beds with cauliflower glauconite pellets from a water depth greater than 100 m and iron oolite from shallower water was arrived at based on evidence from distant areas, but it fits well into the context of the sediments investigated. Provided that the paleobathymetric boundary between environments with formation of cauliflower pellets of pure glauconite and iron ooid accretion was indeed at the depth of around 100 m, water depth would have increased in Canton Aargau from 80 m at the beginning of the Oxfordian to 100 m in the *Densiplicatum* Chron. This net increment of 20 m included some endogenic subsidence. The pertinent *relative* sea level rise occurred in the time span of five ammonite chrons, or probably during more than approximately 1.8 million years. Microbialites with the measured total thickness of between 1 and 15 *millimeters* were deposited in this long period of time in the sections RG 226 at Auenstein/Veltheim (Figs. 28 and 29 in GYGI 1992) and RG 276 in the quarry of Chalch at Holderbank in Canton Aargau (MANGOLD and GYGI 1997, Fig. 2), and sediments with the thickness of *decimeters* at Ueken near Herznach (GYGI and MARCHAND 1982, Fig. 2, section RG 208, beds no. 7a-10). This is additional evidence that eustatic sea level fluctuations were modest in the Late Jurassic, and that they were certainly less than what was concluded by SAHAGIAN et al. (1996).

11.7 The Potential of Ammonites in Sedimentary Geology

A reliable biochronology based on ammonite evolution and time stratigraphy results when a strictly morphological study of ammonite taxonomy is made based on a sufficient number of well-preserved specimens collected from *in situ* in a set of sections measured bed by bed. The sections must be correlated in detail with different, independent methods and then be assembled into a regional frame of reference. The shortest time interval which can be discriminated in ammonite biochronology is the ammonite chron. The average duration of such a chron in Oxfordian and Kimmeridgian time is approximately 370,000 years. The ammonite subchrons and the pertinent subzones which are currently used by European stratigraphers should be upgraded to chrons and zones, respectively, according to the International Stratigraphic Guide, second edition by SALVADOR (1994). This was done in Fig. 1.6 of this study.

Refined taxonomy of representatives of the ammonite genus *Gregoryceras,* mainly from Switzerland by GYGI (1977), provided for the possibility to make an *intercontinental* correlation with an error of approximately ±0.2% of the radiometric age of the pertinent ammonite fossils between northern Switzerland and northern Chile (GYGI and VON HILLEBRANDT 1991). The same could probably be done by comparing, for instance, the work by KNIAZEV (1975) on *Cardioceras* in northeastern Siberia with that by MARCHAND in GYGI and MARCHAND (1982) on *Cardioceras* from northern Switzerland. A point is made with the text given above about time stratigraphy based on ammonites that ammonites are useful in sedimentary geology of the Late Jurassic on a regional, international, and intercontinental scale. Ammonites are not only good markers of relative age in biochronology because of their comparatively high rate of evolution, but are also indicators of facies, especially of water paleodepth.

Regional history of sedimentation based on ammonite biochronology alone can only be reconstructed in deposits from deeper water. Time correlation of sediments laid down in the deeper part of the epicontinental basin with deposits from shallow water requires a combination of ammonite biochronology with mineral stratigraphy based on the comparatively stable clay mineral kaolinite, and a check with sequence stratigraphy. Marker beds provide for the possibility of regional time correlation with great precision, if they are calibrated with ammonite biochronology. The measured and averaged thickness of succession no. 1 in the proximal, thick part in northwestern Switzerland is 185 m, and 0.5 m or 370 times less in the distal, starved part of the Rhodano-Swabian Epicontinental Sea in Canton Schaffhausen. A thinning of approximately 1,300 times was calculated to have occurred between the measured sediment thickness of about 105 m which was laid down by aggradation and at a normal rate during the *Densiplicatum* and the *Antecedens* Chron in the proximal part of succession no. 1 near Liesberg in northwestern Switzerland (GYGI 2000a, Pl. 31) and the coeval, stratigraphically condensed, lowermost bed of the Birmenstorf Member with an average thickness of 8 cm in Canton Aargau. This bed was sedimented in deeper water in the Rhodano-Swabian Epicontinental Basin (GYGI 1977, Pl. 11, section RG 208, bed no. 10 in an excavation near Ueken).

Sedimentation rates which were calculated using a combination of ammonite biochronology with radiochronology were found to be much less than the maximal *possible* rate of isostatic equilibration of the lithosphere. The lithosphere below the sediments investigated was consequently never off isostatic equilibrium during the Late Jurassic Epoch.

The isopachs in the map of "middle Oxfordian marls and limestones" in the upper part of Fig. 7 by WETZEL et al. (2003) are similar to those in the paleogeographic map of the same region which was previously published by GYGI (1969a, Fig. 7 on p. 108). GYGI's map was drawn according to an inappropriate model of sedimentation. The map had later to be thoroughly revised by GYGI and PERSOZ (1986, Pl. 1A), when more ammonites collected from *in situ* were at hand. The revised model of sedimentation published by GYGI and PERSOZ (1986) was based mainly on ammonite nos. 14, 15, and 16 listed in their Tab. 3, and according to the mineral stratigraphic correlations E and F which are represented in their Pl. 1A. All of the three ammonites mentioned were collected from *in situ* and were listed with the corresponding numbers in Tab. 3 by GYGI and PERSOZ (1986). Ammonite no. 14, *Perisphinctes (Dichotomosphinctes) bifurcatus* (QUENSTEDT), which was found and identified by R. ÉNAY (Lyon) in 1964 in the type section of the Günsberg Formation at the head of the landslide called Gschlief, above the village of Günsberg in Canton Solothurn, was figured by GYGI (1995, Fig. 17/2). Ammonite no. 15, *Perisphinctes (Perisphinctes) panthieri* ÉNAY, was found by a farmer in a small, temporary outcrop of the Günsberg Formation beside his homestead at Seewen in Canton Solothurn. The ammonite was given to the geologist P. BITTERLI-BRUNNER, who was mapping the geology of the area at that time. R. ÉNAY identified the specimen, and it was figured by GYGI (1995, Figs. 11 and 17/1). Ammonite no. 16, *Amoeboceras* cf. *serratum* (SOWERBY), was collected by J. HALLER and R. TRÜMPY from the Gerstenhübel Beds, a limestone unit which was then quarried at locality Steinacher north of Mönthal, a village northwest of Brugg in Canton Aargau. This ammonite was given to R. GYGI. It was figured for the first time by ATROPS et al. (1993, Pl. 1, Fig. 6), and again by GYGI (2000a, Pl. 8, Fig. 4).

The ammonite nos. 14, 15, and 16, and the mineral stratigraphic correlations E and F represented in Pl. 1A by GYGI and PERSOZ (1986), are evidence that the Gerstenhübel Beds (abbreviated GER within the Effingen Member), the Grüne Mumienbank Bed in the middle Günsberg Formation (GÜN), and the intercalation of calcareous oolite in the middle of what was called "*Natica* Member" by GYGI and PERSOZ (1986, NAT in Pl. 1A), are coeval sediments of *Bifurcatus* age. The *Natica* Member of GYGI and PERSOZ (1986) is equivalent to the modern Röschenz Member, which is abbreviated RÖS in Fig. 1.5 of this study. The position of ammonite nos. 18, 19, and 21, which are shown in Fig. 1.5 and that are listed in Fig. 1.6 of this study, and the photographs of the pertinent ammonites which were published for the first time elsewhere, confirm that the units mentioned above were all sedimented during the *Bifurcatus* Chron. The juxtaposition of the three coeval lithostratigraphic units Röschenz Member (marl and limestone), Günsberg Formation (mainly calcareous oolite with coral bioherms), and the mixed calcareous and argillaceous Effingen Member with intercalated limestone successions, was represented in the revised paleogeographic map with isopachs by GYGI (1990c, Fig. 6). The isopachs in this map were drawn according to the model of sedimentation which was elaborated by GYGI and PERSOZ (1986, Pl. 1A). In spite of these previously published and well-documented data, WETZEL et al. (2003) alleged in the caption to their Fig. 7 that the stratigraphic resolution in the region shown in the northwestern part of their outdated isopach map was "poor".

The mode of life of almost all of the ammonites fossilized in investigated sediments of Late Jurassic age was necto-benthic. Exceptions are rare phylloceratids and very rare lytoceratids. Most of the ammonite casts investigated are, therefore, indicative of the environment the pertinent specimens were living in, and mainly of water depth. This is summarized in Fig. 4.9. A new result of the investigation of Late Jurassic sediments and of macrofossil assemblages from the epicontinental, Rhodano-Swabian Basin in the Swiss Jura Mountains is that the paleobathymetric boundary at approximately 30–35 m between biofacies with mainly bivalves and facies with mostly ammonites, both vertical and lateral, is probably just as clear-cut and practically without overlap as the bathymetric boundary is between lithofacies with iron ooids and with cauliflower glauconite pellets at the water depth of around 100 m (see main conclusion no. 24). Dwarfing of ammonites living in an environment of dysaerobic bottom water is difficult to prove, but it probably occurred. Rare ammonite taxa, like for instance *Gregoryceras transversarium* (QUENSTEDT) or *Mirosphinctes frickensis* (MOESCH), are well represented among the few taxa with an exceptionally wide geographic range. Ammonites of rare taxa probably lived under greater pressure to migrate than common forms. Presence of the taxa *transversarium* and *frickensis* in northern Chile, which were defined in northern Switzerland, is indicative of *migration* of the pertinent ammonites rather than of *post-mortem drift* of empty shells. Giant ammonite taxa are overrepresented in sediments from comparatively shallow water. This tendency is even more pronounced in nautilids. Giant *Paracenoceras ingens* TINTANT in TINTANT et al. (2002) were found only in deposits from very shallow water.

The composition of ammonite assemblages varies *laterally* in coeval sediments with increasing depth of deposition. This is exemplified in Fig. 4.9 by three Oxfordian

beds of *Cordatum* age: the unnamed, fossil-rich bed in the distal part of the middle Sornetan Member, the Schellenbrücke Bed, and the Glaukonitsandmergel Bed. A similar lateral transition is from the Reuchenette Formation over the Baden Member to the lowermost Schwarzbach Formation of the Kimmeridgian. Variation in the composition of ammonite assemblages in a *vertical,* shallowing-upward succession of sediments is corresponding (see main conclusion no. 36 and WALTHER's "law"). This is evidence that most of the empty ammonite shells that sank to the sea floor and there molded casts in the sediment belonged to animals living in the habitat more or less directly above the sediment the casts are embedded in. *Post-mortem* floating of empty ammonite shells up to the water surface and drift over long horizontal distances was consequently exceptional.

WANI et al. (2005, p. 852) experimented with Recent, living *Nautilus pompilius* LINNAEUS in the Philippines. They found that after death of the animals, most of the shells sank within a few days, even if the shells once floated. The authors stated on p. 849 that the time lag between death of the investigated *Nautilus* and waterlogging of the gas chambers in the phragmocone of the shell depends on the size of the shells. Large shells could float longer than small ones (p. 851), whereas shells with a diameter of less than 135 mm did not float at all. If it were permissible to conclude from this of the behavior of ammonite shells of Late Jurassic age after death of the animals within, this would be one explanation of the close correlation between the depth of deposition of a given sediment and the composition of the ammonite assemblage that is fossilized in it. Caution is mandatory in such comparisons between Recent nautilids and ammonites of the Late Jurassic, because the ratio between casts of nautilids and ammonite casts in the nearly 10,000 cephalopod fossils which were recently collected from sediments of Late Jurassic age in northern Switzerland is only slightly greater than 1:1,000. SCOTT (1940, p. 315) found that nautilids are not abundant and have a low diversity in the sediments with a rich and diverse ammonite assemblage from the Cretaceous epicontinental sea in Texas, USA. This is evidence of a significant difference between the paleoecology of ammonites and that of contemporaneous nautilids. The degree of difference in ecology is surprising between the two groups of cephalopods with shells which are morphologically and functionally so similar.

Many giant casts of the nautilid genus *Paracenoceras* were recently found in the lower Reuchenette Formation in Ajoie region in northern Canton Jura during construction of the Transjurane superhighway. The specimens were probably all found in the boundary beds between the *Evolutum* and the *Tenuicostatum* Zone. Abundant bivalves and absence of ammonites in these beds indicate sedimentation in very shallow water, in spite of the presence of some very fine-grained glauconite, as for instance in bed no. 23 of section RG 340 near Porrentruy (GYGI 2000a, Pl. 17), or in a coeval bed in the old quarry beside the cemetery of Fontenais near Porrentruy, where the holotype of *Paracenoceras ingens* Tintant in Tintant et al. was found (Tintant et al. 2002, Pl. 5, Fig. 1). All of the specimens of this new nautilid taxon with a maximal diameter of complete adults approaching 600 mm, which were figured or mentioned by Tintant et al. (2002), were found in sediments from very shallow water like the uppermost Günsberg Formation north of Grandval in Canton Bern (unpublished section RG 414, bed no. 38), the Laufen Member near Liesberg, or the lower Reuchenette Formation in the region of Ajoie in Canton Jura. No *Paracenoceras ingens* were ever found in sediments from deeper water beyond the shoals of calcareous ooid sand with coral biostromes and small bioherms at Balsthal, which rimmed the lagoonal, shallow-water realm of northwestern Switzerland against the adjacent, deeper Rhodano-Swabian Basin.

The glauconitic, lower Baden Member in Canton Aargau was sedimented in the epicontinental, Rhodano-Swabian Basin at a water depth which is estimated to have been somewhat more than 50 m (Figs. 1.5 and 4.9). The best outcrop of the Baden Member is in the large quarry near Mellikon in northeastern Canton Aargau (section RG 70, bed nos. 124–125, see Gygi 1969a, Pl. 17). The age of the uppermost part of the glauconitic, lower Baden Member (bed no. 124) in this quarry is transitional between the *Evolutum* and the *Tenuicostatum* Chron (Gygi 2003, Fig. 170). Only much smaller nautilids than in Ajoie region were found in the coeval sediments from deeper water of the Rhodano-Swabian Basin, in the rich cephalopod assemblage which was recently collected or excavated from the glauconitic, lower Baden Member near Mellikon. The diameter of the largest adult nautilids at Mellikon is less than 200 mm as for instance that of *Pseudaganides royeri* (De Loriol), which was figured by Tintant et al. (2002, Pl. 1, Fig. 1). One of the possible causes for the absence of giant *Paracenoceras* near Mellikon is the existence of the Kimmeridgian shoal of calcareous ooid sand mentioned above and represented in Fig. 1.5, which separated the landward shallow-water realm from deeper water in the epicontinental basin. A sand bank of calcareous ooids existed between the two facies realms from the Oxfordian *Transversarium* Chron on until at least the *Tenuicostatum* Chron in the Kimmeridgian (Fig. 1.5). The shoal that separated the lagoonal, shallow-water realm with giant *Paracenoceras* in the northwest in Kimmeridgian time from the epicontinental basin near Mellikon must have

restricted water exchange between the two facies realms. The sand bank probably prevented drift of empty shells of giant *Paracenoceras* from the lagoonal realm over the shoal and beyond into the open, epicontinental Rhodano-Swabian Basin.

Drift of empty shells of large perisphinctacean ammonites in the opposite direction, from the epicontinental basin toward and over shoals of calcareous sand in northwestern Switzerland, is documented to have occurred at several occasions. Examples of partial fossilization of large ammonite shells as incomplete casts within calcareous sand on shoals are, for instance, the *Perisphinctes* which was figured by Gygi (1995, Fig. 11) from the Günsberg Formation at Seewen in Canton Solothurn (no. 18 in Figs. 1.5 and 1.6), the *Lithacosphinctes* figured in Gygi (1995, Fig. 20) from the Laufen Member near Dittingen in Canton Basel-Landschaft, or the *Lithacosphinctes* that B. Martin found in the uppermost Balsthal Formation near Balsthal (no. 29 in Fig. 1.5), and which was figured in Gygi (1995, Fig. 19). The three cases mentioned are exceptional. The several casts of large perisphinctids which were found in the lagoonal sediment of the Buix Member with coral reefs are possibly fossils of ammonites which were *living* at least temporarily in the lagoon. No casts of drifted empty shells of large perisphinctacean ammonites were found in Oxfordian sediments younger than the *Transversarium* Zone landward of ooid sand banks in the lagoonal and in the peritidal realm adjacent to land in the northwest (Fig. 1.2). The only exception to this is the specimen mentioned above from the Laufen Member. Most of the empty shells of ammonites which lived in the Rhodano-Swabian Basin later than in the *Transversarium* Chron apparently could only drift upon, but not across the shoals of calcareous sand and beyond into the lagoonal or peritidal environment. This condition changed from the *Acanthicum* Chron on (Fig. 1.5). All of these data are evidence that empty shells of large cephalopods did not normally drift over a significant horizontal distance after death of the animals in the shell during the Late Jurassic in the region investigated. The very few, large casts of ammonite shells which were found in the *Renggeri* Member along with abundant dwarf ammonite casts of iron sulfide must have been molded by empty ammonite shells which drifted in the uppermost part of the water column into the part of the basin where the *Renggeri* Member was sedimented, and which sank there to the sea floor.

List of main conclusions

1. The succession of sediments and ammonite zones investigated is uninterrupted
2. Marl-limestone cycles compared with MILANKOVITCH-type cycles
3. Subsidence: endogenic and exogenic
4. Cause and effect of *exogenic* subsidence
5. Isostatic equilibrium of the lithosphere in the area investigated
6. Areally restricted, major endogenic subsidence and local, minor uplift
7. Paleoclimate
8. Sea level variations
9. Glacio-eustasy in the Late Jurassic?
10. Hermatypic corals: paleoenvironmental conditions
11. Carbonate platforms investigated were rimmed by carbonate sand banks, not coral reefs
12. Drowning of carbonate platforms
13. Calcareous mud: provenance and quantity
14. Sediment transport
15. Transportation of iron from land to sea
16. Widespread, thin iron oolites including mainly ammonites were formed on the basin floor
17. Paleobathymetry
18. Paleobathymetric boundary between facies with iron ooids and cauliflower glauconite pellets
19. Aggradation and progradation: problems unsolved
20. Stromatolites
21. Spherical, centimeter-size calcareous oncoids grew at a widely varying water depth
22. Storm wave base had a great paleobathymetric variability
23. The mode of life of most of the ammonites investigated was necto-benthic
24. Abundance and composition of ammonite assemblage depending on water depth and rate of sedimentation
25. Normal size distribution in a fossilized ammonite assemblage
26. Dwarf size in ammonites: genetically predetermined versus environmentally controlled
27. Ammonites could tolerate less clay mineral sedimentation than hermatypic corals
28. Habitat of giant perisphinctacean ammonites compared with habitat of giant nautilids
29. Time calibration in sediments
30. Time correlation of sediments
31. Rates of sedimentation
32. Interrelation between sea level variation, sedimentation rate, and marker beds
33. Geometry of sedimentation depended mainly on the pattern of sea level variations
34. Sediments of the Late Jurassic in northern Switzerland are a gauge of eustatism
35. GRESSLY's two "laws" of facies
36. WALTHER's "law"

1. The thick sediments from deeper water in the three principal Oxfordian successions shown in Fig. 1.5 are an *uninterrupted succession of uncondensed sediments and ammonite zones* of the type defined in this study. The sediments at the base of the Upper Jurassic Series between Canton Jura and Canton Aargau were laid down at a particularly low rate in iron-oolitic facies in deeper water with a macrofossil assemblage of mainly ammonites (Fig. 8.2A), but the iron-oolitic boundary beds between the Middle and the Upper Jurassic Series which are represented in Figs. 4.14 and 8.2 are not stratigraphically condensed. Sediments which were laid down at an average, "normal" rate in the deep subtidal zone succeed each other in both *vertical* and *lateral*, basinward succession, which is visualized in Fig. 1.5. These sediments include well-preserved ammonite casts. Almost all of the ammonites investigated were collected from *in situ*. The succession of ammonite chrons in the Subtethyan Faunal Province in northern Switzerland from the beginning of the Oxfordian Age to the end of the Kimmeridgian Age is shown in Fig. 1.6. The sedimentary rocks of these ages and the *entire* macrofossil assemblage in them could be studied in excellent natural or in artificial outcrops during more than three

decades. The strata represented in Fig. 1.5 are a detailed frame of reference for ammonite biochronology. Ammonites typical of both the Tethyan and the Boreal Faunal Province occur in the strata investigated. The ammonite succession of the Late Jurassic in northern Switzerland can, therefore, be used as a link between the Tethyan and the Boreal Faunal Province.

2. The *three principal marl-limestone cycles* of the Oxfordian Stage in northern Switzerland were shown for the first time in Pl. 1A by GYGI and PERSOZ (1986). The thick parts of the cycles are uncondensed sediment successions without significant hiati. The time span represented by each of the main three cycles and mainly the time of deposition of *elementary* marl-limestone cycles varies widely. No MILANKOVITCH-type cycles cold consequently be discerned.

3. *Endogenic subsidence* of the basement is caused by non-sedimentary processes deep in the earth interior. *Exogenic* basement subsidence is caused by processes like loading with sediment or with additional water because of a net sea level rise. A special case is displacement along a growth fault in unconsolidated sediments. Fig. 4.12 is evidence that shear in a growth fault can occur in soft deposits along a clear-cut plane like in tectonism, but movement is driven by non-tectonic processes going on within unlithified sediments underneath. The term *tectonic subsidence* as it is currently used by most authors is therefore ambiguous, and it is recommended to substitute it by *endogenic subsidence*. The lithosphere reacts to loading, and even to subtle loads, with isostatic equilibration by *exogenic* subsidence. The rate and the amount of endogenic subsidence and mainly the amount of exogenic subsidence varied in time and space during the Late Jurassic in the area investigated. This is visualized in Fig. 4.15.

4. *Cause and effect of subsidence. Endogenic* subsidence going on in a vast area at an equable rate and during an extended time span is the prerequisite of lasting marginal marine sedimentation like for instance of the lower Vellerat Formation or of the Reuchenette Formation in northwestern Switzerland. Conversely, a thick and shallowing-upward succession of sediments being filled into a pre-existing basin up to sea level causes *exogenic* subsidence of the basement by isostatic equilibration of the lithosphere under the additional sediment load. The weight of the sediments thereby provides for additional accommodation space of 2/3 of the initial depth of the basin.

5. The *lithosphere* reacted to loading with prompt isostatic subsidence, and it probably reacted without delay to even subtle loads. The lithosphere in the region investigated was consequently in the Late Jurassic probably *never off isostatic equilibrium.*

6. The history of *endogenic* and mainly that of *exogenic*, isostatic basement subsidence under the growing load of sediments, or of basement subsidence caused by the weight of additional water brought into the area investigated by rising sea level, could be refined. *Endogenic subsidence in restricted areas* was documented in two cases to have been of the amount of tens of meters in a short time span. *Local uplift* of the order of meters by a process unknown could be proved to have raised the top of shallow marine carbonate sediments above sea level in at least two areas, one near Péry in section RG 307 in the *Planula* Chron of the late Oxfordian (see Fig. 5.7) and the other in the region around Balsthal in the earliest Kimmeridgian Age in sections RG 438, 439, and 450 (see Fig. 5.9 and GYGI 2000a, Fig. 37 on p. 64).

7. *Paleoclimate*: The branching coral with distinctly delimited *annual* accretion bands from a small patch reef in the uppermost St-Ursanne Formation near Courtételle shown in Fig. 3.1 is proof that the climate in the region investigated was in the Late Jurassic *subtropical* and obviously *seasonal*. The climate grew *certainly* drier and *probably* warmer in the course of the Oxfordian Age. The three *principal* marl-limestone successions of the Oxfordian Stage in northern Switzerland which are represented in Fig. 1.5 are the effect of long-term variation between humid and drier climate. Sedimentation of proximal, thick succession no. 1 began in humid climate with homogenous, clayey marl of the *Renggeri* Member. Deposition of succession no. 2 began with marl of the Effingen Member, and marl of the Bure Member is the lowermost part of proximal succession no. 3. Sediments with a high content of clay minerals and with siliciclastic silt are evidence of a climate with much rainfall on neighboring land. The upper part of each succession is limestone, like the St-Ursanne Formation in succession no. 1, or in succession no. 2 the coeval Hauptmumienbank, the Steinibach, and the Geissberg Member, and in succession no. 3 the coeval Courgenay, Balsthal, and Villigen Formation. The limestone units are evidence of a drier climate. The total time of sedimentation was about three million years in succession no. 1, approximately two million years in succession no. 2, and around 1.7 million years in succession no. 3. The proportion of time between sedimentation with a high percentage of terrigenous material in the lower part and deposition of more or less pure carbonates in the upper part was substantially different within each of the three successions.

A climatic variation throughout the Oxfordian Age between times with more and times with less rainfall is indicated by the fact that sedimentation of succession no. 1 began with clayey marl, and that deposition of succession no. 2 started with marl with a higher carbonate content. The tendency of increasingly dry climate continued in succession no. 2 with the first appearance of dolomite in the Günsberg Formation, and in succession no. 3 with pervasive dolomitization and early diagenetic transformation back to dedolomite in the upper Balsthal Formation (Fig. 3.3). The earliest calcite pseudomorphs after sulfate minerals were found in the Hauptmumienbank Member of the *Berrense* Chron (Fig. 5.5C). Such pseudomorphs are common in the Verena Member (Fig. 3.3) and some gypsum occurs in the same member. The skeletal part of the hermatypic coral colony with distinctly delimited, annual accretion bands which is shown in Fig. 3.1 is evidence that climate in the middle Oxfordian Age was subtropically warm with temperature in shallow water varying substantially between summer and winter months. Measurements of paleotemperature that were recently made in the area investigated did not confirm warming of climate in the Late Jurassic for a reason unknown. Drowning of the Late Jurassic carbonate platforms investigated was not caused by climatic change. The smaller amount of rapid relative sea level rises during the Late Jurassic compared with the rapid relative rises in the Middle Jurassic (Fig. 5.1) is an indication that paleoclimate was more equable and possibly somewhat warmer during the Late Jurassic than in the Middle Jurassic.

8. *Sea level variations*. No evidence of a significant tidal range was found. The discrete sea level rises shown in Fig. 5.1 were rapid and essentially eustatic. Eustatic sea level variations in the Late Jurassic amounted to far less than in the Pleistocene. Most of the individual variations in the Late Jurassic were asymmetric with *certainly* rapid rises and *probably* slow falls, where falls occurred at all and left visible traces. Overall tendency of sea level change was transgressive from the latest part of the early Callovian Age until the latest *Transversarium* Chron in the middle Oxfordian Age. From then on until the latest *Berrense* Chron, the tendency was regressive except for a brief, but pronounced transgression of a small amount mainly in the *Hypselum* Chron. Transgression resumed in the *Bimammatum* Chron and prevailed well into the Kimmeridgian Age. A gray band outlines transgressive and regressive tendencies in Fig. 1.5.

A relative *sea level rise of around 80 m* occurred when the thin and very widespread iron oolitic Herznach Formation was laid down between the beginning of the early Callovian *Enodatum* Chron and the earliest part of the *Scarburgense* Chron of the Oxfordian. This can be read from sections RG 307 at Péry and RG 280 at Liesberg in northwestern Switzerland. The relative *sea level rise of approximately 100 m* between the end of

deposition of the Spatkalk Member at the end of the early Bathonian and the beginning of sedimentation of the Birmenstorf Member in the *Densiplicatum* Chron of the middle Oxfordian is documented in sections RG 226 at Auenstein/Veltheim and RG 276 at Holderbank in Canton Aargau. This rise occurred in Canton Aargau during deposition of a thin and widespread, Bathonian iron oolite and subsequent non-deposition in the Callovian and in the early Oxfordian Age. The importance of these probably global processes is not evident from Fig. 5.1. The individual, rapid relative sea level rises that are documented to have occurred in the Late Jurassic and that are shown in Fig. 5.1 amounted each to at most 10 m. Falls of sea level produced either hummocky or planed erosion surfaces on carbonate platforms. No eroded incisions diagnostic of rapid and substantially falling sea level were found at the basinward margin of carbonate platforms. This does not mean that there are none of them. Their apparent absence is possibly the consequence of the paucity and the small areal extent of outcrops. The following five conclusions concerning eustasy could be drawn from Oxfordian sediments in northwestern Switzerland.

(a) The pattern of rapid sea level rises and probably slow falls resembles *qualitatively* eustatic sea level variation during the Quaternary. This is an indication that a minor quantity of ice existed in the Late Jurassic somewhere *on land*. The existence of such ice in very high paleolatitudes could not be proved. The volume of ice waxing and waning on high mountain ranges in lower latitudes in the Oxfordian Age was probably too small to cause significant variation in sea level.

(b) No substantial erosion of Oxfordian sediments out of very shallow water occurred in the region of northwestern Switzerland from the late *Antecedens* Chron on. The record of rapid relative sea level rises in the pertinent sediments is, therefore, probably complete from the middle Oxfordian on, and in that region possibly to the end the Kimmeridgian Age.

(c) Relative sea level rise no. 5 in Fig. 5.1 was rapid, eustatically driven, and amounted to a total of approximately 10 m. The rate of the rise was sufficient to cut short calcareous ooid accretion in the *interior* of the carbonate platform of the St-Ursanne Formation. The average rate of rise no. 5 was less than the maximal possible rate of production and sedimentation of calcareous ooid sand in the Tiergarten Member in the *marginal*, basinward part of the St-Ursanne Formation. A rapid relative sea level rise of 10 m was consequently insufficient to drown an *entire* carbonate platform in the sediments investigated. Rapid relative sea level rise no. 7 in the *Hypselum* Chron created in very shallow water above the uppermost part of the carbonate platform of the Günsberg Formation accommodation space of several meters. This was filled in by the oolitic tidal delta shown in the photograph of Fig. 5.2 and by the large, pelsparitic sand wave shown in Fig. 5.5A. The top of both the tidal delta and the sand wave figured, and of additional, probably coeval calcarenitic units in the formation, is an erosion surface which was presumably caused by a eustatic sea level fall.

(d) Sea level rise no. 10 (in Fig. 5.1) at the end of the *Gredingensis* Chron was rapid. The total amount of the relative rise was less than the 6 m, which is the maximal water depth at which calcareous ooids can be accreted. Rise no. 10, therefore, did not interrupt ooid accretion in the Holzflue Member of the Balsthal Formation and left no trace in the member. Neither could an effect of rise no. 10 be discerned in the well-bedded micritic limestone of the La May Member from the very shallow lagoon of the Courgenay Formation. This can best be judged from the undisturbed section RG 443 of the La May Member in an exploration well near St-Ursanne (GYGI 2000a, Pl. 20). Conversely, rise no. 10 with the small total amount of a few meters had a far-reaching effect in the deeper part of the Rhodano-Swabian Epicontinental Basin. It can be read from Fig. 12 in GYGI et al. (1998) that omission (nondeposition) and subsequent subsolution of early lithified micritic limestone occurred after the rise in the Villigen Formation, for instance, in section RG 62 near Villigen, when export of calcareous mud from the lagoon of the La May Member was for a short time either interrupted, or mud did not reach the area of the Villigen Formation. The pertinent hiatus is sequence boundary O8 in GYGI et al. (1998, Fig. 2). The thin and very widespread, slightly glauconitic marker bed called Knollen Bed was subsequently sedimented directly above the hiatus in a very widespread area that extends from Canton Aargau far into southern Germany, when supply of calcareous mud from the lagoon of the La May Member to the Villigen Formation gradually resumed at an initially low rate.

(e) No coastal onlap or offlap can be studied in northern Switzerland. Only transgressive and regressive tendencies of sedimentation could be concluded from the paleogeographic map of Fig. 1 in GYGI (1986) and from Figs. 1.4 and 1.5 in this study. These tendencies are diagnostic of *eustatic* sea level variation, because sedimentation during the Oxfordian and the early Kimmeridgian Age occurred on a comparatively stable shelf.

9. *Glacio-eustasy*, provided that the process was going on at all in the Late Jurassic, could explain the occurrence of individual, rapid relative sea level rises. When the entire pattern of sea level variation represented in Fig. 5.1 is considered, it appears to be unlikely that these sea level changes were caused by waxing and waning of a major volume of ice on land. No land existed at this time at both the geographic poles. The eustatically driven sea level variation that was documented in northern Switzerland to have occurred in the Late Jurassic cannot be the effect of MILANKOVITCH-type cycles. In case that they were, the rises would have periodically occurred in *regular* time intervals, and their amount would have been approximately equal.

10. *Hermatypic corals* in the Rhodano-Swabian Epicontinental Basin *lived in a paleobathymetric range with the lower limit at the water depth of approximately 20 m.* Plate-like colonies prevailed in the coral assemblage in the depth range or between 10 and 20 m. There is a mass occurrence of plate-like coral colonies in marl of the Liesberg Member that was sedimented in this depth range. *Oxfordian corals had to adapt to rapid, relative sea level rises of at most 10 m,* whereas hermatypic corals of the Quaternary had to cope with rapid sea level rises of as much as 120–130 m. *Corals were more abundant in environments with a high concentration of terrigenous material and nutrients* in suspension in the water *than in water with pure carbonate sedimentation.* Hermatypic corals fossilized in the sediments investigated could tolerate a much higher sedimentation rate of clay minerals than ammonites.

11. The *carbonate platforms investigated were rimmed by a belt of calcareous sand banks,* either ooidal or skeletal. No coral barrier reefs like in the Recent were found. Oxfordian assemblages of hermatypic corals on the basinward side of the marginal belt of sand banks grew in somewhat deeper water on the uppermost part of the basinward slope. Coral assemblages off the sand banks either formed individual bioherms like in the most proximal Pichoux Formation and in the boundary beds between the uppermost Effingen Member and the base of the Günsberg Formation, or biostromes like beyond the outer rim of the carbonate platform of the St-Ursanne Formation for instance near Mervelier and off the carbonate platform of the Balsthal Formation near Olten (Olten Member). Growth of coral biostromes or bioherms in a shallowing-upward succession began as soon as ongoing, aggradational sedimentation reduced

water depth to approximately 20 m. The domal top of reef knobs or patch reefs in the lagoon of the Buix Member (Fig. 4.19B), in the Tiergarten Member, in the uppermost, proximal Pichoux Formation, and in the boundary beds between the Effingen Member and the lowermost Günsberg Formation, grew to an elevation of at most 5 m above the surface of coeval sediment around the reefs. The slope between the outer rim of the carbonate platform of the St-Ursanne Formation and the level floor of the adjacent epicontinental basin had a *primary* declivity of between less than 0.5° and a maximum of 4°.

12. *Drowning of a carbonate platform investigated* occurred when sea level rose rapidly and by an amount of more than 10 m. A borderline case is the Oxfordian carbonate platform of the St-Ursanne Formation, which was subject to rapid relative sea level rise no. 5 of approximately 10 m late in *Antecedens* time (Fig. 5.1). The rate of rise no. 5 was sufficient to cut short accretion of calcareous ooids on the sand bank of Chestel Member in the platform *interior*. The production rate of ooid sand in the Tiergarten Member in the *external*, basinward part of the platform was higher and such that accumulation of calcareous ooid sand could keep up with the rise. The rate of carbonate precipitation was consequently higher near the basinward rim of this carbonate platform than in the interior. When rise no. 5 drowned the oolitic sand bank in the platform interior, a rapidly deepening lagoon with coral patch reefs evolved directly above the sand bank. Rapid relative sea level rise no. 3 of about 15 m was sufficient to drown the carbonate platform of the Dalle nacrée Member during the *Enodatum* Chron, just before the end of the early Callovian (Fig. 5.1).

The carbonate platforms investigated of the Late Jurassic did not drown when a pronounced change from a relatively dry to a humid climate occurred. Such a climatic change is documented at the upper boundary of the wide carbonate platform of the St-Ursanne Formation, which evolved in a period of relatively dry climate with minimal supply of terrigenous material. Rainfall and weathering then substantially increased in the *Schilli* Chron on land in the north, and running water began to transport a great quantity of clay minerals along with some silt-grade, detrital quartz and feldspar from land into the adjacent, marginal marine realm of the lower Vellerat Formation (Fig. 1.2). The coeval carbonate platform of the Günsberg Formation evolved in the time span from the *Schilli* to the *Hypselum* Chron, when sediment transport across the platform by tidal and occasionally by storm-driven currents was at a maximum. The ample siliciclastic material passing in turbulent water mainly in suspension across the Günsberg

carbonate platform reduced the rate of carbonate production on the platform substantially, but did not inhibit it altogether. The effect was that the width of the carbonate platform of the Günsberg Formation is much less than that of the preceding carbonate platform of the St-Ursanne Formation. The very shallow water above the carbonate platform of the Günsberg Formation was near-permanently turbulent mainly in tidal currents. Proof of this is the fact that by far most of the great amount of calcareous mud which was produced in very shallow water in the landward area of the lower Vellerat Formation bypassed the platform along with fine-grained terrigenous material and was then sedimented in deeper water in the Effingen Member. Close to nothing of the fine-grained, allochthonous sediment remained on the Günsberg platform.

13. *Calcareous mud was quantitatively the principal rock-forming constituent in the total volume of Oxfordian sediments investigated.* The contribution by calcareous nannoplankton to sedimentation of mud in the deeper part of the epicontinental basin *in the Oxfordian Age* was insignificant. *Whitings* in very shallow water in vast areas landward of calcareous platforms were consequently *the quantitatively most important process supplying calcareous mud* to the deeper part of the epicontinental Rhodano-Swabian Basin. Whitings are a suspension of calcareous mud produced by inorganic-chemical precipitation from water oversaturated with $Ca(CO)_3$. Schools of fish can stir up sedimented mud when feeding on the shallow bottom. Figure 4.18 is possibly the first photograph ever made showing both types of whitings close together.

14. *Sediment transport:* Argillaceous mud and silt-size, angular detrital quartz as well as some feldspar (Fig. 3.2) were washed by rainfall out of soils on land in the northwest and north. Running water transported the material into the adjacent marginal marine realm in the southeast. Mixed siliciclastic and calcareous sediments laid down in this mostly peritidal environment reach as far as to what is now northwestern Switzerland (Fig. 1.2). Tidal and occasionally storm-driven currents carried fine-grained terrigenous material and calcareous mud in suspension from marginal marine environments across carbonate platforms into deeper water above the upper part of depositional slopes. Water on the slope with mainly mud-grade particles in suspension flowed because of its somewhat greater specific weight compared with that of purer water above downslope to the basin floor in a multiphase process. The declivity of depositional slopes in the proximal Effingen Member of the Wildegg Formation was very slight, according to the scales in the lower left of

Figs. 1.5 and 4.15. Nevertheless, the inclination was sufficient that small debris flows could be triggered probably by earthquakes on the upper slope (Fig. 4.12 in this study and Fig. 13 on Pl. 4 in GYGI 1969a). Debris flows transported and plastically deformed large, semi-consolidated clasts of fine-grained sediment (Fig. 4.13). The debris flows probably graded toward deeper water into turbidity currents. Silt-size detrital quartz and some feldspar in the turbidity currents became concentrated in the lowermost part of turbidites. Such turbiditic beds with a high content of siliciclastic *silt* have as a rule a thickness of a few centimeters. Small turbidites in the Effingen Member can follow one directly above the preceding one (GYGI 1969, Pl. 4, Fig. 12). *No sand-size particles* like calcareous or ferriferous ooids derived from the shallow-water realm were ever found in debris flows or in turbidites.

Water above the basin floor was normally quiet enough that pure mud could be sedimented out of suspension. Bottom water was quasi-stagnant in the area where the *Renggeri* Member was sedimented except during a strong storm, although the environment was unrestricted (Fig. 4.15A). Dysaerobic conditions evolved in bottom water in the *Renggeri* Member because of oxygen consumption by decaying dead plankton settling from near the water surface. A low oxygen content of bottom water became permanent during the entire time of sedimentation of blue-gray marl-clay in the thick *Renggeri* Member. Casts of dwarfed ammonites and thin burrows, both filled with iron sulfide, indicate dysaerobic bottom water even where the member is only 3.3 m thick in its most distal part in section RG 307 near Péry (GYGI 2000a, Pl. 22). The *Renggeri* Member pinches out entirely in the distal direction. The Herznach Formation below with iron ooids of goethite and with a fossil assemblage diagnostic of well-aerated bottom water extends further out into the basin. Dysaerobic conditions in bottom water were temporary in parts of the Effingen Member, where blue-gray marl was laid down out of suspension in an unrestricted basin like that of the *Renggeri* Member.

15. *Iron supply from land to sea* occurred at a large scale in the form of coatings of ferric oxide at the surface of clay minerals and other particles up to sand grade. The process of production of ferric iron oxide on land and transportation to the floor of the epicontinental basin is described in the caption to Fig. 4.11.

16. The *thin and very widespread iron oolite* of the Herznach Formation was sedimented in deeper water on the level floor of the epicontinental, Rhodano-Swabian Basin. The deep-water iron oolites of the Herznach Formation can be followed from eastern France over northern Switzerland to Swabia and to Franconia in southern Germany. Ammonites with a normal size distribution are abundant and prevail everywhere in the macrofossil assemblage in the formation. This is evidence that depth of deposition of the iron oolitic formation was at least 35 m (Fig. 4.9). The ammonite assemblage and numerous, diverse benthic macrofossils in the formation are proof that iron ooids in the Herznach Formation were accreted in well-oxygenated bottom water. The matrix of the iron oolitic sediment is mud, either argillaceous or calcareous. Water above the sediment surface was consequently normally quiet. The iron ooids in the Herznach Formation were episodically accreted at the sediment surface by weak, oscillating bottom currents driven by storm waves. The process of *selective* accretion of iron hydroxide at the surface of ooid cortices is not understood. Neither can the diagenetic transformation of brown, primarily goethitic cortices of iron ooids to green berthierine with ferrous iron *in the same environment* in well-oxygenated bottom water be satisfactorily explained.

It is evident from Fig. 8.2 that bottom currents in the environment of iron ooid accretion could occasionally be unidirectional and strong enough to erode loose sediment and to redeposit it elsewhere. Nevertheless, an origin of iron ooids in the Herznach Formation in very shallow-marine water or even on land as ferriferous spherulites the size of ooids in a soil, and subsequent transport of the particles by a unidirectional bottom current over a long distance to the basin floor must be ruled out. The view that the oolitic iron orebody between Herznach and Wölflinswil in Canton Aargau was formed on a "swell" is incompatible with the fact, that the Herznach Formation is continuous over a great horizontal distance, and that ammonite casts prevail everywhere in the macrofossil assemblage in the formation. The ammonites indicate a more or less uniform depth of sedimentation of more than 35 m everywhere in the formation. No explanation could be found why the normal thickness of decimeters of the Herznach Formation grew locally to meters near Herznach (Fig. 4.10), and how the on average moderate content of iron ooids in the muddy matrix of the sediment could increase to packstone concentration in the upper part of the iron orebody near Herznach.

17. Marks of *paleobathymetry* were found departing from mean sea level down to a water depth of approximately 125 m. Mean sea level could be indicated to an accuracy of probably a few decimeters, because there is no evidence of a significant tidal range. Indicators of the intertidal zone are laminated stromatolites, some of

them crinkled, some with bird's-eye pores, and some with prism cracks. Hermatypic corals began to colonize the sea floor of mixed calcareous and argillaceous mud abruptly with a mass occurrence of platy colonies in biostromes of the Liesberg Member, as soon as aggradational, shallowing-upward sedimentation reduced water depth to approximately the remarkably shallow 20 m, which were calculated semiquantitatively in Chap. 6. The inclined boundary between the Olten Member near Olten and the Geissberg Member (Fig. 1.5) is evidence that the calculated figure of 20 m is of the right order of magnitude. The paleobathymetric range of between 20 and 30 m is characterized by the occurrence of abundant, large bivalves of the genus *Pholadomya* and of many large ostreids. *Pholadomya* are most common in the lower part of their paleobathymetric range. The composition of the macrofossil assemblage in the Bathonian iron oolite from the depth of deposition of between 20 and 30 m in Canton Aargau is very similar to the macrofossil assemblage in the much younger, Oxfordian calcilutite of the Geissberg Member from the same bathymetric range (Sect. 4.15.1). The thin, regionally widespread iron oolite was sedimented at a minimal average rate, whereas the calcilutite of the Geissberg Member was laid down at a substantially higher, normal rate. The similar composition of the bathymetrically well-defined macrofossil assemblages in the two units did not vary through the considerable time span of approximately ten million years. *The close correlation between the depth of deposition of a given sediment and the composition of the macrofossil assemblage in the pertinent unit is visualized in Fig. 4.9. Conversely, the composition of a macrofossil assemblage was largely independent of the sedimentation rate, facies, and age of sediments.*

Ammonites begin to prevail in the macrofossil assemblage without a transition from a paleodepth of approximately 35 m downward. Their percentage can jump up to more than 80% of the macrofossil assemblage following a moderate increase of water depth beyond 35 m. The paleobathymetric boundary between iron-oolitic facies and facies with cauliflower glauconite pellets was at the water depth of about 100 m *in the paleoenvironments investigated*. Water paleodepth of around 125 m in Canton Schaffhausen (Fig. 4.9) at the end of the *Transversarium* Chron can be concluded on the assumption that sea floor topography which existed at the beginning of the Late Jurassic between Canton Aargau and Canton Schaffhausen did not change until the *Transversarium* Chron. The lower paleobathymetric limit to the habitat of hermatypic corals and the minimal water depth from which ammonites begin to prevail in the macrofossil assemblage are clear-cut.

18. The *bathymetric facies boundary*, both vertical and lateral, between the range of water paleodepth in which *iron ooid* accretion was possible and the greater depth range, where *cauliflower pellets of pure glauconite* were formed, is well-defined at the paleodepth of approximately 100 m. The overlap of the two depth ranges was insignificant, provided that it existed at all. The vertical facies boundary between iron ooids in the Herznach Formation below and cauliflower glauconite pellets in the lowermost bed of the Birmenstorf Member directly above is isochronous over a horizontal distance of tens of kilometers. This is evidence that the sea floor was level and around 100 m below sea level, when sedimentation of the Birmenstorf Member began. The widespread oolitic ironstones of the Herznach Formation with a macrofossil assemblage of mostly ammonites were consequently sedimented on a level basin floor as well and at an average water depth of more than 35 m. These thin, deep-water iron oolites can be followed from France over northern Switzerland to Swabia and even to Franconia in southern Germany.

19. *Aggradation and progradation*: No answer to the question could be found, why sedimentation in the proximal, thick part of the partially argillaceous Bärschwil Formation was *aggradational*, whereas sedimentation in the proximal, thick part of the partially argillaceous Effingen Member of the Wildegg Formation was *progradational*. For a reason unknown, sedimentation of the Effingen Member in Canton Aargau turned to be aggradational in the region around the city of Aarau. The micritic limestone of the Villigen Formation directly above was laid down by aggradation between Aarau and Villigen (Fig. 1.5).

20. *Stromatolites* are most abundant in the upper intertidal zone. A crinkled upper surface (GYGI 1992, Fig. 12), prism cracks (this study, Fig. 4.4), and bird's-eye pores (this study, Fig. 4.6), occur only in intertidal stromatolites. Prism cracks in soft marl can be visualized indirectly by their infilling with calcareous mud. Such infillings are preserved at the underside of the limestone bed no. 57 of the unpublished section RG 320 at Undervelier in Canton Jura. Stromatolites from deeper water are much less abundant. Columnar stromatolites on hermatypic corals in a bioherm of the St-Ursanne Formation were shown by GYGI (1992, Fig. 18). Such stromatolites were also found on a coral skeleton in a reef in the boundary beds between the Effingen Member and the Günsberg Formation (GYGI 1992, Fig. 20).

A domal stromatolite from the water depth of close to 100 m in the Schellenbrücke Bed with iron ooids and with some glauconite pellets in section RG 73 near Dangstetten was figured by GYGI (1969a, Pl. 1, Fig. 2).

Mammillate crusts of micrite, phosphate, berthierine, and of goethite with a lustrous surface cover the top of the Schellenbrücke Bed in many sections. The crusts can be the only sediment of the late *Cordatum* Chron in many sections. Such crusts are figured in Gygi (1969a, Pl. 1, Fig. 3) and mainly in Gygi (1992, Figs. 29–31). A stromatolitic crust on the upper side of a dish-shaped siliceous sponge *Discophyma* from the Birmenstorf Member at the water depth of 100 m is shown in Fig. 32 in Gygi (1992). A stromatolite including cauliflower pellets of glauconite on lithified substrate at the water depth of approximately 120 m in the Mumienkalk Bed of section RG 212 at Siblingen in Canton Schaffhausen is represented in the present study in Fig. 8.3. Transitional between stromatolites on stationary substrate and spherical oncoids are stromatolitic crusts which grew on the upper side of ammonite casts lying with the plane of coiling flat on the sediment surface of mud after the casts were washed clear of embedding mud. Casts of perisphinctid ammonites with a diameter of up to 30 cm in the Mumienmergel Bed were occasionally overturned, probably by fish. Stromatolitic crusts could therefore grow all around such casts. Overturning of a perisphinctid cast in the Mumienmergel Bed is documented in Fig. 8.4 by partial geopetal infilling of calcareous mud in empty chambers of the phragmocone.

21. *Spherical, primarily hard oncoids of calcium carbonate* with a diameter of up to several centimeters could grow not only in the very shallow water of widespread lagoons like for instance the one into which the Hauptmumienbank Member was sedimented. Such oncoids grew also at a water paleodepth of about 40 m in the distal part of the lowermost Reuchenette Formation. They even occur in the thin and stratigraphically condensed Mumienkalk Bed in Canton Schaffhausen (Fig. 8.3). This bed was sedimented in probably particularly clear water at a depth of approximately 120 m. The most plausible assumption that can be made about the mode of formation of the centimeter-size, *spherical* oncoids in the Mumienkalk Bed is, that the oncoids were occasionally rolled on the sediment surface by bottom currents that were driven by waves. Such waves must have been great, and they were generated during exceptionally violent storms which probably occurred at rare occasions. Provided that this assumption can be corroborated, storm wave base would be at a widely variable water depth. The fact that the cortices of the spherical oncoids in the Mumienkalk Bed are almost concentric and of similar thickness indicates that time intervals were of limited length when the oncoids were not rolled in periods of quiet bottom water. The flat

mega-oncoids in the Mumienmergel Bed and the successive Mumienkalk Bed were not rolled. Fish in search of food possibly overturned them from time to time.

22. *Storm wave base* is probably at a widely variable water depth. Fine-grained, loose sediment could be eroded by episodic, unidirectional bottom currents driven by storms in the Schellenbrücke Bed at a water depth of close to 100 m (Figs. 4.9 and 8.2C, D). Bottom currents driven by particularly strong storms could roll spherical, calcareous oncoids with a diameter of several centimeters at the surface of fine-grained sediment 120 m below sea level. Conversely, the discoidal limestone bodies with a diameter of as much as 20 cm, which were overturned from time to time and that are covered all around with a crust of goethite with a *shining* surface in the Schellenbrücke Bed, no. 7 in section RG 210 near Gansingen (Fig. 8.3), were probably not overturned by bottom currents at a water depth of up to more than 90 m. Provided that bottom currents had indeed eroded and rounded these bodies in the course of transport, the cobbles would not be encrusted all around with a veneer of goethite with a lustrous surface. Storm wave base at a water depth greater than 100 m is documented in Sect. 4.3.1. Overturning of perisphinctid ammonite casts with a diameter of as much as 30 cm and with a thick, oncolitic or rather stromatolitic crust in the Mumienmergel Bed by bottom currents in water about 120 m deep is highly improbable. The shining surface of the veneer of goethite around some of such mega-oncoids in the Mumienmergel Bed excludes abrasion by erosion in a turbulent bottom current. The term storm wave base and above all distinction between fair weather and storm wave base is avoided in this study.

23. The *mode of life of almost all of the ammonites* which became fossilized in the region investigated was necto-benthic. Possible exceptions are phylloceratid and lytoceratid ammonites. *Post-mortem* drift of empty ammonite shells out of the habitat of the animals and over a long horizontal distance could be documented to have occurred, but it was exceptional. The abundance of ammonites and the composition of ammonite assemblages, the size distribution of complete casts of adult ammonites in an assemblage, and the predominantly calcareous or iron sulfide substance of casts, respectively, are therefore diagnostic of the paleoenvironment of the sediment in which the ammonite casts were embedded. Especially depth of sedimentation and oxygenation of bottom water can be concluded from the ammonite assemblage in a sediment.

24. The abundance and composition of an ammonite assemblage depends on the *depth of deposition* and the *sedimentation rate of* the encasing deposit of calcareous or

argillaceous mud. The abundance of ammonite casts is as a rule at a maximum in sediments from the deep subtidal zone which were laid down at a very low rate. No ammonite casts were ever found in sediments from a peritidal environment. They are absent as well in sediments from very shallow lagoons with spherical, centimeter-size calcareous oncoids like for instance the Hauptmumienbank Member. Ammonites were probably primarily rare mainly because of shallow water in the coral biostromes of the Liesberg Member which grew in a paleobathymetric range of between 6 and 20 m. Some large perisphinctid ammonites could possibly live permanently in water which was no deeper than about 10 m in the lagoon into which the Buix Member was sedimented. Ammonite casts in the sediments investigated predominate in the macrofossil assemblage downward from a minimal paleodepth of deposition of 35 m. The vertical and lateral paleobathymetric boundary between facies with mainly bivalves from a water depth of 30 m or less and facies with a macrofossil assemblage of mainly ammonites from water deeper than approximately 35 m is clear-cut. The percentage of haploceratacean ammonites in an ammonite assemblage has a *tendency* to grow with increasing water depth. It can be read from Fig. 4.9 that this percentage did not increase smoothly with growing water depth. Conversely, the percentage of perisphinctaceans and Cardioceratinae in an ammonite assemblage has a tendency to diminish with increasing water paleodepth. This tendency in boreal *Cardioceras* is unexpected, because water temperature probably diminished with growing water depth. The abundance of ammonite casts rapidly diminishes upward in a succession of fine-grained sediment concomitant with a progressively growing rate of deposition. This is exemplified in the *Renggeri* Member and in the lowermost part of the Effingen Member.

Ammonite casts are as a rule most abundant in beds from deeper water which were sedimented at a particularly low rate, but there is no simple relation between ammonite abundance and a low sedimentation rate. Ammonite casts are particularly abundant in the stratigraphically condensed beds with cauliflower glauconite pellets of the Mumienmergel and the Mumienkalk Bed in Canton Schaffhausen from the water depth of approximately 120 m. The cause could not be found, why casts of ammonites are uncommon in the coeval, stratigraphically condensed lowermost bed of the Birmenstorf Member in Canton Aargau. This bed was laid down at an only slightly lesser water depth of around 100 m and has a similar facies. Few ammonite casts were found when thin iron oolite of the

Scarburgense Zone from a depth of deposition of approximately 80 m was excavated in the upper Herznach Formation on a surface of 20 m^2 in bed no. 20 of section RG 307 at Péry (GYGI 1990a, Fig. 4), on 20 m^2 in bed no. 6 of section RG 280 at Liesberg (GYGI 1990a, Fig. 2), and on 50 m^2 in bed no. 7 of section RG 208 at Ueken near Herznach (this study, Fig. 8.2A). Conversely, ammonite casts are abundant and in many cases particularly well-preserved in iron oolite of the thin Schellenbrücke Bed in Canton Aargau. The bed is the youngest sediment in the Herznach Formation. It was laid down at a very low rate in the *Cordatum* Chron at the depth of about 95 m (Fig. 8.2E).

25. A *normal size distribution in an ammonite assemblage* varied over the entire range between dwarf and giant taxa (classification after GYGI 2001, p. 12). The size of dwarfs in a normal assemblage was predetermined *genetically*. Ammonite assemblages with a normal size distribution lived in the deep subtidal zone, downward from a minimal water depth of 35 m, and in well-oxygenated bottom water with an oxygen content of up to 5 milliliters of dissolved oxygen per liter of water. Such environments with a rich and diverse ammonite assemblage existed when the iron-oolitic Schellenbrücke Bed (depth of deposition about 95 m) and the uncondensed part of the Birmenstorf Member (deposited at a water depth of approximately 105 m) were sedimented, and when the stratigraphically condensed Mumienmergel and the Mumienkalk Bed were laid down at a water depth of around 120 m.

26. *Dwarf size of ammonites* living in well-oxygenated water was *predetermined genetically. Ammonites* living permanently in dysaerobic water were *dwarfed or stunted because of the environment*. The low oxygen content of water in the habitat of such ammonites reduced the *size* of adults, but it apparently did not noticeably reduce ammonite *abundance*.

27. *Ammonites could tolerate a lesser rate of clay mineral sedimentation in their habitat than hermatypic corals.* Ammonite casts were found to be particularly abundant in thin, stratigraphically condensed beds sedimented at a water depth of approximately 120 m like the Mumienmergel Bed and the Mumienkalk Bed. For a reason unknown, ammonites are rare in the coeval, thin and stratigraphically condensed lowermost bed of the Birmenstorf Member with a similar facies from a water depth of about 100 m. Ammonites are very abundant in the thin iron oolite with a calcareous matrix of the Schellenbrücke Bed of the *Cordatum* Chron from a water depth of around 95 m. Conversely, they are rare in the thin iron oolite with a matrix of marl-clay, but with an otherwise similar facies deposited in a water depth of

approximately 80 m at the base of the Oxfordian Stage in the *Scarburgense* Chron between northwestern Switzerland and Canton Aargau. Few ammonite casts were found when this iron oolite, bed no. 20 of section RG 307 at Péry, was excavated on a surface of 20 m^2 (GYGI 1990a, Fig. 4), on 20 m^2 in bed no. 6 of section RG 280 at Liesberg (GYGI 1990a, Fig. 2), and on more than 50 m ^{2} in bed no. 7a of section RG 208 at Ueken (this study, Fig. 8.2E). No simple relation consequently exists between a very low sedimentation rate and ammonite abundance.

Ammonites are fairly abundant in the blue-gray marl-clay of the thin *Scarburgense* Zone with a thickness of close to 3 m in the lowermost *Renggeri* Member at Liesberg (GYGI 1990a, Figs. 2, 3). This part of the member was sedimented at the water depth of approximately 80 m and at a low rate. The rate of sedimentation of pure clay minerals at this locality was calculated in Sect. 4.6 to have been of the order of compacted 6.5 cm per 10,000 years. Ammonite casts are rare in marl of the *Bifurcatus* Zone in the lower Effingen Member below the Gerstenhübel Beds in section RG 37 at Auenstein. This marl from a water depth of around 80 m has a carbonate content of the order of 60%, to judge from the results of chemical analyses represented in GYGI (1969a, Pl. 17). Sedimentation of the order of a compacted thickness of 44 cm of pure clay minerals per 10,000 years was calculated in Sect. 4.6 for the pertinent marl in the lower Effingen Member at Auenstein. This rate is almost seven times the pertinent rate in the lowermost *Renggeri* Member at Liesberg with numerous ammonites.

Hermatypic corals with predominantly plate-like colony shape are approximately 30% of the rock volume of bed no. 105 in the Liesberg Member in section RG 306 at Liesberg. A close-up photograph of the bed is Fig. 32 on p. 54 in GYGI (2000a). The matrix of bed no. 105 in that section is marl with a carbonate content that is estimated at about 70%. The rate of sedimentation of the Liesberg Member was calculated in Sect. 8.1 to have been at least 2.7 m (compacted) per 10,000 years. Pure clay minerals (compacted) were consequently sedimented in the Liesberg Member at a rate of much more than 1 m per 10,000 years, or possibly of as much as about 20 times the pertinent rate of 6.5 cm in the lowermost *Renggeri* Member at Liesberg with numerous ammonites. Bed no. 105 in the upper part of the Liesberg Member at Liesberg with a mass occurrence of plate-like hermatypic corals was laid down at a water depth of between 10 and 15 m. Calcareous oolite from a water depth of less than 6 m in the lower St-Ursanne Formation in section RG 306 follows 10 m above the top of bed no.

105. This is shown on Pl. 31 in GYGI (2000a). The very high sedimentation rate of clay minerals in the Liesberg Member was obviously no detrimental condition to hermatypic corals, but it was probably one reason why ammonite casts are almost absent in the Liesberg Member. The high rate of clay mineral sedimentation and the slight water depth kept living ammonites probably almost entirely out of the environment of the Liesberg Member.

Plate-like colony shape in Recent hermatypic corals is generally assumed to be an adaptation to a low level of illumination. Plate-like, very thin colonies of *Agaricia fragilis* LAMARCK were found to be living in Bermuda in the protected waters of Harrington Sound (map in GYGI 1970, Fig. 17d) in the shade under the overhang of a vertical wall of calcareous eolianite rock that is the shore in parts of Harrington Sound. The author dived below the overhang that has a level ceiling directly below sea level and is horizontally about 4 m deep. A cross-section of the pertinent rocky shore of Harrington Sound is shown in GYGI (1970, Fig. 17e). NEUMANN (1966) documented how vertical walls of eolianite rock on the shore of Harrington Sound are undermined and eroded in deep notches by the boring sponge *Cliona lampa* DE LAUBENFELS. It is noteworthy that plate-like hermatypic corals were found to be living in very clear water of the Red Sea at a water depth of up to 160 m, and that living colonies of *Agaricia fragilis* were collected by the author in Bermuda's Harrington Sound in the deep shade below a wide overhang of rock in water less than 1 m deep.

28. *Casts of giant perisphinctaceans are overrepresented* in ammonite assemblages in sediments from a water paleodepth range of between approximately 30 and 50 m. Examples are the lower Villigen Formation and the lower Baden Member in section RG 70 near Mellikon, both from a water paleodepth of about 50 m, and the distal part of the lower Reuchenette Formation in section RG 21 in the quarry on Mt. Born west of Olten from a water paleodepth approaching 30 m. *Giant nautilids* lived only in the lagoonal, very shallow water of environments like in the most distal part of the upper Röschenz Member, for instance, north of Grandval east of Moutier, in the Laufen Member near Liesberg, and in the platform interior of the lower Reuchenette Formation like in Ajoie region of Canton Jura.

29. *Time calibration* of the sediments investigated was made relatively with ammonite biochronology, and it was attempted to be done directly with radiochronology. Radiometric measurements of age of cauliflower glauconite pellets that were formed together with well-preserved ammonite casts (steinkerns), gave figures

nearly 7% younger than the pertinent ages as published by GRADSTEIN et al. (2004) or OGG et al. (2010). An ammonite chron, which is the time represented by an ammonite zone in the sense of the International Stratigraphic Guide, Second Edition (SALVADOR, ed., 1994, Fig. 10 on p. 62), lasted on average approximately 370,000 years according to Fig. 1.6 in this study, and according to the time scale by GRADSTEIN et al. (2004, p. 310). The ammonite chrons distinguished in Fig. 1.6 are equivalent to the subchrons in previous publications by the author. The ammonite chron as it is conceived in this study is the shortest biochronologic unit that could be discerned.

30. *Time correlation* of sediments from deeper water on a *regional scale* was made with ammonites, or where possible, more precisely with isochronous marker beds like the Knollen Bed (see below). Correlation of deposits from deeper water with units from shallow water, as well as correlation within sediments from shallow water, could be made mainly by means of the vertical variation in the abundance of the clay mineral kaolinite. The curve of vertical variation of kaolinite content in a succession was initially calibrated in the epicontinental basin with ammonite biochronology. Correlations made with kaolinite were checked with sequence stratigraphy and could be confirmed. Another possibility to check time correlations based on kaolinite was use of ammonites that were found in sediments from shallow water, and that are indicated in Fig. 1.5. The few ammonites occurring in sediments from shallow water were found where it was to be expected when using kaolinite correlations. The transgression surfaces above an almost perfectly flat topography in extremely shallow water, which evolved during sedimentation of the middle and again of the upper Röschenz Member, can be rated to be close to isochronous. Rapid transgression over beds from extremely shallow water could cause formation of an isochronous marker bed in deeper water like the Knollen Bed. *Intercontinental time correlation* based on ammonite biochronology was made between northern Switzerland and northern Chile on the level of one ammonite chron, with an error of approximately ±0.2% of the radiometric age assigned to the pertinent ammonites of the genus *Gregoryceras*. Prerequisite of accuracy to such a degree is the time-consuming work of measuring numerous detailed sections, collecting and mainly excavating a great number of ammonites systematically from bed to bed, preparing the well-preserved specimens in the laboratory, devising a practicable method of taxonomy, and identifying specimens.

31. *Sedimentation rates* were calculated based on ammonite biochronology and on the radiochronology which was

published by GRADSTEIN et al. (2004). Rates varied between a maximum of probably not more than approximately 5 m of compacted sediment per 10,000 years, zero (nondeposition or omission) which could last several millions of years without emersion, and negative because of subsolution (corrosion of early lithified limestone), erosion of loose sediment by bottom currents, or subaerial erosion. The highest rates of sedimentation were by far less than the probably maximal possible rate of isostatic equilibration of the lithosphere to loading. The lithosphere was consequently never off isostatic equilibrium in the area investigated. Isostatic equilibration of the lithosphere following even small increments of sediment load, or of water caused by rising sea level, was probably prompt.

32. The interrelation between small-scale variation of sea level, variation of the sedimentation rate, and formation of *marker beds* was different in times of humid climate from that in drier periods. The later, rapid part of sea level *rise no. 6* (in Fig. 5.1) in the *Bifurcatus* Chron occurred when climate on adjacent land was humid. The rise shifted the shoreline and thereby the source of terrigenous sediment landward and thereby interrupted for a short time supply of terrigenous sediment to the realm of very shallow water of the Röschenz Member. The effect was formation of a widespread, thin sheet of calcareous oolite. This bed separates the lower half from the upper half of the marly Röschenz Member (Fig. 1.5). This is additional evidence, that the sea level rises recorded in Fig. 5.1 were rapid. Most of the great quantity of calcareous mud, which was produced at the same time in the realm of the Röschenz Member, was exported across the carbonate platform of the Günsberg Formation and beyond into deeper water, where the lime mud formed the intercalation of micritic limestone of the Gerstenhübel Beds within marl of the Effingen Member. Rapid *rise no. 10* occurred a short time before the end of the *Gredingensis* Chron when climate in the region investigated was comparatively dry. The minor rise substantially reduced or cut short for some time export of lime mud out of the lagoon of the La May Member in the Courgenay Formation landward of the sand bank of the Balsthal Formation. No traces of the rise were found in the shallow-water realm, neither in the La May Member nor in the sand bank of calcareous ooids of the Holzflue Member in the Balsthal Formation. The break in supply of lime mud from the lagoon caused in deeper water some time of omission (nondeposition) and locally subsolution in the Villigen Formation. When supply of lime mud across the vast sand bank of calcareous ooids of the Balsthal Formation gradually resumed after rise no. 10, the thin and very widespread, slightly glauconitic marker bed of the Knollen Bed was laid

down at a low rate in deeper water in the Villigen Formation. The bed is at some localities above a deeply corroded surface (GYGI et al. 1998, Fig. 12). The eustatic component of rapid rise nos. 6 and 10 was not more than a few meters. Nevertheless, the effect of the rises was widespread.

33. The *geometry of sedimentary units* was closely interrelated with variations in sea level. The distal margin of proximal, thick succession no. 1 was essentially stationary during the entire time of sedimentation of the succession (Fig. 1.5). This was the effect of a coeval, net sea level rise which was most of the time gradual and was probably eustatically driven. Supply of mainly mudgrade sediment and the average rate of relative sea level rise were then in equilibrium. Sea level was approximately constant most of the time when succession no. 2 was sedimented. The result was net progradation of the Günsberg Formation and the Steinibach Member (Figs. 1.5 and 4.15B, C). Progradation of the Günsberg Formation was halted when sea level began to rise gradually in the *Bifurcatus* Chron. Rapid and essentially eustatic, but small-scale sea level rises and concomitant events of transgression occurred in the later part of the *Bifurcatus* Chron and in the early *Hypselum* Chron. Strong progradation resumed during sedimentation of the youngest part of the Effingen Member. Progradation slowed down when the carbonate end-member of succession no. 2 (Steinibach and lowermost Olten Member) was laid down during the *Berrense* Chron, and came to a stillstand in the *Bimammatum* and mainly in the *Gredingensis* Chron. This is documented by the geometry of the Olten Member. Demise of the coral biostrome of the Olten Member was brought about by the progressively increasing rate of sea level rise during the *Planula* Chron. Continuing, gradual sea level rise at a high rate caused backstepping during deposition of the uppermost part of succession no. 3. Backstepping continued until coral biostromes and small bioherms appeared at Balsthal in the Kimmeridgian Age (Fig. 1.5).

34. *Sediments of the Late Jurassic in the Swiss part of the Jura Mountains can probably be used as a gauge of global, eustatic sea level variation* during the pertinent time span. This is because of the uninterrupted succession of sediments, which were laid down mostly at a normal rate. The strata are a complete and well-documented lithostratigraphic frame of reference. These sediments include a probably complete ammonite succession of the Subtethyan Faunal Province and ammonites of the Tethyan and the Boreal Faunal Province.

35. GRESSLY'S *first "law" of facies* is based on the lateral exclusion of hermatypic coral/calcareous oolite facies and coeval facies with siliceous sponges and ammonites. His second "law" is open to interpretation (see Chap. 1, "Previous work"). It probably anticipated progradation.

36. WALTHER'S *"law"* is well exemplified by succession no. 1. The *vertical* facies transition in the proximal thick, aggradational and shallowing-upward succession no. 1 in northwestern Switzerland begins with a thin bed of iron oolite including a macrofossil assemblage with mostly ammonites. The bed was sedimented at a very low rate at the water depth of approximately 80 m. Hermatypic corals are abundant in the Liesberg Member of the uppermost Bärschwil Formation and in the Tiergarten Member of the upper St-Ursanne Formation. Sedimentation of proximal succession no. 1 filled the basin in northwestern Switzerland to close below sea level for the first time at the top of the calcareous-oolitic sand bank of Chestel Member in the internal part of the St-Ursanne Formation, and later at the top of Tiergarten Member in the marginal, basinward part of the formation. Above is the peritidal Vorbourg Member in lowermost succession no. 2 with a stromatolite from the upper intertidal zone, which is shown in Figs. 1.5 and 4.6.

The *lateral* facies variation in coeval sediments in the uppermost part of succession no. 1 can be studied in a continuous succession between proximal deposits from shallowest water down to the basin floor in the following units. The lagoon of the Buix Member with coral patch reefs and few ammonites was rimmed by the sand bank of the calcareous-oolitic Tiergarten Member with small coral reefs in the marginal part of the carbonate platform of the St-Ursanne Formation. The adjacent, micritic limestone of the Pichoux Formation was sedimented upon the depositional slope extending to the basin floor. The unit there grades into the Birmenstorf Member with biostromes of siliceous sponges and with abundant ammonites. This member was sedimented on the level basin floor at the depth of somewhat more than 100 m. The Birmenstorf Member thins on a minor slope which is the transition to the Mumienmergel Bed, the Mumienkalk Bed, and the glauconitic, unnamed bed directly above in the area of Canton Schaffhausen (Figs. 1.5 and 8.3). The thin beds in Canton Schaffhausen were sedimented on a level sea floor at the water depth of 120–125 m. The macrofossil assemblage in these beds is practically the same as that in the Birmenstorf Member. The correlation line in Fig. 1 in GYGI (2001) between the top of the Birmenstorf Member and the top of the coeval, thin beds in Canton Schaffhausen was drawn inappropriately. Time equivalence of the sediments mentioned is documented mainly with figured specimens of the ammonite taxon *Perisphinctes (Perisphinctes) alatus* ÉNAY.

Acknowledgments

Fieldwork was funded by the Swiss National Science Foundation, grants no. 2.211.69 and 2.165.78, as well as by several private foundations. Thanks are owed to the many persons who helped to gather the material this book is based on. The names of these persons are mentioned in the acknowledgments in earlier papers by the author. Careful reviews of preliminary versions of this text were made by R. BURKHALTER, A. HALLAM, and A. STRASSER. A. HALLAM and A. STRASSER suggested further reading and gave exact references. K. B. FÖLLMI and A. D. MIALL contributed some remarks. All of this and comments by anonymous reviewers made printing of this book possible. This is gratefully acknowledged. Most of all, the author is grateful to JESUS for his guidance and help.

References

ABBINK, O., TARGARONA, J., BRINKHUIS, H. and VISSCHER, H., 2001, Late Jurassic to earliest Cretaceous palaeoclimatic evolution of the southern North Sea: Global and Planetary Change, v. 30, p. 231-256.

ALDINGER, H., 1945, Zur Stratigraphie des Weissen Jura Delta in Württemberg: Jahresberichte und Mitteilungen des oberrheinischen geologischen Vereins, Neue Folge, v. 31, p. 111-152.

ALDINGER, H., 1957a, Summary of the opinions about the mode of formation of oolitic ironstones, which were read before the annual convention of the Deutsche Geologische Gesellschaft on May 10 and11 at Stuttgart, Germany: Zeitschrift der Deutschen Geologischen Gesellschaft, vol. 109, no. 1, p. 2-6 (in German).

ALDINGER, H., 1957b, Zur Entstehung der Eisenoolithe im Schwäbischen Jura: Zeitschrift der Geologischen Gesellschaft, v. 109, no. 1, p. 7-9.

ALDINGER, H., and FRANK, M., 1943, Vorkommen und Entstehung der südwestdeutschen jurassischen Eisenerze: Neues Jahrbuch für Mineralogie, Geologie und Paläontologie, Abteilung B, Geologie Paläontologie, vol. 88, no. 2, p. 293-336.

ALLENBACH, R., 2002, The ups and downs of "tectonic quiescence" - recognizing differential subsidence in the epicontinental sea of the Oxfordian in the Swiss Jura Mountains: Sedimentary Geology, v. 150, p. 323-342.

ALLENBACH, R., and van KONIJNENBURG-van CITTERT, J. H. A., 1997, On a small flora with araucariaceous conifers from the Röschenz Beds of Court, Jura Mountains, Switzerland: Eclogae geologicae Helvetiae, v. 90, p. 571-579.

ARKELL, W. J., 1933, The Jurassic System in Great Britain: Clarendon Press, Oxford, 681 p.

ARKELL, W. J., 1935-48, Ammonites of the English Corallian Beds: Monographs of the Palaeontographical Society London, v. 92, lxxiv and 420 p.

ARKELL, W. J., 1947, The geology of Oxford: Clarendon Press, Oxford, VIII and 267 p.

ARKELL, W. J., 1956, Jurassic geology of the world: Oliver and Boyd, Edinburgh and London, 806 p.

ATROPS, F., GYGI, R., MATYJA, B. A., and WIERZBOWSKI, A., 1993, The Amoeboceras faunas in the Middle Oxfordian-lowermost Kimmeridgian, Submediterranean succession, and their correlation value: Acta geologica polonica, v. 43, no. 3-4, p. 213-227.

BARD, E., HAMELIN, B., and FAIRBANKS, R. G., 1990, U-Th ages obtained by mass spectrometry in corals from Barbados: sea level during the past 130 000 years: Nature, v. 346, p. 456-458.

BATHURST, R. G. C., 1959, Diagenesis in Mississippian calcilutites and pseudobreccias: Journal of Sedimentary Petrology, v. 29, no. 3, p. 365-376.

BERGER, J.-P., 1986, Dinoflagellates of the Callovian-Oxfordian boundary of the "Liesberg-Dorf" quarry (Berner Jura), Switzerland: Neues Jahrbuch für Geologie und Paläontologie, Abhandlungen, v. 172, no. 3, p. 331-355.

BERT, D., 2004, Révision, étude systématique et évolution du genre Gregoryceras SPATH, 1924 (Ammonoidea, Oxfordien): Annales du Muséum d'Histoire naturelle de Nice, tome XIX, p. 1-182.

BERT, D., ÉNAY, R. and ATROPS, F., 2009, Les Gregoryceras (Ammonitina) de l'Oxfordien moyen terminal et supérieur téthysien: révision systématique, biostratigraphie et évolution: Geobios, v. 42, p. 451-493.

BIRKHÄUSER, P., ROTH, P., MEIER, B., and NAEF, H., 2001, 3D-Seismik: Räumliche Erkundung der mesozoischen Sedimentschichten im Zürcher Weinland: NAGRA Technischer Bericht 00-03, text and appendix.

BITTERLI-DREHER, P., 1979, Cyclic sedimentation in the upper Bathonian-Callovian of the Swiss Jura Mountains: Association des Sédimentologues Français, Publication spéciale no. 1, p. 99-109.

BÖHLKE, J. and CHAPLIN, C. C. G., 1970 (second printing), Fishes of the Bahamas and adjacent tropical waters: Livingston Publishing Company, Wynnewood, Pennsylvania, 771 p.

BOLLIGER, W., and BURRI, P., 1967, Versuch einer Zeitkorrelation zwischen Plattformcarbonaten und tiefermarinen Sedimenten mit Hilfe von Quarz-Feldspat-Schüttungen (mittlerer Malm des Schweizer Jura): Eclogae geologicae Helvetiae, v. 60, no. 2, p. 491-507.

BOLLIGER, W., and BURRI, P., 1970, Sedimentologie von Schelf-Carbonaten und Beckenablagerungen im Oxfordien des zentralen Schweizer Jura: Beiträge zur geologischen Karte der Schweiz, Neue Folge, v. 140, p. 1-96.

BONNOT, A. and GYGI, R. A., 1998, Les Euaspidoceratinae (Ammonitina, Aspidoceratidae) d'Herznach (Suisse septentrionale) à la fin de la zone à Cordatum (Oxfordien inférieur): Eclogae geologicae Helvetiae, v. 91, p. 493-512.

BONNOT, A., and GYGI, R. A., 2001, Les Euaspidoceratinae (Ammonitina, Aspidoceratidae) de la zone à Transversarium (Oxfordien moyen) de Suisse septentrionale (cantons d'Argovie et de Schaffhouse): Eclogae geologicae Helvetiae, v. 94, no. 3, p. 427-445.

BRÄM, H., 1965, Die Schildkröten aus dem oberen Jura (Malm) der Gegend von Solothurn: Schweizerische Paläontologische Abhandlungen, v. 83, p. 1-190.

BRETT, C. E. and BAIRD, G. C., 1986, Comparative taphonomy: A key to paleoenvironmental interpretation based on fossil preservation: Palaios, v. 1, p. 207-227.

BRINDLEY, G. W., BAILEY, S. W., FAUST, G. T., FORMAN, S. A., and RICH, C. I., 1968, Report of the Nomenclature Committee of the Clay Minerals Society: Clays and Clay Minerals, v. 16, p. 322-324.

BÜCHI, U. P., LEMCKE, K., WIENER, G., and ZIMDARS, J., 1965, Geologische Ergebnisse der Erdölexploration auf das Mesozoikum im Untergrund des schweizerischen Molassebeckens: Bulletin des Vereins schweizerischer Petroleum-Geologen und -Ingenieure, v. 32, no. 82, p. 7-38.

BURKHALTER, R. M., 1995, Ooidal ironstones and ferruginous microbialites: origin and relation to sequence stratigraphy (Aalenian and Bajocian, Swiss Jura mountains): Sedimentology, v. 42, p. 57-74.

CALLOMON, J. H., 1960, New sections in the Corallian Beds around Oxford, and the subzones of the *Plicatilis* Zone. Proceedings of the Geologist's Association, London, v. 71, part 2, p.177-208.

CALLOMON, J. H., 1964, Notes on the Callovian and Oxfordian Stages: Colloque du Jurassique Luxembourg 1962, volume des comptes rendus et mémoires de l'institut grand-ducal, section des sciences naturelles, physiques et mathématiques, p. 269-291. St. Paul, Luxembourg.

CARIOU, E., and HANTZPERGUE, P., coord., 1997, Biostratigraphie du Jurassique ouest-européen et méditerranéen. Zonations parallèles et distribution des invertébrés et microfossiles: Bulletin du Centre de Recherches Elf Exploration et Production, Pau, Mémoire 17, 422 p.

CARROLL, D., 1958, Role of clay minerals in the transportation of iron: Geochimica et Cosmochimica Acta, v. 14, p. 1-28.

CARTE GÉOLOGIQUE DE LA FRANCE à 1/50 000, no. 474 Montbéliard, avec note explicative, 1973.

CAVELIER, C., and ROGER, J. (coord.), 1980, Les étages français et leurs stratotypes: Mémoires du Bureau des Recherches géologiques et minières, v. 109, p. 1-295.

CECIL, C. B., 1990, Paleoclimate control on stratigraphic repetition of chemical and siliciclastic rocks: Geology, v. 18, p. 533-536.

CHARLTON, D. S., 1969, Intertidal zonation of Bermuda's rocky shores as an indicator of tide range and wave energy: Bermuda Biological Station for Research, Special Publication no. 2, p. 27-34.

CHAUVE, P., and PERRIAUX, J., 1974, Le Jura. In: DEBELMAS, J. (ed.): Géologie de la France. Doin, Paris, p. 443-464.

CHUMAKOV, N. M., and FRAKES, L. A., 1997, Mode of origin of dispersed clasts in Jurassic shales, southern part of the Yana-Kolyma fold belt, North East Asia: Palaeogeography, Palaeoclimatology, Palaeoecology, v. 128, p. 77-85.

CLOUD, P. E., 1955, Physical limits of glauconite formation: American Association of Petroleum Geologists Bulletin, v. 39, no. 4, p. 484-492.

COE, A. L. (ed.), BOSENCE, D. W., CHURCH, K. D., FLINT, S. S., HOWELL, J. A., and WILSON, R. C. L., 2003, The sedimentary record of sea-level change: Cambridge University Press, Cambridge, 287 p.

COLLIN, P. Y., LOREAU, J. P., and COURVILLE, P., 2005, Depositional environments and iron ooid formation in condensed sections (Callovian-Oxfordian, south-eastern Paris basin, France): Sedimentology, v. 52, p. 969-985.

COLLOQUE DU JURASSIQUE LUXEMBOURG 1962, 1964, Recommendations: Volume des comptes rendus et mémoires de l'Institut grand-ducal, section des sciences naturelles, physiques et mathématiques p. 84-86.

CONTINI, D., 1976, Évolution des milieux de sédimentation au cours de l'Oxfordien en Franche-Comté: Bulletin de la Fédération de la Société d'Histoire naturelle de Franche-Comté, v. 77, p. 43-54.

CREVELLO, P. D., and SCHLAGER, W., 1980, Carbonate debris sheets and turbidites, Exuma Sound, Bahamas: Journal of Sedimentary Petrology, v. 50, no. 4, p. 1121-1148.

CROSS, T. A., and HOMEWOOD, P. W., 1997, Amanz Gressly's role in founding modern stratigraphy: Geological Society of America Bulletin, v. 109, no. 12, p. 1617-1630.

DERCOURT, J. et al. (eds.), 2000, Atlas Peri-Tethys, with 24 palaeogeographical maps and explanatory notes, 268 p.: Commission for the Geological Map of the World, Paris.

DIEBOLD, P., 1988, Der Nordschweizer Permokarbon-Trog und die Steinkohlenfrage der Nordschweiz: Vierteljahrsschrift der Naturforschenden Gesellschaft in Zürich, v. 133, no. 1, p. 143-174.

DIEBOLD, P., NAEF, H., and AMMANN, M., 1991, Zur Tektonik der zentralen Nordschweiz: NAGRA Technischer Bericht 90-04, Text and appendix.

DIEBOLD, P., and NOACK, T., 1997, Late Paleozoic troughs and Tertiary structures in the eastern folded Jura. In: PFIFFNER, O. A., LEHNER. P., HEITZMANN, P., MUELLER, ST., and STECK, A. (ed.): Deep structure of the Swiss Alps. Birkhäuser, Basel, p. 59-63.

DIETL, G., and GYGI, R., 1998, Die Basis des Callovian (Mittlerer Jura) bei Liesberg BL, Nordschweiz: Eclogae geologicae Helvetiae, v. 91, p. 247-260.

DREYFUSS, M., ROLLET, A., and ROLLET, M., 1964, Remarques sur les types d'étage définis par Marcou en Franche-Comté: Argovien et Séquanien: Colloque du Jurassique Luxembourg 1962, Volume des comptes rendus et mémoires, Institut grand-ducal, section des sciences naturelles, physiques et mathématiques, p. 301-306.

DROMART, G., GARCIA, J.-P., PICARD, S., ATROPS, F., LÉCUYER, C., and SHEPPARD, S. M. F., 2003, Ice age at the Middle-Late Jurassic transition?: Earth and Planetary Science Letters, v. 213, p. 205-220.

DUNHAM, R. J., 1962, Classification of carbonate rocks according to depositional texture. In: HAM, W. E., ed., Classification of carbonate rocks: American Association of Petroleum Geologists, Memoir 1, p. 108-121.

DUPRAZ, C., and STRASSER, A., 2002, Nutritional modes in coral-microbialite reefs (Jurassic, Oxfordian, Switzerland): Evolution of trophic structure as a response to environmental change: Palaios, v. 17, p. 449-471.

EARDLEY, A. J., 1938, Sediments of Great Salt Lake, Utah: American Association of Petroleum Geologists Bulletin, v. 22, no. 10, p. 1305-1411.

EKDALE, A. A., and MASON, T. R., 1988, Characteristic trace-fossil associations in oxygen-poor sedimentary environments: Geology, v. 16, no. 8, p. 720-723.

EL ALBANI, A., MEUNIER, A., and FÜRSICH, F., 2005, Unusual occurrence of glauconite in a shallow lagoonal environment (Lower Cretaceous, northern Aquitaine Basin, SW France): Terra Nova, v. 17, no. 6, p. 537-544.

ÉNAY, R., 1964, Les faunes d'ammonites et la zonation de l'Oxfordien supérieur du Jura méridional: Colloque du Jurassique Luxembourg 1962, volume des comptes rendus et mémoires de l'institut grand-ducal, section des sciences naturelles, physiques et mathématiques, p. 487-501. St. Paul, Luxembourg.

ÉNAY, R., 1966, L'Oxfordien dans la moitié sud du Jura français: Nouvelles Archives du Musée d'Histoire naturelle de Lyon, v. 8, no. 1-2, p. 1-624.

ÉNAY, R., and GYGI, R., 2001, Les ammonites de la zone à Bifurcatus (Jurassique supérieur, Oxfordien) de Hinterstein, près de Oberehrendingen (canton d'Argovie, Suisse): Eclogae geologicae Helvetiae, v. 94, p. 447-487.

ÉTALLON, M. A., 1862, Etudes paléontologiques sur le Haut-Jura. Monographie du Callovien: Mémoires de la Société d'Emulation du Doubs, sér. 3, v. 6, p. 53-260.

FAVRE, E., 1876, Description des fossiles du terrain oxfordien des Alpes fribourgeoises: Mémoires de la société paléontologique suisse, v. 3, p. 1-76.

FIRSTBROOK, P. L., FUNNELL, B. M., HURLEY, A. M., and SMITH, A. G., 1979, Paleoceanic reconstructions 160-0 Ma.: Scripps Institution of Oceanography, 41 p.

FISCHER, H., 1965, Oberer Dogger und unterer Malm des Berner Jura: Tongruben von Liesberg. In: SCHAUB. H., and LUTERBACHER, H. (ed.): Proceedings of the 9th European congress on micropaleontology: Bulletin der Vereinigung Schweizerischer Petrol-Geologen und -Ingenieuren, v. 31, p. 25-36.

FISCHER, H., and GYGI, R., 1989, Numerical and biochronological time scales correlated at the ammonite subzone level; K-Ar, Rb-Sr ages, and Sr, Nd, and Pb sea-water isotopes in an Oxfordian (Late Jurassic) succession of northern Switzerland: Geological Society of America Bulletin, v. 101, p. 1584-1597.

GALLIHER, E. W., 1935, Geology of glauconite: American Association of Petroleum Geologists Bulletin, v. 19, p. 1569-1601.

GARRETT, P., SMITH, D. L., WILSON, A. O., and PATRIQUIN, D., 1971, Physiography, ecology, and sediments of two Bermuda patch reefs: Journal of Geology, v. 79, no. 6, p. 647-668.

GEIKIE, A., 1905, The founders of geology: Dover Publications, New York, reprint (undated), second edition, first published in 1962, 486 p., unabridged an unaltered republication of the second (1905) edition of the work originally published by Macmillan and Co. in 1897.

GHASEMI-NEJAD, E., SARJEANT, W. A. S., and GYGI, R. A., 1999, Palynology and paleoenvironment of the uppermost Bathonian and Oxfordian (Jurassic) of the northern Switzerland sedimentary basin: Schweizerische paläontologische Abhandlungen, v. 119, p. 1-69.

GRADSTEIN, F. M., OGG, J. G., and SMITH, A. G., 2004, A geologic time scale 2004: Cambridge University Press, Cambridge, 589 p.

GREPPIN, E., 1893, Etude sur les mollusques des couches coralligènes des environs d'Oberbuchsiten: Mémoires de la Société paléontologique suisse, v. 20, p. 1-109.

GRESSLY, A., 1838-41, Observations géologiques sur le Jura soleurois: Nouvelles Mémoires de la Société helvétique des Sciences naturelles, v. 3, 4, and 5, 349 p.

GRESSLY, A., 1864, Rapport géologique sur les terrains parcourus par les lignes du réseau des chemins de fer jurassiens par le Jura bernois. In: Rapports concernant le réseau des chemins de fer du Jura bernois: Annexe 3, p. 87-105. Rieder and Simmen, Bern.

GRÜN, W., and ZWEILI, F., 1980, Das kalkige Nannoplankton der Dogger/Malm-Grenze im Berner Jura bei Liesberg (Schweiz): Jahrbuch der Geologischen Bundesanstalt (Wien), v. 123, no. 1, 231-341.

GUGGENBÜHL, P., 1980, 3rd edition, Unsere einheimischen Nutzhölzer: Stocker-Schmid, Dietikon, 406 p.

GWINNER, M. P., 1962, Geologie des Weissen Jura der Albhochfläche (Württemberg): Neues Jahrbuch für Geologie und Paläontologie, Abhandlungen, v. 115, no. 2, p. 137-221.

GYGI, R. A., 1966, Über das zeitliche Verhältnis zwischen der transversarium-Zone in der Schweiz und der plicatilis-Zone in England (Unt. Malm, Jura): Eclogae geologicae Helvetiae, v. 59, no. 2, p. 935-942.

GYGI, R. A., 1969a, Zur Stratigraphie der Oxford-Stufe (oberes Jura-System) der Nordschweiz und des süddeutschen Grenzgebietes: Beiträge zur geologischen Karte der Schweiz, Neue Folge, v. 136, p. 1-123.

GYGI, R. A., 1969b, An estimate of the erosional effect of *Sparisoma viride* (Bonnaterre), the Green Parrotfish, on some Bermuda Reefs. In: Reports of Research of the 1968 Seminar on Organism-Sediment Interrelationships (edited by GINSBURG, R. N., and GARRET, P.). Bermuda Biological Station for Research, Special Publication No. 2, p. 137–143.

GYGI, R. A., 1970, Coral reefs in Bermuda today, and in the Jura Mountains 140 million years ago: Sandoz Bulletin, v. 16, p. 21-40.

GYGI, R. A., 1973, Tektonik des Tafel- und Faltenjura vom Rhein bei Koblenz bis nach Wildegg. Schichtfolge von der Trias bis ins Tertiär: Jahresberichte und Mitteilungen des oberrheinischen geologischen Vereins, Neue Folge, v. 55, p. 13-22.

GYGI, R. A., 1975, *Sparisoma viride* (Bonnaterre), the Stoplight Parrotfish, a major sediment producer on coral reefs of Bermuda?: Eclogae geologicae Helvetiae, v. 68, no. 2, p. 327-359.

GYGI, R. A., 1977, Revision der Ammonitengattung *Gregoryceras* (Aspidoceratidae) aus dem Oxfordian (Oberer Jura) der Nordschweiz und von Süddeutschland. Taxonomie, Phylogenie, Stratigraphie: Eclogae geologicae Helvetiae, v. 70, no. 2, p. 435-542.

GYGI, R. A., 1981, Oolitic iron formations: marine or not marine?: Eclogae geologicae Helvetiae, v. 74, no. 1, p. 233-254.

GYGI, R. A., 1986, Eustatic sea level changes of the Oxfordian (Late Jurassic) and their effect documented in sediments and fossil assemblages of an epicontinental sea: Eclogae geologicae Helvetiae, v. 79, no. 2, p. 455-491.

GYGI, R. A., 1990a, The Oxfordian ammonite succession near Liesberg BE and Péry BE, northern Switzerland: Eclogae geologicae Helvetiae, v. 83, no. 1, p. 177-199.

GYGI, R. A., 1990b, The Oxfordian in northern Switzerland: International Subcommission on Jurassic Stratigraphy, Oxfordian Working Group, 2nd Oxfordian Working Group Meeting, Basel and Jura range of northern Switzerland, 70 p.

GYGI, R. A., 1990c, Die Paläogeographie im Oxfordium und frühesten Kimmeridgium in der Nordschweiz: Jahreshefte des geologischen Landesamts Baden-Württemberg, v. 32, p. 207-222.

GYGI, R. A., 1990d, The ammonoid genus *Gregoryceras* (Oxfordian, Late Jurassic) in the Monti Lessini, Province of Verona, Italy: Eclogae geologicae Helvetiae, v. 83, no. 3, p. 799-812.

GYGI, R. A., 1991a, Proposed validation of the specific names *crenatus* BRUGUIERE, 1798, and *renggeri* OPPEL, 1863 (Class Cephalopoda, Order Ammonoidea) from the Oxfordian Stage of the Upper Jurassic: Paläontologische Zeitschrift, v. 65, no. 1/2, p. 119-125.

GYGI, R. A., 1991b, Die vertikale Verbreitung der Ammonitengattungen *Glochiceras, Creniceras* und *Bukowskites* im Späten Jura der Nordschweiz und im angrenzenden Süddeutschland: Stuttgarter Beiträge zur Naturkunde, Serie B (Geologie und Paläontologie), no. 179, p. 1-41.

GYGI, R. A., 1992, Structure, pattern of distribution and paleobathymetry of Late Jurassic microbialites (stromatolites and oncoids) in northern Switzerland: Eclogae geologicae Helvetiae, v. 85, no. 3, p. 799-842.

GYGI, R. A., 1995, Datierung von Seichtwassersedimenten des Späten Jura in der Nordwestschweiz mit Ammoniten: Eclogae geologicae Helvetiae, v. 88, no. 1, p. 1-58.

GYGI, R. A., 1998, Taxonomy of perisphinctid ammonites of the early Oxfordian (Late Jurassic) from near Herznach, Canton Aargau, Switzerland: Palaeontographica, Abteilung A, v. 251, p. 1-37.

GYGI, R. A., 1999, Ammonite ecology in Late Jurassic time in northern Switzerland: Eclogae geologicae Helvetiae, v. 92, no. 1, p. 129-137.

GYGI, R. A., 2000a, Integrated stratigraphy of the Oxfordian and Kimmeridgian (Late Jurassic) in northern Switzerland and adjacent southern Germany: Memoirs of the Swiss Academy of Sciences, v. 104, 151 p.

GYGI, R. A., 2000b, Annotated index of lithostratigraphic units currently used in the Upper Jurassic of northern Switzerland: Eclogae geologicae Helvetiae, v. 93, no. 1, p. 125-146.

GYGI, R. A., 2000c, Zone boundaries and subzones of the Transversarium Ammonite Zone (Oxfordian, Late Jurassic) in the reference section of the zone, northern Switzerland. In: Advances in Jurassic research 2000 (edited by Hall, R. L., and Smith, P. L.), Proceedings of the 5th International Symposium on the Jurassic System, Vancouver 1998. GeoResearch Forum, v. 6, p. 77-83.

GYGI, R. A., 2001, Perisphinctacean ammonites of the type Transversarium Zone (Middle Oxfordian, Late Jurassic) in northern Switzerland: Schweizerische Paläontologische Abhandlungen, v. 122, p. 1-169.

GYGI, R. A., 2003, Perisphinctacean ammonites of the Late Jurassic in northern Switzerland: a versatile tool to investigate the sedimentary geology of an epicontinental sea: Schweizerische Paläontologische Abhandlungen, v. 123, v. p. 1-232.

GYGI, R. A., and MCDOWELL, F., 1970, Potassium-argon ages of glauconites from a biochronologically dated Upper Jurassic sequence of northern Switzerland: Eclogae geologicae Helvetiae, v. 63, no. 1, p. 111-118.

GYGI, R. A., SADATI, S.-M., and ZEISS, A., 1979, Neue Funde von *Paraspidoceras* (Ammonoidea) aus dem Oberen Jura von Mitteleuropa - Taxonomie, Ökologie, Stratigraphie: Eclogae geologicae Helvetiae, v. 72, no. 3, p. 897-952.

Gygi, R. A., and Marchand, D., 1982, Les faunes de *Cardioceratinae (Ammonoidea)* du Callovien terminal et de l'Oxfordien inférieur et moyen (Jurassique) de la Suisse septentrionale: Stratigraphie, paléoécologie, taxonomie préliminaire: Geobios, v. 15, no. 4, p. 517-571.

Gygi, R. A., and Persoz, F., 1986, Mineralostratigraphy, litho- and biostratigraphy combined in correlation of the Oxfordian (Late Jurassic) formations of the Swiss Jura range: Eclogae geologicae Helvetiae, v. 79, no. 2, p. 385-454.

Gygi, R. A., and Persoz, f. 1987, The epicontinental sea of Swabia (southern Germany) in the Late Jurassic - factors controlling sedimentation: Neues Jahrbuch für Geologie und Paläontologie, Abhandlungen, v. 176, no. 1, p. 49-65.

Gygi, R. A., and Hillebrandt, A. von, 1991, Ammonites (mainly *Gregoryceras*) of the Oxfordian (Late Jurassic) in northern Chile and time-correlation with Europe: Schweizerische Paläontologische Abhandlungen, v. 113, p. 135-185.

Gygi, R. A., and Marchand, D., 1993, An early Oxfordian ammonite bed in the Terrain à Chailles Member of northern Switzerland and its sequence stratigraphical interpretation: Eclogae geologicae Helvetiae, v. 86, no. 3, p. 997-1013.

Gygi, R. A., Coe, A. L., and Vail, P. R., 1998, Sequence stratigraphy of the Oxfordian and Kimmeridgian (Late Jurassic) in northern Switzerland. In: Mesozoic-Cenozoic sequence stratigraphy of European basins (edited by De Graciansky, P. C., Hardenbol, J., Jacquin, T., and Vail, P. R.). SEPM (Society of sedimentary Geology) Special Publication no. 60, p. 527-544.

Hallam, A., 1964, Origin of the limestone-shale rhythm in the Blue Lias of England: a composite theory: Journal of Geology, v. 72, p. 157-169.

Hallam, A., 1965, Environmental causes of stunting in living and fossil marine benthonic invertebrates: Palaeontology, v. 8, part 1, p. 132-155.

Hallam, A., 1985, A review of Mesozoic climates: Journal of the Geological Society, London, v. 142, p. 433-445.

Hallam, A., 1999, Evidence of sea-level fall in sequence stratigraphy: Examples from the Jurassic: Geology, v. 27, no. 4, p. 343-346.

Hallam, A., 2001, A review of the broad pattern of Jurassic sea-level changes and their possible causes in the light of current knowledge: Palaeogeography, Palaeoclimatology, Palaeoecology, v. 167, p. 23-37.

Hallock, P., and Schlager, W., 1986, Nutrient excess and the demise of coral reefs and carbonate platforms: Palaios, v. 1, p. 389-398.

Haq, B. U., Hardenbol, J., and Vail, P. R., 1988, Mesozoic and Cenozoic chronostratigraphy and cycles of sea-level change: *In:* Wilgus, C. K. et al. (ed.), Sea-level changes: an integrated approach: SEPM (Society of Sedimentary Geology), Special Publication no. 42, p. 71-108.

Hardenbol, J., Thierry, J., Farley, M. B., Jacquin, T., and Vail, P. R., 1998, Jurassic sequence chronostratigraphy, chart 6. *In:* De Graciansky, P.-C., Hardenbol, J., Jacquin, T., and Vail, P. R. (ed.), Mesozoic and Cenozoic sequence stratigraphy of European Basins, SEPM (Society of Sedimentary Geology), Special Publication no. 60.

Harms, J. C., Southard, J. B, Spearing, D. R., and Walker, R. G, 1975, Depositional environments as interpreted from primary sedimentary structures and stratification sequences: Society of Economic Paleontologists and Mineralogists, Short Course 2, 161 p.

Hauber, L., 1971, Zur Geologie des Salzfeldes Schweizerhalle-Zinggibrunn (Kt. Baselland): Eclogae geologicae Helvetiae, v. 64, no. 1, p. 163-183.

Hauber, L., 1980, Geology of the salt field Rheinfelden-Riburg, Switzerland: 5th Symposium on Salt, Cleveland 1980, Proceedings v. 1, p. 83-90.

Hauerstein, G., 1966, Perisphinctes (Arisphinctes) aus der Plicatilis-Zone (Mittel-Oxfordium) von Blumberg/Südbaden (Taxionomie; Stratigraphie). Ph. D. thesis, University of München, 112 p.

Hedberg, H. D., ed., 1976, International Stratigraphic Guide: Wiley, New York, 200 p.

Heer, O., 1865, Die Urwelt der Schweiz: Schulthess, Zürich, 622 p.

Heim, Alb., 1919, Geologie der Schweiz, Band 1, Molasse und Juragebirge: Tauchnitz, Leipzig, 704 p.

Heim, Alb., 1932, Bergsturz und Menschenleben: Fretz und Wasmuth, Zürich, 218 p.

Heim, Arn., 1934, Stratigraphische Kondensation: Eclogae geologicae Helvetiae, v. 27, no. 2, p. 372-383.

Heim, Arn., 1958, Oceanic sedimentation and submarine discontinuities: Eclogae geologicae Helvetiae, v. 51, no. 3, p. 642-649.

Hess, H., 1968, Ein neuer Seestern (*Pentasteria longispina* n. sp.) aus den Effinger Schichten des Weissensteins (Kt. Solothurn): Eclogae geologicae Helvetiae, v. 61, no. 2, p. 607-614.

Hjulström, F., 1935, Studies of the morphological activity of rivers as illustrated by the River Fyris: Bulletin of the Geological Institution of the University of Upsala, vol. 25, p. 221-527.

Hostettler, B., 2006, Die fossilen regulären Echiniden der Günsberg-Formation: Diplomarbeit University of Bern.

Iams, W. J., 1969, New methods for studying the growth rates of reef-building organisms: Bermuda Biological Station for Research, Special Publication no. 2, p. 65-77.

Insalaco, E., 1996a, Upper Jurassic microsolenid biostromes of northern and central Europe: facies and depositional environment: Palaeogeography, Palaeoclimatology, Palaeoecology, v. 121, p. 169-194.

Insalaco, E., 1996b, The use of Late Jurassic coral growth bands as palaeoenvironmental indicators: Palaeontology, v. 39, no. 2, p. 413-431.

James, N. P., and Ginsburg, R. N., 1979, The seaward margin of Belize barrier and atoll reefs: International Association of Sedimentologists, Special Publication no. 3, 191 p.

Jank, M., 2004, New insights into the development of the Late Jurassic Reuchenette Formation of NW Switzerland (late Oxfordian to late Kimmeridgian, Jura Mountains). Ph. D. thesis, University of Basel, 122 p.

Jank, M., Wetzel, A., and Meyer, C. A., 2006, The Late Jurassic sea-level fluctuations in NW Switzerland (Late Oxfordian to Late Kimmeridgian): closing the gap between the Boreal and Tethyan realm in Western Europe: Facies, vol. 52, p. 487-519.

Jeannet, A., 1951, Stratigraphie und Palaeontologie des oolithischen Eisenerzlagers von Herznach und seiner Umgebung (1. Teil): Beiträge zur Geologie der Schweiz, geotechnische Serie, v. 13, no. 5, p. 1-240.

Keller, O., and Krayss, E., 2005, Der Rhein-Linth-Gletscher im letzten Hochglazial: Vierteljahrsschrift der Naturforschenden Gesellschaft in Zürich, v. 150, no. 3/4, p. 69-85.

Kelts, K., and Hsü, K. J., 1980, Resedimented facies of 1875 Horgen slumps in Lake Zürich and a process model of longitudinal transport of turbidity currents: Eclogae geologicae Helvetiae, v. 73, no. 1, p. 271-281.

Kemmerling, G. L. L., 1911, Geologische Beschreibung der Ketten von Vellerat und Moutier: Ph. D. thesis University Freiburg im Breisgau, 42 p.

Kemper, E., and Schmitz, H. H., 1981, Glendonite - Indikatoren des polarmarinen Ablagerungsmilieus: Geologische Rundschau, v. 70, no. 2, p. 759-773.

Kendall, C. G. St. C., and Alsharhan, A. S., 2011, Holocene geomorphology and recent carbonate-evaporite sedimentation of the coastal region of Abu Dhabi, United Arab Emirates. *In:* Kendall, C. G. St. C. and Alsharhan, A. S., ed., Quaternary carbonate and evaporite sedimentary facies and their ancient analogues: International Association of Sedimentologists, Special Publication Nr. 43, p. 45-88.

Kimberley, M. M., 1979, Origin of oolitic iron formations: Journal of Sedimentary Petrology, v. 49, no. 1, p. 111-132.

KNIAZEV, V. G., 1975, Ammonites and zonal stratigraphy of the lower Oxfordian in northern Siberia: Academy of Science of the USSR, Siberian Branch, Transactions of the Institute of Geology and Geophysics, "Nauka" edition, Moscow, v. 275, p. 1-139 (in Russian).

KUENEN, P. H., 1960, Marine Geology: 3rd printing, Wiley, New York, 551 p.

LAMBECK, K., and CHAPPELL, J., 2001, Sea level change through the last glacial cycle: Science, v. 292, p. 679-686.

LATHUILIÈRE, B., GAILLARD, C., HABRANT, N., BODEUR, Y., BOULLIER, A., ÉNAY, R., HANZO, M., MARCHAND, D., THIERRY, J., and WERNER, W., 2005, Coral zonation of an Oxfordian reef tract in the northern French Jura: Facies, v. 50, p. 545-559.

MACINTYRE, I. G., and REID, R. P., 1992, Comment on the origin of aragonite needle mud: a picture is worth a thousand words: Journal of Sedimentary Petrology, v. 62, no. 6, p. 1095-1097.

MAGNÉ, J., and MASCLE, G., 1964, L'Argovien d'Andelot-en-Montagne (Jura). Révision du stratotype: Colloque du Jurassique Luxembourg 1962, Volume des comptes rendus et mémoires de l'institut grand-ducal, section des sciences naturelles, physiques et mathématiques, p. 307-332. St. Paul, Luxembourg.

MANGOLD, C., and GYGI, R. A., 1997, Bathonian ammonites from Canton Aargau, northern Switzerland: Stratigraphy, taxonomy, and biogeography: Geobios, v. 30, no. 4, p. 497-518.

MARCHAND, D., 1979, Un nouvel horizon paléontologique: l'horizon à Paucicostatum (Oxfordien inférieur, zone à Mariae, base de la sous-zone à Scarburgense) : Comptes Rendus sommaires de la Société géologique de France, fascicule 3, p. 122-124.

MARCHAND, D., and TARKOWSKI, R., 1992, Les ammonites du niveau vert de Zalas (Oxfordien inférieur, Pologne du Sud) : condensation ou concentration des faunes: Bulletin of the Polish Academy of Sciences, Earth Sciences, v. 40, no. 1, p. 55-65.

MARCHAND, D., GYGI, R. A., BONNOT, A., and FORTWENGLER, D., 2000, Les ammonites du Callovien terminal (zone à lamberti) et de l'Oxfordien basal (zone à mariae) d'Argovie (Suisse septentrionale) : Revue de Paléobiologie, Genève, v. 19, no. 1, p. 179-189.

MARCOU, J., 1848, Recherches géologiques sur le Jura salinois. Première partie: Mémoires de la Société géologique de France, sér. 2, v. 3, no. 1, p. 1-151.

MARTIN, B., 1984, Zur Geologie der Weissenstein-Kette zwischen Matzendorf (SO) und Wiedlisbach (BE). Unpublished licentiate thesis, University of Bern, 283 p.

MCMANUS, J. F., 2004, A great grand-daddy of ice cores: Nature, v. 429, p. 611-612.

MERIAN, P., 1821, Uebersicht der Beschaffenheit der Gebirgsbildungen in den Umgebungen von Basel, mit besonderer Hinsicht auf das Juragebirge im Allgemeinen: Beiträge zur Geognosie, v. 1, p. 1-156.

MEYER, C. A., 1984, Palökologie und Sedimentologie der Echinodermenlagerstätte Schofgraben (mittleres Oxfordian, Weissenstein, Kt. Solothurn): Eclogae geologicae Helvetiae, v. 77, no. 3, p. 649-673.

MIALL, A. D., 1997, The geology of stratigraphic sequences: Springer, Berlin, 433 p.

MILANKOVITCH, M., 1941, Kanon der Erdbestrahlung und seine Anwendung auf das Eiszeitenproblem: Königlich Serbische Akademie, Editions Spéciales, Section des Sciences Mathématiques et Naturelles, tome 33, Belgrad, 633 p.

MILLIMAN, J. D., FREILE, D., STEINEN, R. P., and WILBER, R. J., 1993, Great Bahama Bank aragonitic muds: mostly inorganically precipitated, mostly exported: Journal of Sedimentary Petrology, v. 63, no. 4, p. 589-595.

MOESCH, C., 1863, Vorläufiger Bericht über die Ergebnisse der im Sommer 1862 ausgeführten Untersuchungen im Weissen Jura der Kantone Solothurn und Bern: Verhandlungen der schweizerischen naturforschenden Gesellschaft, Luzern 1862, p. 156-168.

MOESCH, C., 1867, Geologische Beschreibung des Aargauer Jura und der nördlichen Gebiete des Kantons Zürich: Beiträge zur geologische Karte der Schweiz, v. 4, p. 1-319.

MOORE, G. T., HAYASHIDA, D. N., ROSS, C. A., and JACOBSON, S. R., 1992, Paleoclimate of the Kimmeridgian/Tithonian (Late Jurassic) world: I. Results using a general circulation model: Palaeogeography, Palaeoclimatology, Palaeoecology, v. 93, p. 113-150.

MÜLLER, W. H., HUBER, M., ISLER, A., and KLEBOTH, P. 1984, Erläuterung zur geologischen Spezialkarte Nr. 121 der zentralen Nordschweiz 1:100'000: Nationale Genossenschaft für die Lagerung radioaktiver Abfälle, NAGRA Technischer Bericht 84-25, p. 1-234.

MUTTI, M., 1994, Association of tepees and paleokarst in the Ladinian Calcare Rosso (Southern Alps, Italy): Sedimentology, v. 41, no. 3, p. 621-641.

NAGRA, 1988, Sondierbohrung Weiach, Geologie, NAGRA Technischer Bericht 86-01, Text- und Beilage-Band.

NAGRA, 1990, Sondierbohrung Riniken, Untersuchungsbericht: NAGRA Technischer Bericht 88-09, 125 p.

NAGRA 2001, Sondierbohrung Benken: NAGRA Technischer Bericht NTB 00.01, volume of text and volume of appendices.

NEUENDORF, K. K. E., MEHL, A. P., Jr., and JACKSON, J. A. (ed.), 2005, Glossary of Geology, 5th edition: American Geological Institute, Alexandria, 779 p.

NEUMANN, A. C., 1966, Observation on coastal erosion in Bermuda and measurements of the boring rate of the sponge Cliona lampa: Limnology and Oceanography, v. 11, p. 92-108.

NORRIS, M. L., and HALLAM, A., 1995, Facies variations across the Middle-Upper Jurassic boundary in Western Europe and the relationship to sea-level changes: Palaeogeography, Palaeoclimatology, Palaeoecology, v. 116, p. 189-245.

NOTZ, R., 1924, Geologische Untersuchungen an der östlichen Lägern. Ph. D. thesis, University of Zürich, 58 p.

ODIN, G. S., and MATTER, A., 1981, De glauconiarum origine: Sedimentology, v. 28, p. 611-641.

ODIN, G. S., KNOX, R. W. O'B, GYGI, R. A., and GUERRAK, S., 1988, Green marine clays from the oolitic ironstone facies: habit, mineralogy, environment: Developments in Sedimentology, v. 45, p. 29-52.

OERTLI, H. J., and ZIEGLER, M. A., 1958, Présence d'un Séquanien lacustre dans la région de Pontarlier (Département Doubs, France): Eclogae geologicae Helvetiae, v. 51, no. 2, p. 385-390.

OGDEN, J. C., 1977, Carbonate-sediment production by parrot fish and sea urchins on Caribbean Reefs. In: FROST, S. H., WEISS, M. P., AND SAUNDERS, J. B. (ed.), Reefs and related carbonate-ecology and sedimentology: American Association of Petroleum Geologists, Studies in Geology, v. 4, p. 281-288.

OGG, J. G., OGG, G., and GRADSTEIN, F. M., 2010, The concise geologic time scale. Reprint of 2008: Cambridge University Press, Cambridge, 177 p.

OLSEN, P. E., and KENT, D. V., 1999, Long-period Milankovitch cycles from the Late Triassic and Early Jurassic of eastern North America and their implications for the calibration of the Early Mesozoic time-scale and the long-term behaviour of the planets: Philosophical Transactions of the Royal Society of London, series A, v. 357, no. 1757, p. 1761-1786.

OPPEL, A., 1863, Palaeontologische Mittheilungen aus dem Museum des koeniglich-bayerischen Staates. III. Ueber jurassische Cephalopoden (Fortsetzung), p. 163-266.

OPPEL, A., and WAAGEN, W., 1866, Über die Zone des Ammonites transversarius: Geognostisch-paläontologische Beiträge, v. 1, no. 2, p. 207-318.

ORBIGNY, A. D', 1842-49, Paléontologie Française. Terrains Oolithiqes ou Jurassiques: A. d'Orbigny, Paris, 642 p.

PERRIER, R., and QUIBLIER, J., 1974, Thickness changes in sedimentary layers during compaction history; methods for quantitative evaluation: American Association of Petroleum Geologists Bulletin, v. 58, no. 3, p. 507-520.

PETTIJOHN, F. J., 1975, Sedimentary rocks, 3rd edition: Harper and Row, New York, 628 p.

PFRUNDER, V. R., and WICKERT, H., 1970, Einige Versuche über den Einfluss der chemischen Zusammensetzung und der Mahlung auf die Sinterung von Zement-Rohmehlen: Zement-Kalk-Gips, v. 23, no. 4, p. 147-152.

PICARD, L., and HIRSCH, F., 1987, The Jurassic stratigraphy in Israel and the adjacent countries: Israel Academy of Sciences and Humanities, Section of Sciences, Jerusalem, 106 p.

PITTET, B., and STRASSER, A., 1998, Depositional sequences in deep-shelf environments formed through carbonate-mud import from the shallow platform (Late Oxfordian, German Swabian Alb and eastern Swiss Jura): Eclogae geologicae Helvetiae, v. 91, no. 1, p. 149-169.

PORRENGA, D. H., 1967, Glauconite and chamosite as depth indicators in the marine environment: Marine Geology, v. 5, p. 495-501.

PRESS, F., and SIEVER, R., 1982, Earth (3rd edition): Freeman, San Francisco, 613 p.

PÜMPIN, V. F., 1965, Riffsedimentologische Untersuchungen im Rauracien von St. Ursanne und Umgebung (zentraler Schweizer Jura): Eclogae geologicae Helvetiae, v. 58, no. 2, p. 799-876.

PURSER, B. H., 1979, Middle Jurassic sedimentation on the Burgundy platform: Publication spéciale de l'Association des Sédimentologistes français, v. 1, p. 75-97.

QUENSTEDT, F. A., 1846-49, Petrefactenkunde Deutschlands, 1. Abteilung, 1: Cephalopoden: Fues, Tübingen.

RAIS, P. S. C., 2007, Evidence for a major paleoceanographic reorganization during the Late Jurassic; insights from sedimentology and geochemistry: Ph. D. dissertation Federal Institute of Technology ETH Zürich, 149 p.

RANKEY, E. C., 2004, On the interpretation of shallow shelf carbonate facies and habitats: How much does water depth matter?: Journal of Sedimentary Research, v. 74, no. 1, p. 2-6.

RANKEY, E. C., and REEDER, S. L., 2009, Holocene ooids of Aitutaki Atoll, Cook Islands, South Pacific: Geology, v. 37, no. 11, p. 971-974.

RICHARDS, F. A., 1957, Oxygen in the ocean. In: HEDGPETH, J. W. (ed.), Treatise on marine ecology and paleoecology, v. 1: Ecology: Geological Society of America Memoir, v. 67, no. 1, p. 185-238.

ROLLIER, L., 1888, Etude stratigraphique sur le Jura bernois. Les faciès du Malm jurassien: Eclogae geologicae Helvetiae, v. 1, p. 1-88.

ROLLIER, L., 1892, Sur la composition et l'extension du Rauracien dans le Jura. Eclogae geologicae Helvetiae, v. 3, no. 3, p. 271-293.

ROLLIER, L., 1898, Deuxième supplément à la description géologique de la partie jurassienne de la feuille VII de la Carte géologique de la Suisse au 1:100,000, Jura bernois et régions adjacentes du Jura neuchâtelois, soleurois, bâlois et du Département du Doubs: Matériaux pour la Carte Géologique de la Suisse, nouvelle série, v. 8, 206 p.

ROLLIER, L., 1911, Les faciès du Dogger ou Oolithique dans le Jura et les régions voisines. Georg, Genève et Bâle, 352 p.

ROWLEY, D. B., and LOTTES, A. L., 1988, Plate-kinematic reconstructions of the North Atlantic and Arctic: Late Jurassic to Present: Tectonophysics, Special Issue, v. 155, p. 73-120.

SAEMANN, H., 1921, Untersuchung der Fricktaler Eisenerze und ihre Verhüttbarkeit. Ph. D. thesis, University of Zürich, Sauerländer, Aarau, 56 p.

SAHAGIAN, D., PINOUS, O., OLFERIEV, A., and ZAKHAROV, V., 1996, Eustatic curve of the Middle Jurassic-Cretaceous based on Russian Platform and Siberian stratigraphy: Zonal resolution. American Association of Petroleum Geologists Bulletin, v. 80, no. 9, p. 1433-1458.

SALVADOR, A., ed., 1994, International stratigraphic guide, second edition: Geological Society of America, Boulder, 214 p.

SAVRDA, C. E., BOTTJER, D. J., and GORSLINE, D. S., 1984, Development of a comprehensive oxygen-deficient marine biofacies model: Evidence from Santa Monica, San Pedro, and Santa Barbara Basins, California Continental Borderland: American Association of Petroleum Geologists Bulletin, v. 86, no. 9, p. 1179-1192.

SAVRDA, C. E., and BOTTJER, D. J., 1986, Trace-fossil model for reconstruction of paleo-oxygenation in bottom waters : Geology, v. 14, p. 3-6.

SCHLAGER, W., 1999, Scaling of sedimentation rates and drowning of reefs and carbonate platforms: Geology, v. 27, no. 2, p. 183-186.

SCHMIDT-KALER, H., 1961, Stratigraphische und tektonische Untersuchungen im Malm des nordöstlichen Riesrahmens (Auszug): Geologische Blätter für Nordost-Bayern und angrenzende Gebiete, v. 11, p. 190-200.

SCHNEIDER, A., 1960, Geologie des Gebietes von Siegfriedblatt Porrentruy (Berner Jura): Beiträge zur Geologischen Karte der Schweiz, Neue Folge, v. 109, p. 1-72.

SCHULZ, M., and SCHÄFER-NETH, C., 1997, Translating Milankovitch climate forcing into eustatic fluctuations via thermal deep water expansion: a conceptual link: Terra Nova, v. 9, no. 5/6, p. 228-231.

SCHWEIGERT, G., and CALLOMON, J. H., 1997, Der bauhini-Faunenhorizont und seine Bedeutung für die Korrelation zwischen tethyalem und subborealem Oberjura: Stuttgarter Beiträge zur Naturkunde, Serie B (Geologie und Paläontologie), no. 247, p. 1-69.

SCOTT, G., 1940, Paleoecological factors controlling the distribution and mode of life of Cretaceous ammonoids in the Texas area: Journal of Paleontology, v. 14, no. 4, p. 299-323.

SEIBOLD, E., 1952, Chemische Untersuchungen zur Bankung im unteren Malm Schwabens: Neues Jahrbuch für Geologie und Paläontologie, Abhandlungen, v. 95, no. 3, p. 337-370.

SHINN, E. A., STEINEN, R. P., LIDZ, B. H., and SWART, P. K., 1989, Perspectives: Whitings, a sedimentological dilemma: Journal of Sedimentary Petrology, v. 59, no. 1, p. 147-161.

SMITH, P. E., EVENSEN, N. M., and YORK, D., 1993, First successful ^{40}Ar-^{39}Ar dating of glauconies: Argon recoil in single grains of cryptocrystalline material: Geology, v. 21, p. 41-44.

SMITH, W., 1817, Stratigraphical system of organized fossils, with reference to the specimens of the original geological collection in the British Museum: explaining their state of preservation and their use in identifying the British strata: Williams, London, 121 p.

STANLEY, G. D., 1981, Early history of scleractinian corals and its geological consequences: Geology, v. 9, p. 507-511.

STÄUBLE, A. J., 1959, Zur Stratigraphie des Callovian im zentralen Schweizer Jura: Eclogae geologicae Helvetiae, v. 52, no. 1, p. 57-176.

STECKLER, M. S., and WATTS, A. B., 1978, Subsidence of the Atlantic-type continental margin off New York: Earth and Planetary Science Letters, v. 41, p. 1-13.

STRASSER, A., HILLGÄRTNER, H., HUG, W., and PITTET, B., 2000, Third-order depositional sequences reflecting Milankovitch cyclicity: Terra Nova, v. 12, p. 303-311.

STRASSER, A., 2007, Astronomical time scale for the Middle Oxfordian to Late Kimmeridgian in the Swiss and French Jura Mountains: Swiss Journal of Geosciences, v. 100, p. 407-429.

TALBOT, M. R., 1973, Major sedimentary cycles in the Corallian Beds (Oxfordian) of southern England: Palaeogeography, Palaeoclimatology, Palaeoecology, v. 14, p. 293-317.

TERZAGHI, R. D., 1940, Compaction of lime mud as a cause of secondary structure: Journal of Sedimentary Petrology, v. 10, no. 2, p. 78-90.

TINTANT, H., 1959, Études sur les ammonites de l'Oxfordien supérieur de Bourgogne. I - Les genres Platysphinctes nov. et Larcheria nov.: Bulletin scientifique de Bourgogne, v. 19, p. 109-144.

TINTANT, H., GYGI, R. A., and MARCHAND, D., 2002, Les nautilidés du Jurassique supérieur de Suisse septentrionale: Eclogae geologicae Helvetiae, v. 95, p. 429-450.

TRÜMPY, R., 1959, Hypothesen über die Ausbildung von Trias, Lias und Dogger im Untergrund des schweizerischen Molassebeckens: Eclogae geologicae Helvetiae, v. 52, no. 2, p. 435-448.

USTASZEWSKI, K., SCHUMACHER, M. E., and SCHMID, S. M., 2005, Simultaneous normal faulting and extensional flexuring during rifting: an example from the southernmost Upper Rhine Graben: International Journal of Earth Science (Geologische Rundschau), v. 94, p. 680-696.

VÉDRINE, S., 2007, High-frequency palaeoenvironmental changes in mixed carbonate-siliciclastic sedimentary systems (Late Oxfordian, Switzerland, France, and southern Germany): GeoFocus, v. 19, p. 1-216.

WALKER, R. G., and PLINT, A. G., 1992, Wave and storm-dominated shallow marine systems. In: WALKER, R. G., and JAMES, N. P., ed., Facies models, response to sea level change: GEO text, v. 1, Geological Association of Canada.

WALTHER, J., 1893-94, Einleitung in die Geologie als historische Wissenschaft: Fischer, Jena, 1055 p.

WANI, R., KASE, T., SHIGETA, Y., and DE OCAMPO, R., 2005, New look at ammonoid taphonomy, based on field experiments with modern chambered nautilus: Geology, v. 33, no. 11, p. 849-852.

WEEDON, G. P., JENKYNS, H. C., COE, A. L., and HESSELBO, S. P., 1999, Astronomical calibration of the Jurassic time-scale from cyclostratigraphy in British mudrock formations: Philosophical Transactions of the Royal Society of London, ser. A, v. 357, no. 1757, p. 1787-1813.

WETZEL, A., ALLENBACH, R., and ALLIA, V., 2003, Reactivated basement structures affecting the sedimentary facies in a tectonically "quiescent" epicontinental basin: an example from NW Switzerland: Sedimentary Geology, v. 157, p. 153-172.

WÜRTENBERGER, F. J., and WÜRTENBERGER, L., 1866, Der Weisse Jura im Klettgau und angrenzenden Randengebirg: Verhandlungen des naturwissenschaftlichen Vereins Karlsruhe, v. 2, p. 11-68.

YOUNG, T. P., 1989, Phanerozoic ironstones: an introduction and review: In: YOUNG, T. P., and GORDON TAYLOR, W. E., ed., Phanerozoic ironstones: Geological Society of London, Special Publication no. 46, p. ix-xxv.

ZEISS, A., 1955, Zur Stratigraphie des Callovien und Unter-Oxfordien bei Blumberg (Südbaden): Jahreshefte des geologischen Landesamtes Baden-Württemberg, v. 1, p. 239-266.

ZIEGLER, B., 1963, Ammoniten als Faziesfossilien: Paläontologische Zeitschrift, v. 31, p. 96-102.

ZIEGLER, B., 1967, Ammoniten-Ökologie am Beispiel des Oberjura: Geologische Rundschau, v. 56, no. 2, p. 439-464.

ZIEGLER, M. A., 1962, Beiträge zur Kenntnis des untern Malm im zentralen Schweizer Jura. Ph. D. thesis, University of Zürich, 51 p.

ZIEGLER, P. A., 1956, Zur Stratigraphie des Séquanien im zentralen Schweizer Jura. Mit einem Beitrag von E. GASCHE: Beiträge zur geologischen Karte der Schweiz, Neue Folge, v. 102, p. 37-101.

ZIEGLER, P. A., 1982, Geological atlas of Western and Central Europe: Shell International Petroleum Maatschapie, Elsevier, Amsterdam.

ZIEGLER, P. A., 1988, Evolution of the Arctic-North Atlantic and the Western Tethys: Memoirs of the American Association of Petroleum Geologists, v. 43, p. 1-198.

Index

A

Acanthicum Chron (and Zone), 16, 144
Accommodation space, 81, 86–88, 97, 99, 111, 113, 116, 118–119, 162, 170, 172, 178, 187
Accretion bands, annual, in hermatypic corals, 33, 35–36
Ahermatypic corals, 130
Algae, calcifying green, 93
    characean, 173
Ammonite aperture (peristome), morphologic differentiation, 66
    assemblage, normal size distribution in the, 58, 63–64, 68, 71, 78
    cast, calcareous, 26, 47, 51, 61, 62, 64, 132, 157
    cast of iron sulfide, 65, 66
    chron, 26, 27
    crowding of septa, 66, 67
    drift of empty shells, 190
    dwarfing, environmental, 66
    dwarfing, genetic, 66
    egression, 66
    intolerance of heavy clay mineral sedimentation, 70
    maturity, 66
    morphologic differentiation of the aperture (peristome), 66
    peristome (aperture), ontogenetically precocious, 66, 143
    sensitiveness of heavy clay mineral sedimentation, 70
    septa, crowding of, 66, 67
    size classification, 66
    stunting, 66
    taxonomy, 132, 173, 187
    tolerance of low oxygenation in bottom water, 70
    zone, 26, 133
Annual accretion bands in coral colonies, 33, 35–36
Antecedens Chron (and Zone), 16, 136
Aragonite needles, 93
Aragonitic ooze, 92
Argovien, 10, 22, 23

B

Backstepping, 80, 124–125, 171
Barrier reefs, 89
Base of the Kimmeridgian Stage, 10, 18, 21, 174
    of the Oxfordian Stage, 17, 21, 154
Basement subsidence, 86
Berrense Chron (and Zone), 16, 141
Berthierine, 78, 98
Bifurcatus Chron (and Zone), 16, 139
Bimammatum Chron (and Zone), 16, 141
Biochronology, 13, 26, 187
Bioerosion, 93
Bioherm, 87, 128, 129
Biostrome, 127, 129
Biotite, glauconitization of, 41
Bird's-eyes, 49, 81

B

Blackened lithoclasts (pebbles), 123, 170
Boreal Faunal Province, 136
Boring sponges, 93
Bottom water undersaturated with $Ca(CO)_3$, 158
Boundary beds of Middle/Upper Jurassic Series, 11, 17
    of Oxfordian/Kimmeridgian Stages, 18, 21, 174
Boundary sections, base of Oxfordian and Kimmeridgian Stages, 21, 154, 174

C

Calcareous cast of ammonite, 64, 157
Calcareous concretions, 109
Calcareous mud, 60, 92, 93
Calcareous ooids, limnic, 71
    marine, 31, 39, 50, 112
    overcrowded, 110
Calcareous oolite, 39, 88, 101, 112
Calcifying green algae, 93
Calcilutite, 59, 65, 67, 98
Calcimetry, 27
Calcium sulfate minerals, 32, 33, 39, 121
Calcrete nodules, 39, 124
Caletanum (or *Eudoxus*) Chron (and Zone), 16, 146
Callovian Age, 65, 99
Carbonate platform, 86, 87, 89, 91, 99, 151
Carbonate ramp, 89
Cause and effect of subsidence, 162, 170, 172, 192
Cement of low-magnesian calcite spar, 87
Chalk, 47, 93
Chamosite, 78
Characean algae, 33, 38, 47, 181
Chert nodules, 93
Chestel Member, 28, 101
Climatic change, long-term, 33, 179
Clinoform, 79, 80
Coal, lignite, 33, 38, 47, 87, 173, 181
Coccolithophorids, 30, 91, 92
Colloque du Jurassique Luxembourg 1962, 11, 18, 20, 23
Compaction, 55, 89, 91, 170, 173
Condensation, stratigraphic, 46, 152
Contour currents, 73, 78
Corals
    ahermatypic, 130
    annual accretion bands, 33, 35–36
    bioherm, 22, 88, 101, 127, 128
    biostrome, 88, 127, 128
    hermatypic, tolerance of heavy clay mineral sedimentation, 68, 194
    reef, 12, 33, 89, 101
Cordatum Chron (and Zone), 16, 134
Corrosion, subaqueous (subsolution), 31, 155, 156
Crenulated stromatolite, 116